# Zero-dimensional Carbon Nanomaterials

## Fundamentals and applications

# Zero-dimensional Carbon Nanomaterials

## Fundamentals and applications

**Edited by**
**Pratima R Solanki**
*Special Centre for Nano Sciences (SCNS), Jawaharlal Nehru University (JNU), Delhi, India*

**Ravindra Pratap Singh**
*Department of Biotechnology, Indira Gandhi National Tribal University, Madhya Pradesh, India*

**IOP** Publishing, Bristol, UK

ISBN 978-0-7503-4048-9 (ebook)
ISBN 978-0-7503-4034-2 (print)
ISBN 978-0-7503-4036-6 (myPrint)
ISBN 978-0-7503-4035-9 (mobi)

DOI 10.1088/978-0-7503-4048-9

Version: 20220601

IOP ebooks

British Library Cataloguing-in-Publication Data: A catalogue record for this book is available from the British Library.

Published by IOP Publishing, wholly owned by The Institute of Physics, London

IOP Publishing, Temple Circus, Temple Way, Bristol, BS1 6HG, UK

US Office: IOP Publishing, Inc., 190 North Independence Mall West, Suite 601, Philadelphia, PA 19106, USA

*Dedicated to those who care, conserve, protect but do not spoil/destroy the beauty and unique characteristics of Kshit (Earth), Jal (Water), Pawak (Fire), Gagan (Sky), and Sameera (Air).*

# Contents

# Preface

Zero-dimensional nanomaterials have been considered as a forerunner of nanotechnology due to their extremely small structures, high surface-to-volume ratios, and unique properties that have made them more promising than bulk materials. Zero-dimensional nanomaterials exhibit either a spherical or quasi-spherical shape with a diameter of less than 100 nm. Due to their excellent quantum and edge confinement effects, zero-dimensional nanomaterials exhibit unique properties such as high chemiluminescence and photoluminescence. Various zero-dimensional nanomaterials such as graphene quantum dots (GQDs), fullerenes, carbon quantum dots (CQDs), inorganic quantum dots (QDs), polymeric dots, etc, have been explored and have also found potential roles in many domains, such as biomedicine, sensing, nanodevices, energy, cosmetics, catalysis, and material sciences. Of the zero-dimensional carbon nanomaterials, CQDs and GQDs have attracted significant interest in both the world of technology and the scientific domain. This interest is also reflected in the almost exponential growth of scientific publications on the subject. CQDs exhibit tunable emission, which paved the way for a revolutionary biocompatible alternative over semiconductor quantum dots. GQDs, on the other hand, have brilliant optical as well as electrochemical properties. Understanding the photophysical and electrical properties of these zero-dimensional carbon nanomaterials will help in designing more useful applications.

This book, *Zero-dimensional Carbon Nanomaterials: Fundamentals and Applications*, is unique as it examines collectively the literature on zero-dimensional nanomaterials, focusing on CQDs and GQDs. This book has fourteen chapters describing every aspect of zero-dimensional nanomaterials required for study and research purposes. This book mainly focuses on the unique zero-dimensional properties of nanomaterials and presents various synthesis methods for zero-dimensional nanomaterials. Further, it also presents the reader with potential applications of zero-dimensional nanomaterials in various domains. Hence, this book provides an extensive review of the fundamentals of zero-dimensional carbon nanomaterials, which is very much needed for budding researchers in this domain. This book's major objective is to focus on the multifaceted applications of these nanomaterials and it can be utilized to help students and professionals understand the importance of zero-dimensional carbon nanomaterials. Further, this book can also serve as a textbook for students, but its major focus will be to become a book of great importance for professionals, scientists, researchers, and academics in this research field.

***Aims and Scope***

This book is suitable for wide range of audiences working in the technology domain and scientific circles as it presents a wide range of literature on zero-dimensional carbon nanomaterials, namely CQDs and GQDs. The major points this book will cover are as follows:

- An overview of zero-dimensional nanomaterials.
- The properties, synthesis, and applications of CQDs and GQDs.

- Understanding the photophysical and electrical properties of these zero-dimensional carbon nanomaterials that will help in developing more useful applications.

Pratima R Solanki
Ravindra Pratap Singh

# Acknowledgement

It gives us immense pleasure to acknowledge the Vice-Chancellor of Jawaharlal Nehru University, Delhi, India, Professor Shri Prakash Mani Tripathi, Honorable Vice-Chancellor of Indira Gandhi National Tribal University, Amarkantak, India, and Dr Jay Singh and Mr Kshitij RB Singh of the Department of Chemistry, Institute of Science, Banaras Hindu University, Varanasi, Uttar Pradesh, India, for providing constant assistance in all possible ways. It is also our great pleasure to acknowledge and express our enormous debt to all the contributors who have provided their quality material to prepare this book. We are grateful to our beloved family members (Mr Karamdev Solanki, Dr Babita Singh, Miss Aparajita RB Singh, Miss Kanush Solanki, and Mr Viven Solanki) who joyfully supported and stood with us in many hours of our absence to finish this book project. Thanks are also due to Caroline Mitchell, Calum Sims, Robert Trevelyan, and the entire publishing team for their patience and extra care in publishing this book.

# Editors' biographies

## Pratima R Solanki

Dr Solanki received her PhD degree in biological sciences from Maharishi Dayanand University, Rohtak, Haryana, India. She worked as a postdoctoral fellow at National Physical Laboratory, New Delhi, Institute of Genomics and Integrative Biology, New Delhi, and Tokyo Institute of Technology (TIT), Japan. Dr Solanki has received prestigious fellowships and awards such as the Fast Track-Young Scientist Award from the Department of Science and Technology, New Delhi, India, and Senior Research Associateship (SRA)/pool Scientist from the Council of Scientific Industrial Research. She is actively engaged in synthesizing nanomaterials (carbon dots, metal and metal oxide nanoparticles, and nanofibers) and their nanocomposites with biopolymers/conducting polymers and self-assembled monolayers. She is the co-author of more than 175 research papers published in reputed international and national journals, and 16 international book chapters, with citations of < 7781 and an H-index of 48 (according to Google Scholar). She has also filed 16 Indian patents (two patents granted). She has successfully completed and is currently running various research projects granted from funding agencies such as DST, BIG-BIRAC, DBT, ICMR, and UGC. She has healthy collaboration with reputed scientists at the national and international levels. She has delivered invited lectures in various national (35) and international (20) platforms such as conferences, seminars, symposiums, etc. She has served as a founder member and life member of various scientific societies. She is actively engaged in the research and development of biosensors utilizing nanostructured materials for healthcare and environmental monitoring. She received the Visitor Award 2019 from the President of India.

## Ravindra Pratap Singh

Dr Singh received his BSc from Allahabad University, India, and his MSc and PhD in Biochemistry from Lucknow University, India. He is currently working as an Assistant Professor in the Department of Biotechnology, Indira Gandhi National Tribal University, Amarkantak MP, India. He has previously worked as a scientist at various esteemed laboratories around the world, such as Sogang University, IGR, Paris, etc. His work and research interests include biochemistry, biosensors, nanobiotechnology, electrochemistry, material sciences, and biosensor applications in biomedicine, environmental monitoring, agriculture, and forensics. He has to his credit several reputed national and international honors/awards. Dr Singh has authored over 50 articles in international peer-reviewed journals, more than 30 book chapters of international repute, and has served as editor for three books. He serves as a reviewer for many reputed international journals and is also a member of many international societies. He is currently also involved in editing various books, which will be published by internationally reputable publication houses, namely IOP Publishing, CRC Press, Elsevier, and Springer Nature. Moreover, he is a book series editor of *Emerging Advances in Bionanotechnology*, CRC Press, Taylor and Francis Group. He has also guest-edited a special issue on *Smart and Intelligent Nanobiosensors: Multidimensional Applications* for *Materials Letters*, Elsevier.

# List of contributors

**Charles Oluwaseun Adetunji**
Applied Microbiology, Biotechnology and Nanotechnology Laboratory, Department of Microbiology, Edo University Iyamho, PMB 04, Auchi, Edo State 312101, Nigeria

**V Akhil**
Toxicology Division, Biomedical Technology Wing, Sree Chitra Tirunal Institute for Medical Sciences and Technology (Govt. of India), Poojapura, Trivandrum 695 012, Kerala, India

**Eduardo E Andrada Suarez**
Planta Piloto de Procesos Industriales Microbiológicos (PROIMI-CONICET), Avenida Belgrano y Pasaje Caseros, T40001MVB, Tucumán, Argentina

**Geanne Matos de Andrade**
Center for Research and Drug Development (NPDM), Federal University of Ceara, Rua Cel. Nunes de Melo 1000, Porangabussu, Fortaleza, Ceará, 60430-270 Brazil

Department of Physiology and Pharmacology, Faculty of Medicine, Federal University of Ceara, Rua Cel. Nunes de Melo 1127, Porangabussu, Fortaleza, Ceará, 60430-270 Brazil

**Geanne Matos de Andrade**
Department of Clinical Medicine, Faculty of Medicine, Federal University of Ceara, Rua Costa Mendes, 1608, Porangabussu, Fortaleza, Ceará, 60430-140, Brazil
Department of Physiology and Pharmacology, Faculty of Medicine, Federal University of Ceara, Rua Cel. Nunes de Melo 1127, Porangabussu, Fortaleza, Ceará, 60430-270 Brazil

Department of Morphology, Institute of Biomedicine, Laboratory of the Biology of Tissue Healing, Ontogeny and Nutrition, School of Medicine, Federal University of Ceara, Rua Cel. Nunes de Melo, 1315, Fortaleza, Ceará, 60430-270, Brazil

**A Arathi**
Toxicology Division, Biomedical Technology Wing, Sree Chitra Tirunal Institute for Medical Sciences and Technology (Govt. of India), Poojapura, Trivandrum 695 012, Kerala, India

**Kamal Arora**
Department of Energy Science and Engineering, IIT Powai, Mumbai, Maharashtra 400076, India

**Reinaldo Barreto Oriá**
Department of Morphology, Institute of Biomedicine, Laboratory of the Biology of Tissue Healing, Ontogeny and Nutrition, School of Medicine, Federal University of Ceara, Rua Cel. Nunes de Melo, 1315, Fortaleza, Ceará, 60430-270, Brazil

**Gaurav Bartwal**
Department of Chemistry, Indian Institute of Technology, Kanpur, Uttar Pradesh 208016, India

**Najla Bentrad**
Faculty of Biological Sciences, Department of Biology and Physiology of Organisms, University of Sciences and Technology Houari Boumediene (USTHB), Algiers, Algeria

**Rosalin Beura**
University School of Basic and Applied Sciences, Guru Gobind Singh Indraprastha University, Sector 16C, Dwarka, New Delhi 110078, India

**Jessica Rabelo Bezerra**
Center for Research and Drug Development, Federal University of Ceara, Rua Cel. Nunes de Melo 1000, Porangabussu, Fortaleza, Ceará, 60430-270, Brazil

Department of Physiology and Pharmacology, Faculty of Medicine, Federal University of Ceara, Rua Cel. Nunes de Melo 1127, Porangabussu, Fortaleza, Ceará, 60430-270, Brazil

**Vikas Chaudhary**
CSIR-Central Scientific Instruments Organisation, Sector-30, Chandigarh 160030, India and Academy of Scientific and Innovative Research (AcSIR), Ghaziabad 201002, India

**Albert Layo Costa de Assis**
Department of Clinical Medicine, Faculty of Medicine, Federal University of Ceara, Rua Costa Mendes, 1608, Porangabussu, Fortaleza, Ceará 60430-140, Brazil

Center for Research and Drug Development, Federal University of Ceara, Rua Cel. Nunes de Melo 1000, Porangabussu, Fortaleza, Ceará 60430-270, Brazil

**Sergio A Cuozzo**
Planta Piloto de Procesos Industriales Microbiológicos (PROIMI-CONICET), Avenida Belgrano y Pasaje Caseros, T40001MVB, Tucumán, Argentina

Facultad de Ciencias Naturales e Instituto Miguel Lillo, Universidad Nacional de Tucumán, Miguel Lillo 205, 4000, Tucumán, Argentina

**Awadesh Kumar Dwivedi**
Department of Chemistry, Kisan College Sohsarai, Patliputra University, Nalanda, Bihar, India

**Martín A Fernández-Baldo**
Instituto de Química de San Luis (INQUISAL), Consejo Nacional de Investigaciones Científicas y Técnicas (CONICET), Universidad Nacional de San Luis (UNSL), Chacabuco 917, D5700BWS, San Luis, Argentina

**Jessica Maria Pessoa Gomes**
Department of Clinical Medicine, Faculty of Medicine, Federal University of Ceara, Rua Costa Mendes, 1608, Porangabussu, Fortaleza, Ceará 60430-140, Brazil

Center for Research and Drug Development, Federal University of Ceara, Rua Cel. Nunes de Melo 1000, Porangabussu, Fortaleza, Ceará 60430-270, Brazil

**Abel Inobeme**
Department of Chemistry, Edo University Iyamho, PMB 04, Auchi, Edo State 312101, Nigeria

**X Joseph**
Toxicology Division, Biomedical Technology Wing, Sree Chitra Tirunal Institute for Medical Sciences and Technology (Govt. of India), Poojapura, Trivandrum 695 012, Kerala, India

**Sumit Kumar**
Department of Chemistry, University of Delhi, Delhi 110007, India

**Sumit Kumar**
School of Electronics and Electrical Engineering, Lovely Professional University, Phagwara, Punjab, India

**John Tsado Mathew**
Department of Chemistry, Ibrahim Badamasi Babangida University Lapai, Niger State 911101, Nigeria

**Elaine Gabutin Mission**
Bioeconomy Institute, Universidad de Valladolid, Valladolid, Spain

**Satyabrata Mohapatra**
University School of Basic and Applied Sciences, Guru Gobind Singh Indraprastha University, Sector 16C, Dwarka, New Delhi 110078, India

**Vanya Nayak**
Department of Chemistry, Institute of Science, Banaras Hindu University, Varanasi, Uttar Pradesh 221005, India

**Charu Negi**
CSIR-Central Scientific Instruments Organisation, Sector-30, Chandigarh 160030, India

**Olugbemi T Olaniyan**
Laboratory for Reproductive Biology and Developmental Programming, Department of Physiology, Edo University Iyamho, PMB 04, Auchi, Edo State 312101, Nigeria

**Pratibha Pandey**
Department of Chemistry, Hansraj College, University of Delhi, Delhi

**Mohanan PV**
Toxicology Division, Biomedical Technology Wing, Sree Chitra Tirunal Institute for Medical Sciences and Technology (Govt. of India), Poojapura, Trivandrum 695012, Kerala, India

**Kushagra Rajendra**
Amity School of Earth and Environmental Science, Amity University Haryana, Gurugram, India

**Garima Rathee**
Special Centre for Nanoscience, Jawaharlal Nehru University, New Delhi 110067, India

**Jyotsna Rathee**
Maharaja Surajmal Institute of Technology, New Delhi 110007, India

**Stephen Rathinaraj Benjamin**
Center for Research and Drug Development (NPDM), Federal University of Ceara, Rua Cel. Nunes de Melo 1000, Porangabussu, Fortaleza, Ceará 60430-270, Brazil

**Eli José Miranda Ribeiro Júnior**
Faculty of Inhumas (FacMais), Avenida Monte Alegre, no. 100 Residencial—Monte Alegre, Inhumas, 75400-000 Goiás-GO, Brazil

**Olga Rubilar**
Departamento de Ingeniería Química, Centro de Excelencia En Investigación Biotecnológica, Aplicada al Medio Ambiente (CIBAMA), Universidad de La Frontera, Temuco, Chile

**Anas Saifi**
CSIR-Central Scientific Instruments Organisation, Sector-30, Chandigarh 160030, India

Academy of Scientific and Innovative Research (AcSIR), Ghaziabad 201002, India

**Jay Singh**
Department of Chemistry, Institute of Science, Banaras Hindu University, Varanasi, Uttar Pradesh 221005, India

**Kshitij RB Singh**
Department of Chemistry, Institute of Science, Banaras Hindu University, Varanasi, Uttar Pradesh 221005, India

**Ravindra Pratap Singh**
Department of Biotechnology, Indira Gandhi National Tribal University, Amarkantak, Madhya Pradesh 484887, India

**Pratima R Solanki**
Special Centre for Nanoscience, Jawaharlal Nehru University, New Delhi, India

**Tyciane de Souza Nascimento**
Department of Clinical Medicine, Faculty of Medicine, Federal University of Ceara, Rua Costa Mendes, 1608, Porangabussu, Fortaleza, Ceará 60430-140, Brazil

Center for Research and Drug Development, Federal University of Ceara, Rua Cel. Nunes de Melo 1000, Porangabussu, Fortaleza, Ceará 60430-270, Brazil

**Smriti Sri**
Special Centre for Nanoscience, Jawaharlal Nehru University, New Delhi, India

**Saurabh Srivastava**
Department of Applied Science and Humanities, Rajkiya Engineering College Ambedkar Nagar (Dr APJ Abdul Kalam Technical University, Lucknow), Uttar Pradesh 224122, India

**Mansi Swami**
Department of Chemistry, Biochemistry and Forensic Science, Amity School of Applied Sciences, Amity University Haryana, Gurugram, India

**Gonzalo Tortella**
Departamento de Ingeniería Química, Centro de Excelencia En Investigación Biotecnológica, Aplicada al Medio Ambiente (CIBAMA), Universidad de La Frontera, Temuco, Chile

**Mansi Tyagi**
University School of Basic and Applied Sciences, Guru Gobind Singh Indraprastha University, Sector 16C, Dwarka, New Delhi 110078, India

**Monika Vats**
Department of Chemistry, Biochemistry and Forensic Science, Amity School of Applied Sciences, Amity University Haryana, Gurugram, India

**Tanvi Vats**
Department of Applied Chemistry, School of Vocational Studies and Applied Sciences, Gautam Buddha University, Greater Noida, India

# Chapter 1

# Introduction: zero-dimensional carbon nanomaterials

**Kshitij RB Singh, Vanya Nayak, Smriti Sri, Jay Singh, Pratima R Solanki and Ravindra Pratap Singh**

Carbon is the fifteenth most abundant element found in the Earth's crust and the fourth most abundant element present in the Universe by mass, following hydrogen, helium, and oxygen. Further, there are various types of carbon nanomaterials, of which zero-dimensional carbon nanomaterials, including quantum dots and nanodiamonds, have gained the attention of researchers worldwide. These materials are being utilized in various domains such as biomedicine, agriculture, environmental monitoring, etc. This chapter discusses the various synthesis routes of zero-dimensional carbon nanomaterials by covering their various properties and their applications in the biomedical, agricultural, and environmental domains.

## 1.1 Introduction

Nanotechnology has emerged as one of the most promising technologies and has immense potential in various domains. Nanotechnology uses materials whose dimensions lie in the nano-range (1–100 nm). Their extremely small size and high surface-to-volume ratio help them exhibit diverse properties such as mechanical, optical, biological, chemical, physical, and catalytic properties, etc, that distinguish them from bulk matter [1]. The nanoparticle-based material can be either organic (carbon nanomaterials, etc) or inorganic (metals, etc). The word 'carbon' is derived from the Latin, meaning coal, and is a non-metal element present in every living organism, and thus we can firmly say that it is one of the fundamental aspects of life. Carbon is one of the most astonishing elements as it can be hard, like diamond, and it can also be soft, like graphite. Carbon is one of the basic elements of life; it is non-toxic and is also used to filter other toxins [2].

doi:10.1088/978-0-7503-4048-9ch1

In addition, carbon exhibits the highest melting point among all the elements and is present in a free form in nature. Carbon nanomaterials are known as the chemical fruits of the mother tree of carbon. Carbon nanomaterials have captured a lot of attention from researchers as they exhibit unique electrical, structural, mechanical, optical, biological, and chemical properties that are studied widely and utilized in various domains. The carbon nanomaterials can be classified according to their dimensionality, which depends on their size (<100 nm) in different dimensions, such as zero-dimensional (0D), one-dimensional (1D), and two-dimensional (2D). Fullerenes, metal conjugated carbon nanoparticles, carbon and graphene quantum dots (CQDs and GQDs), and nanodiamonds fall under the 0D carbon nanomaterials, whereas carbon nanotubes and carbon nanowalls come under the 1D and 2D carbon nanomaterial categories, respectively [3].

This chapter aims to provide a detailed description of 0D carbon nanomaterials and their potential biomedical, environmental, and agricultural applications. Moreover, this chapter also highlights the various synthesis routes, properties, and toxicological assessments of carbon-based nanomaterials, as illustrated in figure 1.1, which is followed by a detailed description of different synthesis routes in chapter 2. Further, chapter 3 elaborates on the various properties of 0D carbon nanomaterials, and is followed by chapter 4, which will briefly discuss a toxicological overview of 0D carbon nanomaterials. Chapter 5 is dedicated to the different utilities of

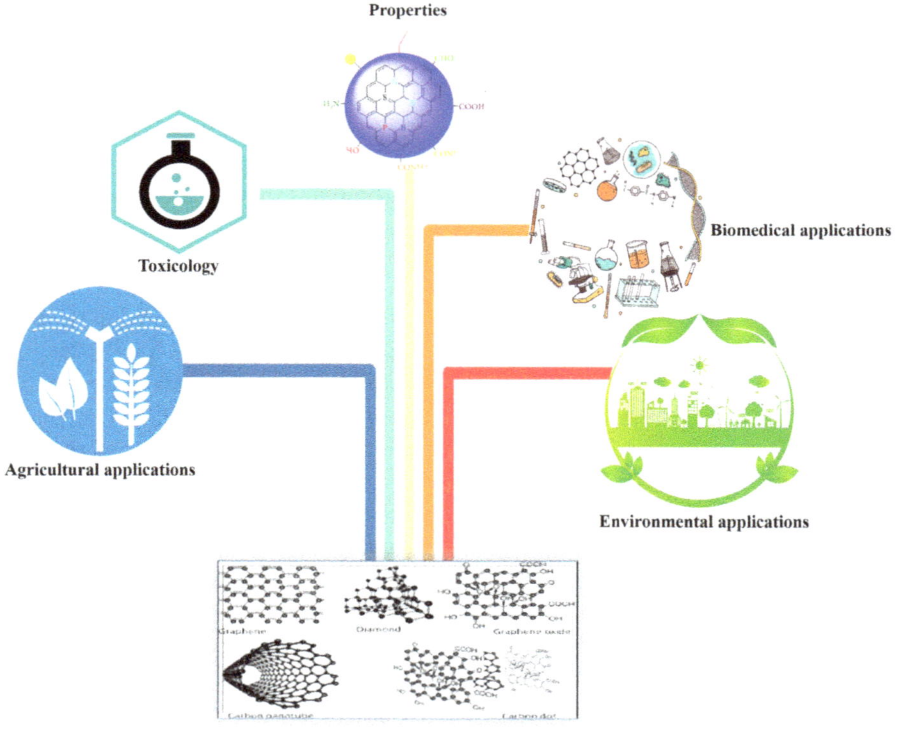


**Figure 1.1.** Schematic illustration of the topics covered in this chapter.

0D carbon nanomaterials in the environmental domain. Chapters 6 and 7 will deal with the application of various 0D carbon nanomaterials used for the treatment and diagnosis of cancer by highlighting their roles in developing biosensors for cancer diagnosis. Further, bio-imaging applications are covered briefly in chapter 8, accompanied by chapter 9, which discusses the utility of 0D carbon nanomaterials for diseases affecting the central nervous system. The potential of 0D carbon nanomaterials in photothermal therapy and targeted drug delivery are discussed in chapter 10, and chapter 11 covers the role of different 0D carbon nanomaterials in tissue engineering. Chapter 12 covers the anti-biofilm potential of 0D carbon nanomaterials. Finally, chapters 13 and 14 deal with the role of 0D carbon nanomaterials in the agriculture domain.

## 1.2 Synthesis

The synthesis approach for any nanomaterial plays a crucial role in deciding its applications as well as its characteristics. There are various methods for synthesizing 0D carbon-based nanomaterials, such as the physical, chemical, and biological routes. The physical routes involve electrospraying, pyrolysis, sputtering, thermal evaporation, etc, whereas the chemical routes include the sol–gel techniques, solvothermal method, chemical vapour deposition, microemulsion, etc. However, the physical and chemical routes require harsh chemicals and are expensive; therefore, these days researchers have shifted their focus to the biological methods for synthesizing 0D carbon nanomaterials. The biological routes use various biological compounds such as plants, algae, microorganisms, biomolecules such as enzymes, proteins, DNA, RNA, etc, which makes the nanomaterials less cytotoxic and more biocompatible. Moreover, these biologically synthesized 0D carbon nanomaterials are economical and sustainable as they are cost-effective and minimize the use of harsh chemicals. Owing to these characteristics, biologically synthesized carbon nanomaterials provide profound benefits in various domains such as biomedicine, agriculture, environmental monitoring, sensing, pharmaceuticals, the food industry, etc. An experiment was performed to synthesize CQDs using the orange peel to detect *Escherichia coli* (*E. coli*) present in milk [4]. The results of the experiment showed that the synthesized green carbon quantum dots had a size of 4.2 nm and displayed bright green fluorescence. Since the prepared CQDs exhibited high fluorescence, they were capable of detecting *E. coli* in milk. This method of synthesis was found to be cheap, effective, and simple [4]. Similarly, by utilizing fenugreek seed extract, CQDs of size $4.25 \pm 0.56$ nm were synthesized through single-step microwave plasma-enhanced decomposition in 5 min. This experiment concluded that natural CQD synthesis through microwave plasma-enhanced decomposition was three times faster and more energy efficient than the traditional thermal decomposition method and can be employed to synthesize diverse and complex bionanomaterials [5].

On the other hand, various chemical routes have also been employed to synthesize CQDs and GQDs. For example, CQDs of size 2 nm were prepared using the hydrothermal method, which was then linked with molecularly imprinted silica for the fluorometric detection of fipronil, a commonly used insecticide that can also generate neurodevelopmental disorders [6]. Similarly, by employing the electrochemical exfoliation method, GQDs and graphene oxide quantum dots (GO-QDs) of different morphologies were synthesized with an average size ranging from 2–3 nm, and it was found that these GQDs and GO-QDs exhibited blue to green luminescence under 365 nm UV irradiation [7]. To avoid the use of harsh chemicals and metal impurities, graphene quantum dots of size 3.0–4.5 nm were synthesized through the hydrothermal method using hydrogen peroxide ($H_2O_2$) and black carbon as the precursor. It was observed that the process was completed in 90 min, and the synthesized GQDs showed excellent light stability, high salt tolerance, high biocompatibility, and decreased cytotoxicity [8]. A further detailed examination of the different synthesis routes of 0D carbon nanomaterials is provided in chapter 2.

## 1.3 Properties

CQDs and GQDs exhibit their optical adsorption range in the UV region at 230–320 nm and 270–390 nm, respectively. This phenomenon is due to the presence of the $\pi$–$\pi$* and n–$\pi$* transition in the C=C and C=O bonds, respectively [9]. However, it has been reported that CQDs show more adsorption of wavelengths than other 0D carbon nanomaterials. Moreover, among all the 0D carbon nanomaterials, CQDs and GQDs show broad-spectrum utilities in various domains owing to their tunable photoluminescence (PL) characteristics. The CQDs have emissive traps on their surfaces; therefore, they show low PL quantum yield (QY), whereas, owing to their layered structure and high crystallinity, GQDs exhibit high PL QY. However, the accurate mechanism of generating PL is still unexplored [10] and is said to be caused by the presence of aromatic structures, quantum confinement effects (QCEs), emissive traps, unbound zigzag sites, oxygen-containing groups, and edge defects [11–13]. Various factors such as pH, temperature, solubility, surface oxidation degree, polarity, size, surface functionalization, heteroatom doping, etc, affect the PL of CQDs and GQDs. An experiment was conducted in which the absorption peak energy of the GQDs was observed to decrease from 6.2 eV to 4.6 eV, as the size of the GQDs increased from 5–35 nm, and the shape of the GQDs changed from circular and elliptical to zigzag and armchair edges to hexagonal and parallelogram-type rectangular with mostly armchair edges. Further, the presence of oxygen-functional groups (OFGs) such as C–O, C=O, and COOH was found to increase the size of GQDs; therefore, it can be concluded that the QCEs strongly affect the PL properties of quantum dots [14].

Further, PL also depends on the pH of the solution, as it was observed that GQDs exhibited strong PL in alkaline conditions in contrast to acidic conditions [15]. However, it was noted that neutral pH emitted the highest PL intensity due to the functional groups and surface states [16]. Apart from pH and size, the solution's temperature and concentration also affect the PL intensity of GQDs. Moreover, it has been observed that the temperature of the reaction influences the functional groups, surface and molecular structure, and PL of GQDs. As the temperature increases the PL QY also increases because of the carbonization process and doping of N atoms in the system [16].

Additionally, QDs exhibit high surface free energy that helps in decreasing the aggregation of the particles because aggregation hinders the optical properties of the QDs, such as the aggregation-induced emission enhancement (AIEE) effect, aggregation-induced red-shift emission (AIRS) effect, and aggregation-induced emission quenching (AIEQ) effect [17], and these aggregation based properties can be utilized in sensing applications [17]. Further, the 0D carbon nanomaterials exhibit electrochemiluminescence (ECL), a type of luminescence that occurs during electrical reactions. In ECL a voltage transfers an electron to an excited state and creates a stable luminescence emission [18]. It has been observed that GQDs show the same ECL behaviour as QDs [19, 20]. Various factors enhance the ECL properties, such as increasing the defects and functional groups on the surface which can increase the solubility and thus enhance the ECL, which can be further utilized for imaging and diagnosis purposes [19–21]. Chapter 3 elaborates further on the various properties of 0D carbon nanomaterials.

## 1.4 Applications

### 1.4.1 Biomedical

The 0D carbon nanomaterials have certainly received a lot of attention from researchers in the biomedical field for cancer therapy, drug delivery, gene therapy, biosensors, etc, as shown in figure 1.2. It has been observed that the conjugation of carbon nanomaterials with other nanomaterials has enhanced their properties and increased their uses in the biomedical domain. Cancer is one of the most lethal diseases whose mortality rate is growing rapidly; therefore, the treatment and diagnosis of cancer have now become the call of the hour. For this aim, carbon nanomaterials were found to be highly efficient in detecting and treating cancer.

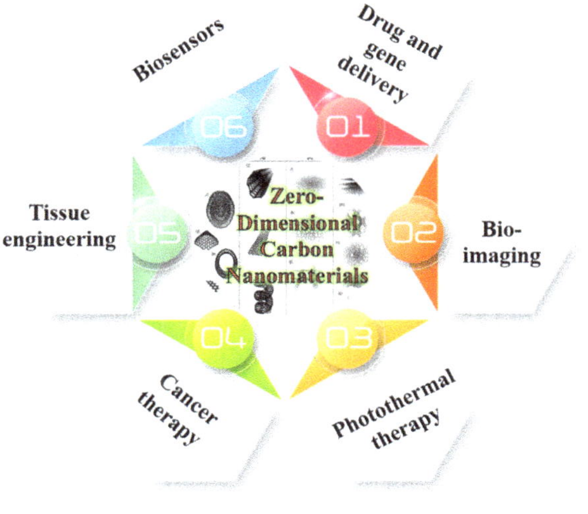


**Figure 1.2.** Schematic illustration of various biomedical utilities of zero-dimensional carbon nanomaterials.

Since CQDs and GQDs exhibit high biocompatibility, low toxicity, and high photoluminescence properties, they are highly acceptable for cancer treatment. Moreover, the high photoluminescence of both GQDs and CQDs helps in tracking the nanoparticles within the body to deeply observe the release of the drug at the targeted organ [22, 23]. Further, it has been observed that by increasing the temperature the proteins of cancer cells start to denature, which kills the cancer cells, therefore cancer-based photothermal therapy (PTT) using 0D carbon nanomaterials is being utilized. PTT is considered one of the best therapies to treat cancer as it is non-invasive, shows low side effects, and has low biotoxicity and high biocompatibility. Although the single-walled carbon nanotubes (SWCNTs) were the first carbon-based nanomaterials to be utilized in PTT, eventually they were replaced by CQDs and GQDs due to their extremely small size [24]. In an experiment, CQDs were synthesized using polythiophene benzoic acid, which showed light absorption from 400–700 nm and emitted a bright red photoluminescence peak between 640–680 nm at various excitations. It was reported that these synthesized CQDs could be utilized for both PTT and photodynamic therapy (PDT). Moreover, it was also noted that the prepared CQDs did not exhibit any cytotoxicity in dark conditions. However, poor near-infrared radiation (NIR) limits the use of CQDs in PTT. Nonetheless, many approaches, such as doping CQDs with metals, preparing CQDs with NIR absorption precursors, conjugating CQDs with certain dyes with NIR absorption, etc, have been developed that help in overcoming this limitation [25].

With advancing research, many techniques have been employed to diagnose various diseases and one eye-catching technique is the use of fluorescent probes. Multiple nanomaterials have been employed in developing fluorescent probes. Carbon-based nanomaterials are considered useful fluorescent probes for cancer detection despite limitations such as low photo-stability and a short life span [26]. Photo-bleached CQDs and GQDs exhibit unique cellular permeability, excellent optical properties, low cytotoxicity, etc; therefore, they are utilized for labelling and bio-imaging [27]. It is a well-known fact that diagnosing a disease is the first step in combating it. Yang *et al* were the first to report the utilization of CQDs as a PL contrast agent. The PEGglytated CQDs were injected inside mice, and various FL images were recorded at different wavelengths. They observed that PEGglytated CQDs exhibited excellent internalization in cancer cells, active addition at the tumour site, effectively delivered the drug at the targeted site, and showed long circulation times. The negative charge of the cell membrane was found to be the reason for easy entry into the cancer cells and excellent efficiency of internalization [28]. Further, CQDs and GQDs were found to be useful in targeted drug and gene delivery. One of the anticancer drugs, doxorubicin, was efficiently delivered by the DNA-CQDs as it showed better drug loading capacity and was stable under different pH conditions. Moreover, it was noted that the presence of N and P in DNA enhanced the photoluminescence of the CQDs; therefore, they can also be utilized for imaging purposes [29].

Further, CQDs have shown their great potential in biosensors by detecting several biomarkers such as glucose, RNA, DNA, proteins, enzymes, etc. For instance, a

CQD-based DNA biosensor was developed to detect the cystic fibrosis transmembrane regulator (CFTR) gene. This developed sensor exhibited high selectivity and sensitivity with a detection limit of 0.16 nM [30]. Moreover, research has started to focus on regenerating injured or damaged tissue and this is considered a significant challenge in tissue engineering. Therefore, an appropriate and permanent solution is required to overcome this challenge. For this, 0D carbon nanomaterials were found to be a promising candidate as they can be incorporated easily into scaffolds owing to their high surface area, tunable properties, high stability, high biocompatibility, etc [31, 32]. A fabricated scaffold should be highly porous, biocompatible, have a high surface area, and control biodegradation for sufficient cell–scaffold interactions. For instance, *p*-phenylenediamine functionalized CQD mediated silk fibroin PLA (CQDs-SF-PLA) nanofibrous scaffolds were developed for cardiac tissue engineering and treatment. These scaffolds were found to exhibit better mechanical activity compared to the non-functionalized scaffolds [32]. Similarly, a surface-modified CQD (bacterial cellulose (BC) or polyvinyl alcohol (PVA))–silica nanoparticle (SiNP)-scaffold (SF) based nanofiber was developed for the regeneration of hair and skin tissue with the help of electrospinning and electrospraying methods. It was observed that the PVA-CQD-SiNP-SF nanofiber showed better elongation properties, high antibacterial activity, increased proliferation of NIH-3T3 cells, and enhanced the speed of *in vivo* skin regeneration, which was found to be 65% higher than for the BC-CQD-SiNP-SF nanofiber when observed on mice. Therefore, the PVA-CQD-SiNP-SF nanofiber is a potential candidate for wound dressing, inhibiting the growth of bacteria, and increasing the healing of skin wounds and hair regeneration [33].

### 1.4.2 Agriculture

Carbon dots have found immense potential in the agricultural field as well. However, certain limitations, such as the requirement for complex equipment, low yield, burdensome synthesis, high temperatures, etc, increase the cost of the application of carbon dots in the agriculture domain to unsatisfactory levels. Carbon dots are the most promising nanomaterials, as they exhibit excellent chemical inertness, low toxicity, high biocompatibility, water-solubility, environmentally friendly synthesis routes, and widely available precursors, which help them to ensure the safety of crops, humans, and the environment. Moreover, the high chemical inertness of carbon dots increases their storability and stability after being added to a nutrient solution. Since carbon dots have extremely small structures, they can be absorbed and transported through plant cells easily. Further, carbon dots exhibit fluorescence properties that track their route after absorption and monitor their action [34]. However, the application of carbon dots in agriculture is still unexplored and is at a very nascent stage, but according to a study conducted by Tripathi and Sarkar *et al* in 2015, carbon dots showed positive impacts when tested on the roots of wheat by adding 150 $\mu$g of CQDs to the plant. However, the paper presented no emission spectra, and the cultivation period of only ten days made their results controversial [35]. Apart from this study, another study was conducted by

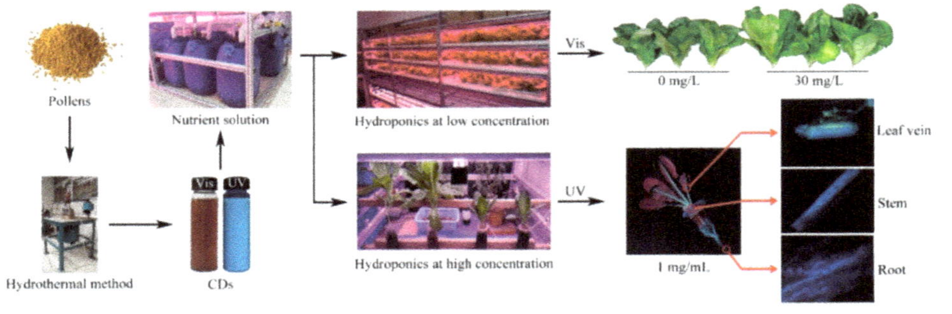


**Figure 1.3.** Schematic illustration of the experiment. (Reproduced with permission from [36]. Copyright 2017 American Chemical Society.)

Zheng *et al* in 2017, in which carbon dots were synthesized with a green method using rapeseed pollen in a hydrothermal approach and were directly added to Hoagland nutrient solution used for hydroponic cultivation. Further, this hydroponically cultivated Rome lettuce was used as a model to study the impact of CQDs on the plant at different concentrations, as shown in figure 1.3. The results of this experiment showed that 20–30 mg $L^{-1}$ CQDs can efficiently increase the production yield by decreasing the nitrate content. Moreover, it was confirmed that the carbon dots are transferred through nutrient solutions to vessels through vascular bundles and cellular pathways through the transpiration stream, therefore rapeseed pollen mediated synthesized carbon dots can act as an excellent material for labelling cells *in vivo*, affecting the physiology of the plant, and increase the yield [36]. Chapters 12 and 13 of this book further highlight the various applications of 0D carbon nanomaterials in agriculture.

### 1.4.3 Environment

The presence of unwanted chemical entities in the environment that prevent or hamper natural processes is known as pollution. With the increase in globalization, population, urbanization, and industrialization, pollution is also increasing at an alarming rate [37, 38]. Therefore, improving and enhancing the quality of water, air, and soil is a great challenge. One of the key approaches in protecting the environment is identifying and investigating environmental pollutants. In combination with material sciences, nanotechnology has exponentially increased the treatment of environment-based pollution. 0D carbon nanomaterials are a class of engineered nanomaterials (ENMs) that show exceptional properties and are widely used to purify heavy metal ion contaminated water. 0D carbon nanomaterials are commonly used in pollution detection methods compared to standard methods because the former exhibit advantages such as simplicity, rapidity, accuracy, and efficiency. Moreover, their well-defined sub-nanometer structure helps them show fluorescence emission that allows them to act as an efficient fluorescent nanoprobe for detecting various environmental pollutants [39–41]. The carbon dot based biosensors work through two mechanisms: detecting contaminants by analyzing the change in fluorescence signals due to direct interaction between carbon dots and analytes, and functionalizing carbon dots with specific recognition molecules to detect the analytes [42].

Heavy metal pollution is considered highly toxic and dangerous for human beings and animals, even at low concentrations [43]. Zero-dimensional carbon nanomaterials have emerged as promising materials in overcoming heavy metal pollution. For example, an experiment was performed in which hydrothermally synthesized CQDs detected heavy metals such as $Cu^{2+}$, $Co^{2+}$, and $Ni^{2+}$. The synthesized CQDs were used as nanoprobes, and the sensing was based on the 'turn-off' approach. With the help of photoinduced electron transfer (PET), it was observed that as the concentration of the metal was increased, the fluorescence of the CQDs decreased. Further, these generated signals were successfully applied to the drinking water sample analysis [44]. Moreover, CQDs exhibit different surface charges, high water stability, high biocompatibility, and low cytotoxicity; therefore, they are also employed to detect various pathogenic bacteria [45–48]. For instance, a turn-off fluorescent sensor array was designed to detect different bacteria, in which six different bacteria were identified with a precision rate of 91.6%, as shown in figure 1.4 [49]. Further, chapter 5 of this book describes the applications of 0D carbon nanomaterials in the environmental domain.

## 1.5 Toxicity

One of the significant limitations of utilizing CQDs and GQDs is their cytotoxic parameters which are still not fully explored. Much deeper research and investigation are still required to explore the different effective parameters. The *in vitro* toxicity depends on dosage, dosage concentration, cell type, etc [50–53]. It was found that the polyethyleneimine (PEI)-coated CQD-based nanohybrids of size $6.5 \pm 2$ nm showed less cytotoxicity at lower concentrations when investigated on kidney epithelial cells of African green monkeys. Further, with the help of MTT (3-(4,5-dimethylthiazol-2-yl)–2,5-diphenyltetrazolium bromide) assays, it was also recorded that these nanohybrids efficiently killed 39% of the cells at 600 $\mu$g mL$^{-1}$ [54]. Further, in another experiment pristine fluorescent carbon quantum dots of size 7 nm were investigated on human bronchial epithelial cells (16HBE). It was noted that the CQDs present on the surface of the cells showed oxidative stress and decreased cell viability [55]. Similarly, owing to the uniquely small size and high oxygen content, the GQDs also demonstrated low cytotoxic effects when investigated on MC3T3 osteoblast cell lines of mouse calvaria [56]. However, a study reported the adverse effects of hydroxyl-modified GQDs (OH-GQDs) of size $10.3 \pm 1.9$ nm when studied on human lung carcinoma cell lines H1299 and A549. These synthesized OH-GQDs at 50 $\mu$g mL$^{-1}$ lowered cell viability and decreased intracellular ROS generation at a significant level. Further, it was also noted that these OH-GQDs generated G0/G1 arrest and cell senescence, and inhibited Rb phosphorylation in both types of cells [57].

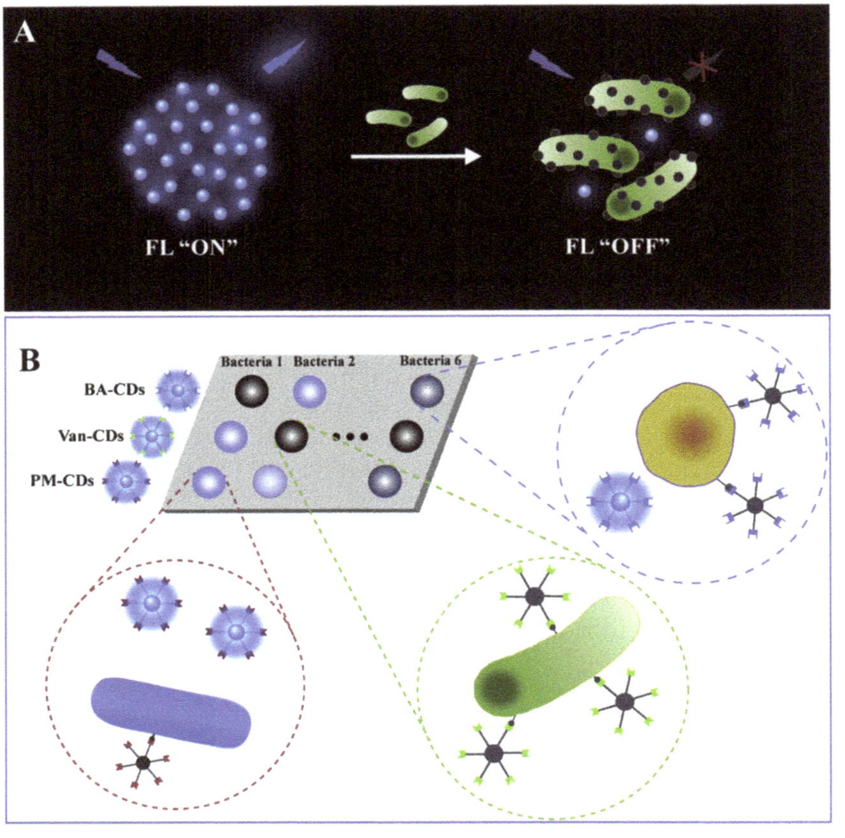


**Figure 1.4.** Schematic representation of pattern based recognition of bacteria by functionalizing three different receptors on carbon dots: (a) a decrease in the fluorescence intensity of carbon dots due to binding of bacteria and (b) the fluorescence pattern produced from the different responses of the carbon dots toward six different bacteria. (Reproduced with permission from [49]. Copyright 2019 Elsevier.)

Moreover, various factors such as size, cell permeability, functionalization, and dispersion of zero-dimensioned carbon nanomaterials play an important role in generating differential toxicity. Although various studies have shown the toxic effects of zero-dimensional carbon nanomaterials in both *in vitro* and *in vivo*, specific issues are still unexplored and need to be addressed as soon as possible. Certain issues lack research, such as showing the comparative toxic effects of two or more 0D carbon nanomaterials on the same cell lines. Similarly, a systematic study is still required to efficiently demonstrate the different toxic effects depending on the nanoparticles' surface charge (positive or negative) [58]. Overall, the correct knowledge of the toxicity of different 0D carbon nanomaterials will help the researchers of interdisciplinary backgrounds deliver more enhanced biocompatible zero-dimensional carbon nanomaterials to society in the future.

## 1.6 Conclusion

Science and technology are advancing day by day. The growth in science and technology has benefitted humans and the environment. Nanotechnology is a domain that has grabbed much attention from researchers, as it manipulates the properties of bulk matter at nanodimensions. Among the many nanomaterials, 0D carbon nanomaterials have successfully been utilized in various domains such as biomedicine, sensors, environmental monitoring, agriculture, robotics, etc. The extremely small size of 0D carbon nanomaterials helps them to exhibit various distinct and tunable properties that allow them to be utilized in various domains. The 0D carbon nanomaterials include the carbon and graphene quantum dots and nanodiamonds. Further, the unique fluorescent and photoluminescent properties of 0D carbon nanomaterials help them develop various types of sensors and can also be used for imaging purposes. This chapter briefly highlights the different properties and the potential of 0D carbon nanomaterials in the biomedical, environmental, and agricultural domains. Moreover, this chapter also highlights the various methods used to synthesize 0D carbon nanomaterials. Further, toxicology studies of 0D carbon nanomaterials have also been discussed that researchers can use to overcome the various limitations.

## Acknowledgments

KRBS and VN are grateful to BHU (IoE grant) for financial support throughout this work and they are also very grateful to Professor Mrituanjay D Pandey of the Department of Chemistry, Banaras Hindu University, India, for his constant support and motivation. SS and PRS are grateful to their respective institutions for providing support to carry out this work. JS expresses their gratitude to the DST-INSPIRE faculty Fellowship, BHU (IoE grant), and UGC New Delhi for financial support. RPS is grateful to VC, IGNTU, Amarkantak, India, for providing constant financial support and for motivating good science.

## References

[1] Singh K R, Nayak V and Singh R P 2021 Introduction to bionanomaterials: an overview *Bionanomaterials* (Bristol: IOP Publishing)

[2] Pantarotto D, Briand J-P, Prato M and Bianco A 2004 Translocation of bioactive peptides across cell membranes by carbon nanotubes *Chem. Commun.* **1** 16

[3] Rauti R, Musto M, Bosi S, Prato M and Ballerini L 2018 Properties and behavior of carbon nanomaterials when interfacing neuronal cells: how far have we come? *Carbon* **143** 430–46

[4] Hu X *et al* 2021 Green one-step synthesis of carbon quantum dots from orange peel for fluorescent detection of *Escherichia coli* in milk *Food Chem.* **339** 127775

[5] Dager A, Baliyan A, Kurosu S, Maekawa T and Tachibana M 2020 Ultrafast synthesis of carbon quantum dots from fenugreek seeds using microwave plasma enhanced decomposition: application of C-QDs to grow fluorescent protein crystals *Sci Rep.* **10** 12333

[6] Yang C, Wang L, Zhang Z, Chen Y, Deng Q and Wang S 2020 Fluorometric determination of fipronil by integrating the advantages of molecularly imprinted silica and carbon quantum dots *Microchim. Acta* **187** 12

[7] Ahirwar S, Mallick S and Bahadur D 2017 Electrochemical method to prepare graphene quantum dots and graphene oxide quantum dots *ACS Omega* **2** 8343–53
[8] Lu Q *et al* 2017 A facile and simple method for synthesis of graphene oxide quantum dots from black carbon *Green Chem.* **19** 900–4
[9] Eda G *et al* 2010 Blue photoluminescence from chemically derived graphene oxide *Adv. Mater.* **22** 505–9
[10] Zhu S *et al* 2012 Surface chemistry routes to modulate the photoluminescence of graphene quantum dots: from fluorescence mechanism to up-conversion bioimaging applications *Adv. Funct. Mater.* **22** 4732–40
[11] Sciortino A, Cannizzo A and Messina F 2018 Carbon nanodots: a review—from the current understanding of the fundamental photophysics to the full control of the optical response *C: J. Carbon Res.* **4** 67
[12] Shen J, Zhu Y, Chen C, Yang X and Li C 2011 Facile preparation and upconversion luminescence of graphene quantum dots *Chem. Commun.* **47** 2580–2
[13] Chen S *et al* 2013 Full-range intracellular pH sensing by an aggregation-induced emission-active two-channel ratiometric fluorogen *J. Am. Chem. Soc.* **135** 4926–9
[14] Kim S *et al* 2012 Anomalous behaviors of visible luminescence from graphene quantum dots: interplay between size and shape *ACS Nano* **6** 8203–8
[15] Pan D, Zhang J, Li Z and Wu M 2010 Hydrothermal route for cutting graphene sheets into blue-luminescent graphene quantum dots *Adv. Mater.* **22** 734–8
[16] Gu J, Zhang X, Pang A and Yang J 2016 Facile synthesis and photoluminescence characteristics of blue-emitting nitrogen-doped graphene quantum dots *Nanotechnology* **27** 165704
[17] Liu M A and Reeves P R 2019 Customizable cloning of whole polysaccharide gene clusters by yeast homologous recombination *Methods Mol. Biol.* **1954** 1–14
[18] Xu Y, Liu J, Gao C and Wang E 2014 Applications of carbon quantum dots in electrochemiluminescence: a mini review *Electrochem. Commun.* **48** 151–4
[19] Zhang R, Adsetts J R, Nie Y, Sun X and Ding Z 2018 Electrochemiluminescence of nitrogen- and sulfur-doped graphene quantum dots *Carbon* **129** 45–53
[20] Lu Q, Wei W, Zhou Z, Zhou Z, Zhang Y and Liu S 2014 Electrochemiluminescence resonance energy transfer between graphene quantum dots and gold nanoparticles for DNA damage detection *Analyst* **139** 2404–10
[21] Wang R, Wu H, Chen R and Chi Y 2019 Strong electrochemiluminescence emission from oxidized multiwalled carbon nanotubes *Small* **15** 1901550
[22] Durán N, Simões M B, de Moraes A C M, Fávaro W J and Seabra A B 2016 Nanobiotechnology of carbon dots: a review *J. Biomed. Nanotechnol.* **12** 1323–47
[23] Li X, Shi L, Li L, Dong C, Li C and Shuang S 2019 Recent advances in carbon nanodots: properties and applications in cancer diagnosis and treatment *J. Anal. Test* **3** 37–49
[24] Shi Kam N W, O'Connell M, Wisdom J A and Dai H 2005 Carbon nanotubes as multifunctional biological transporters and near-infrared agents for selective cancer cell destruction *Proc. Natl Acad. Sci.* **102** 11600–5
[25] Kang E B, Phuong P T M, Lee G, Lee S, In I and Park S Y 2019 pH-selective fluorescent probe with photothermal ablation of bacteria based NIR dye-embedded zwitterionic carbon dots *Macromol. Res.* **27** 720–8

[26] Pedram P, Mahani M, Torkzadeh-Mahani M, Hasani Z and Ju H 2016 Cadmium sulfide quantum dots modified with the human transferrin protein siderophiline for targeted imaging of breast cancer cells *Microchim. Acta* **183** 67–71
[27] Geng B *et al* 2018 NIR-responsive carbon dots for efficient photothermal cancer therapy at low power densities *Carbon* **134** 153–62
[28] Yang S T *et al* 2009 Carbon dots as nontoxic and high-performance fluorescence imaging agents *J. Phys. Chem.* C **113** 18110–4
[29] Ding H, Du F, Liu P, Chen Z and Shen J 2015 DNA-carbon dots function as fluorescent vehicles for drug delivery *ACS Appl. Mater. Interfaces* **7** 6889–97
[30] García-Mendiola T *et al* 2018 Carbon nanodots based biosensors for gene mutation detection *Sens. Actuators* B **256** 226–33
[31] Dehghani A, Ardekani S M, Hassan M and Gomes V G 2018 Collagen derived carbon quantum dots for cell imaging in 3D scaffolds via two-photon spectroscopy *Carbon* **131** 238–45
[32] Yan C *et al* 2020 Photoluminescent functionalized carbon quantum dots loaded electroactive silk fibroin/PLA nanofibrous bioactive scaffolds for cardiac tissue engineering *J. Photochem. Photobiol.* B **202** 111680
[33] Abolghasemzade S, Pourmadadi M, Rashedi H, Yazdian F, Kianbakht S and Navaei-Nigjeh M 2021 PVA based nanofiber containing CQDs modified with silica NPs and silk fibroin accelerates wound healing in a rat model *J. Mater. Chem.* B **9** 658–76
[34] Khodakovskaya M V *et al* 2013 Carbon nanotubes as plant growth regulators: effects on tomato growth, reproductive system, and soil microbial community *Small* **9** 115–23
[35] Tripathi S and Sarkar S 2015 Influence of water soluble carbon dots on the growth of wheat plant *Appl. Nanosci.* **5** 609–16
[36] Zheng Y *et al* 2017 Bioimaging application and growth-promoting behavior of carbon dots from pollen on hydroponically cultivated rome lettuce *ACS Omega* **2** 3958–65
[37] Mehndiratta P, Jain A, Srivastava S and Gupta N 2013 Environmental pollution and nanotechnology *Environ. Pollut.* **2** 49–58
[38] Zhang X *et al* 2017 Mussel-inspired fabrication of functional materials and their environmental applications: progress and prospects. *Appl. Mater. Today* **7** 222–38
[39] Devi P, Rajput P, Thakur A, Kim K-H and Kumar P 2019 Recent advances in carbon quantum dot-based sensing of heavy metals in water *Trends Anal. Chem.* **114** 171–95
[40] Sun X and Lei Y 2017 Fluorescent carbon dots and their sensing applications *Trends Anal. Chem.* **89** 163–80
[41] Wang Y and Hu A 2014 Carbon quantum dots: synthesis, properties and applications *J. Mater. Chem.* C **2** 6921–39
[42] Long C, Jiang Z, Shangguan J, Qing T, Zhang P and Feng B 2021 Applications of carbon dots in environmental pollution control: a review *Chem. Eng. J.* **406** 126848
[43] Jaishankar M, Tseten T, Anbalagan N, Mathew B B and Beeregowda K N 2014 Toxicity, mechanism and health effects of some heavy metals *Interdiscip. Toxicol.* **7** 60–72
[44] Mintz K J, Guerrero B and Leblanc R M 2018 Photoinduced electron transfer in carbon dots with long-wavelength photoluminescence *J. Phys. Chem.* C **122** 29507–15
[45] Bhaisare M L, Gedda G, Khan M S and Wu H-F 2016 Fluorimetric detection of pathogenic bacteria using magnetic carbon dots *Anal. Chim. Acta* **920** 63–71

[46] Zhong D, Zhuo Y, Feng Y and Yang X 2015 Employing carbon dots modified with vancomycin for assaying gram-positive bacteria like *Staphylococcus aureus Biosens. Bioelectron.* **74** 546–53
[47] Yang L, Deng W, Cheng C, Tan Y, Xie Q and Yao S 2018 Fluorescent immunoassay for the detection of pathogenic bacteria at the single-cell level using carbon dots-encapsulated breakable organosilica nanocapsule as labels *ACS Appl. Mater. Interfaces* **10** 3441–8
[48] Wang H *et al* 2018 Development of a fluorescence assay for highly sensitive detection of *Pseudomonas aeruginosa* based on an aptamer-carbon dots/graphene oxide system *RSC Adv.* **8** 32454–60
[49] Zheng L, Qi P and Zhang D 2019 Identification of bacteria by a fluorescence sensor array based on three kinds of receptors functionalized carbon dots *Sens. Actuators* B **286** 206–13
[50] Ko N R, Nafiujjaman M, Lee J S, Lim H-N, Lee Y-K and Kwon I K 2017 Graphene quantum dot-based theranostic agents for active targeting of breast cancer *RSC Adv.* **7** 11420–7
[51] Bagheri Z, Ehtesabi H, Hallaji Z, Latifi H and Behroodi E 2018 Investigation the cytotoxicity and photo-induced toxicity of carbon dot on yeast cell *Ecotoxicol. Environ. Saf.* **161** 245–50
[52] Wu C, Wang C, Han T, Zhou X, Guo S and Zhang J 2013 Insight into the cellular internalization and cytotoxicity of graphene quantum dots *Adv. Healthc. Mater.* **2** 1613–9
[53] Şenel B, Demir N, Büyükköroğlu G and Yıldız M 2019 Graphene quantum dots: synthesis, characterization, cell viability, genotoxicity for biomedical applications *Saudi Pharm. J.* **27** 846–58
[54] Emam A N, Loutfy S A, Mostafa A A, Awad H and Mohamed M B 2017 Cyto-toxicity, biocompatibility and cellular response of carbon dots–plasmonic based nano-hybrids for bioimaging *RSC Adv.* **7** 23502–14
[55] Zhang X *et al* 2013 A cytotoxicity study of fluorescent carbon nanodots using human bronchial epithelial cells *J. Nanosci. Nanotechnol.* **13** 5254–9
[56] Zhu S *et al* 2011 Strongly green-photoluminescent graphene quantum dots for bioimaging applications *Chem. Commun.* **47** 6858–60
[57] Tian X *et al* 2016 Hydroxylated-graphene quantum dots induce cells senescence in both p53-dependent and -independent manner *Toxicol. Res.* **5** 1639–48
[58] Kumar A, Singh K R B, Ghate M D, Lalhlenmawia H, Kumar D and Singh J 2022 Bioinspired quantum dots for cancer therapy: a mini-review *Mater. Lett.* **313** 131742

**IOP** Publishing

Zero-dimensional Carbon Nanomaterials
Fundamentals and applications
**Pratima R Solanki and Ravindra Pratap Singh**

# Chapter 2

# The chemical and biological synthesis of zero-dimensional carbon nanomaterials

**Tanvi Vats, Sumit Kumar and Pratibha Pandey**

Zero-dimensional nanomaterials have attracted great interest in the research community ever since their development. Their unique properties are due to their ultra-small size and high surface-to-volume ratio. The synthesis of nanomaterials plays a huge role in determining their properties; this is because the size, mono-dispersity, and synthesis conditions are significant in defining the characteristics of nanomaterials and thus determining the role of nanomaterials. In the current chapter we have described the synthesis of zero-dimensional carbon nanomaterials using chemical routes in addition to the more benign and eco-friendly biological route.

## 2.1 Introduction

The synthesis of nanomaterials is one of the key aspects of nanotechnology as it is the first step in translating theoretical knowledge into practical applications. As the unique properties of nanomaterials are ascribed to their size, the quest to utilize nanomaterials in a plethora of fields begins with obtaining nanoparticles of desired shape, size, morphology, crystal structure, and chemical composition. The synthesis of nanoparticles aims at the optimization of certain parameters, namely particle size, particle shape, crystalline or amorphous morphology, and size distribution

Nanoparticle synthesis can be achieved through two broad pathways, namely the top-down approach and bottom-up approach. The former mainly consists of simplifying bulk material into smaller nanoparticles. It is brought about by consecutive removal of crystal planes of bulk materials whereas the latter builds up nanoparticles from atoms to molecules to crystals (figure 2.1) [1–3].

Both of the aforementioned synthesis approaches play an important role in material fabrication. Although the top-down method is accompanied by non-uniform size and inherent crystal defects, such approaches still remain the preferred choice for bulk

doi:10.1088/978-0-7503-4048-9ch2

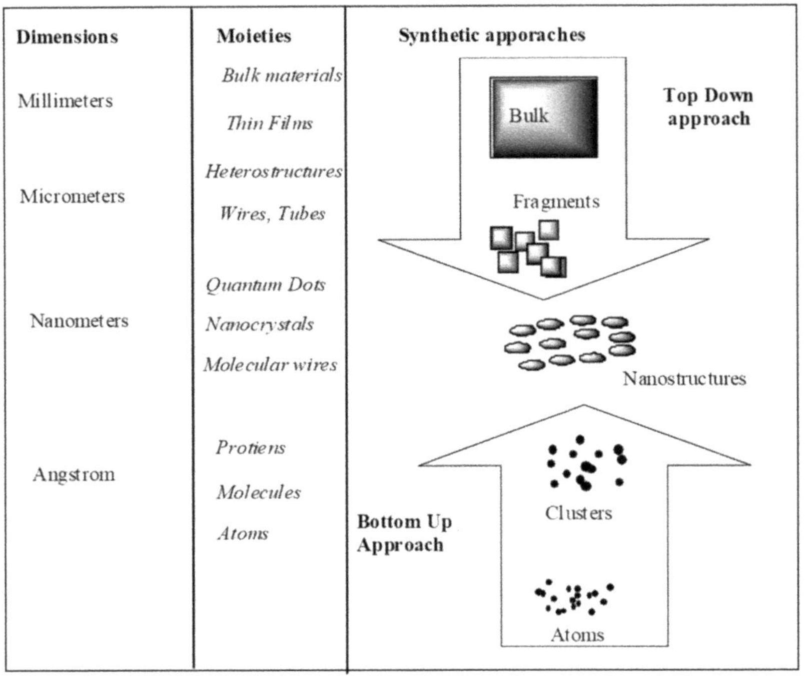


**Figure 2.1.** Synthesis approaches for nanomaterial synthesis.

production of materials. The bottom-up approach gives better uniformity and monodispersed nanoparticles and is the preferred approach for device fabrication. It is pertinent to note that the size and hence the properties of nanomaterials depend upon the synthesis route selected, therefore the desired application is a major factor in selecting the fabrication method. Synthesis can be achieved by four methods, namely [1–5] physical, chemical, mechanical, and biological synthesis (figure 2.2).

Physical methods include physical vapour deposition, laser vapourization, laser pyrolysis, ion beam techniques, nanolithography, microwave assisted synthesis, and plasma synthesis. Chemical methods comprise chemical vapour deposition, epitaxial growth, colloidal dispersion, the sol–gel method, the hydrothermal route, micro-emulsions, and the polymer route. Mechanical methods include mechanical alloying, high energy ball milling, mechanochemical synthesis, mechanochemical activation synthesis, and severe plastic deformation [1, 6]. It is important to note that while physical and chemical methods are bottom-up approaches, mechanical methods are top-down approaches.

Biological methods are a relatively new entry in the arena of nanomaterial synthesis. Biological methods are environmentally benign as well as economical. The reaction conditions are milder compared to other methods as the synthesis is performed at physiological pH, temperature, and pressure. Apart from the removal of extreme reaction conditions and the complex paraphernalia of reaction step-up

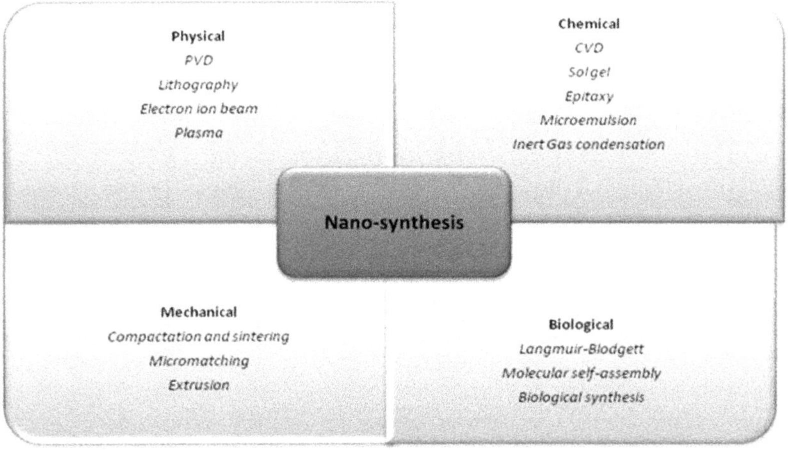


**Figure 2.2.** Different synthesis approaches for nanomaterial synthesis.

used in physical and chemical synthesis, a major advantage of biological methods is that the nanoparticles obtained have good polydispersity, dimensionality, and stability [7]. In the current chapter we will be dealing exclusively with the chemical and biological synthesis of zero-dimensional carbon nanomaterials. Although the chapter will elaborate on the various chemical and biological methods used, we will centre our discussion on carbon quantum dots and fullerenes.

## 2.2 Chemical routes of synthesis

Chemical synthesis is also called wet synthesis. It includes mainly chemical vapour deposition, epitaxial growth, colloidal dispersion, the sol–gel method, the hydrothermal route, microemulsions, and the polymer route. The common principle behind chemical synthesis is the mixing of quantified solutions under controlled conditions of temperature, pressure, and pH, which results in the formation of an insoluble precipitate which can be separated from the reaction mixture. Particle shape and size can be controlled easily using these methods [1].

### 2.2.1 Chemical vapour deposition (CVD)

CVD is a synthesis route where a solid is deposited on a heated surface by a chemical reaction from the vapour phase. The activation energy required for the reaction to proceed can be supplied by various sources, as depicted in figure 2.3.

Typically, the CVD set up is comprised of a gas supply system, a reaction chamber where deposition takes place, and finally an exhaust system. The precursor gases flow into the reaction chamber which in turn is loaded with a substrate on which the product is deposited [1, 8]. The system is represented schematically in figure 2.4. The advantages and disadvantages of this popular method are listed in table 2.1.

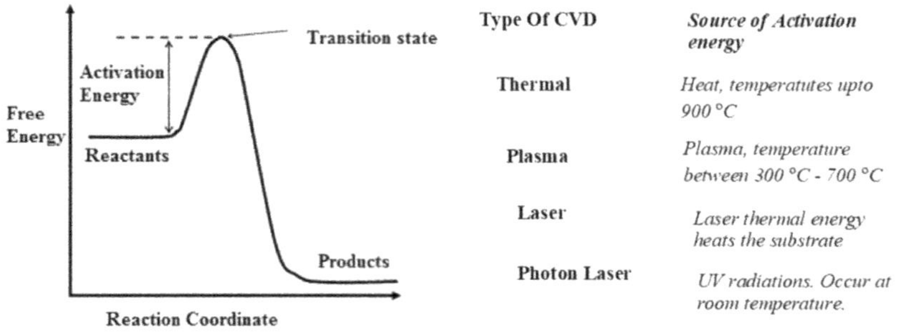


**Figure 2.3.** Types of CVD based on activation energy.

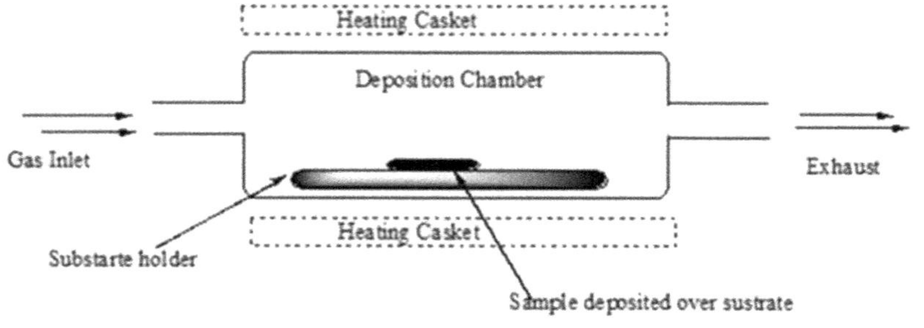


**Figure 2.4.** Schematic representation of CVD.

**Table 2.1.** Advantages and disadvantages of the CVD technique.

| Advantages | Disadvantages |
|---|---|
| The films produced are quite symmetrical. The conformal nature of films is a significant aspect as properties at the nano-level are highly sensitive to shape and size. | Material safety is a major concern as the deposited material could be corrosive, toxic, or pyrophoric. |
| The deposited materials are of high purity. | |
| The deposition rate is fast. | |

## 2.2.2 Laser ablation

The laser ablation method is a top-down approach for nanoparticle synthesis. This uses a laser as an energy source for removal or ablation of surface atoms of a solid target material. The process involves concentrating a high energy laser beam at a certain point on the target and evaporating the surface atoms. The collision between evaporated species leads to the formation of excited electronic states which in turn forms a plasma plume containing the nanoparticles. The advantages include high purity materials while the process falls short on control over particle size [9, 10]. The process is depicted in figure 2.5.

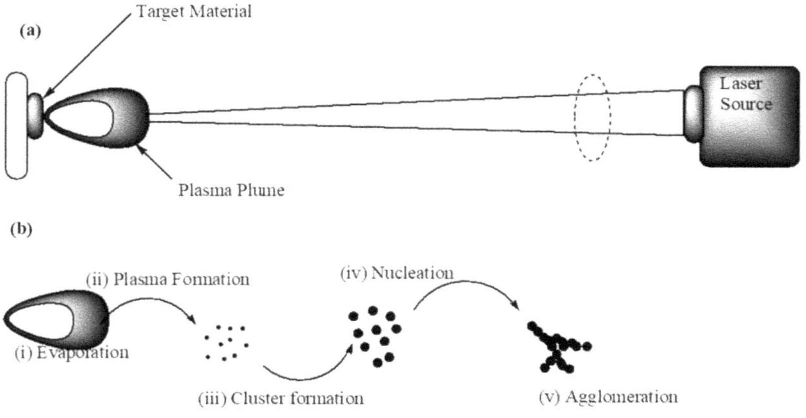


**Figure 2.5.** (a) Schematic assembly of a laser ablation process. (b) Formation of nanoparticles in the plasma plume.

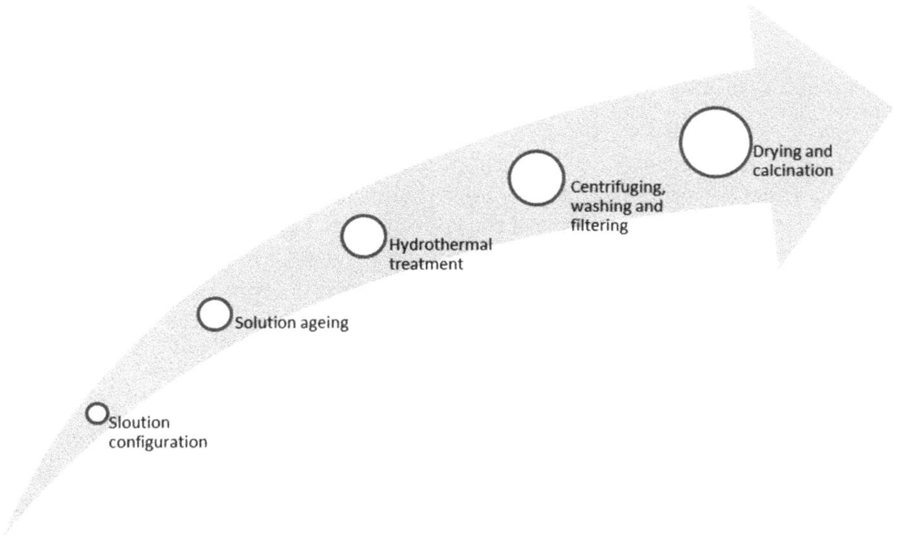


**Figure 2.6.** Steps leading to the formation of product in hydrothermal synthesis.

### 2.2.3 Hydrothermal/solvothermal route

The solvothermal process brings about a chemical reaction in sealed vessels in which the temperature can be brought close to its critical points though continuous heating with the help of an autoclave [11]. The process is called hydrothermal when water is used as the solvent and at times to describe the ambient conditions as well. This high temperature and high pressure process is helpful in the synthesis of those substances which are insoluble at ordinary temperature and pressure [1]. The reaction proceeds in a closed vessel creating a high temperature and pressure reaction by heating the reaction vessel and the vapour pressure generated is the source of high pressure. The various steps involved in the process are depicted in figures 2.6 [12, 13] and 2.7.

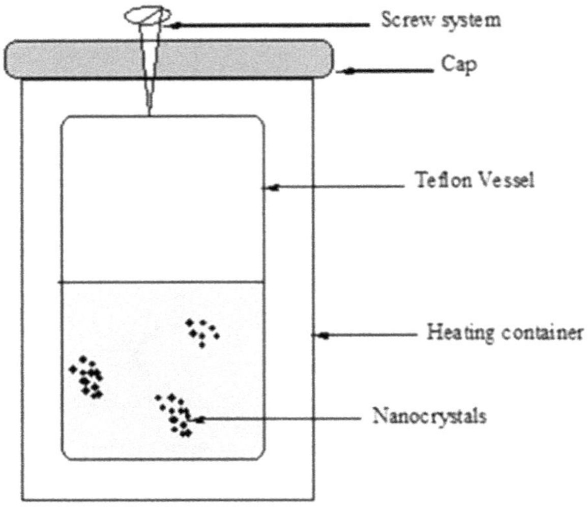


**Figure 2.7.** Hydrothermal system: basic set-up.

### 2.2.4 Sol–gel synthesis

Sol–gel is a wet chemical process which uses a chemical solution or colloidal particles, sol, to generate an integrated network, gel. The precursors undergo hydrolysis and polycondensation to form colloidal nanoparticles dispersed in the solvent. This sol over time gives rise to a continuous network which on calcinations give rise to the requisite nanoparticles [14]. The process can be used in multiple ways to generate powders, thin films, or dense ceramics. It is represented schematically in figure 2.8.

## 2.3 Biological synthesis of nanoparticles

Biological synthesis refers to the synthesis of nanoparticles using plant extracts and micro-organisms such as bacteria and fungi. This is at times referred to as 'phytonanotechnology' and has paved the way for an environmentally benign, less complicated, and more economical synthesis of nanoparticles. Among the numerous advantages of biological synthesis, its scalability, the use of an aqueous medium, and biocompatibility are ranked top-most [15–18].

The biosynthesis of nanoparticles utilizes different parts of plants, including the roots, shoots, leaves, and seeds of fruits. While the exact synthesis route employed in phytonanotechnology is yet to be deciphered, several research papers indicate that organic acids, proteins, vitamins, and secondary metabolites such as alkaloids, flavonoids, terpenoids, polysaccharides, and heterocyclic compounds lead to the synthesis of various types of nanoparticles.

In addition to plant products, micro-organisms also hold great potential to be used as a tool for nanotechnology. Common micro-organisms are yeast, fungi, and bacteria. Micro-organisms have many reductase enzymes as an inherent part of their biological structure. These enzymes lend micro-organisms their ability to accumulate

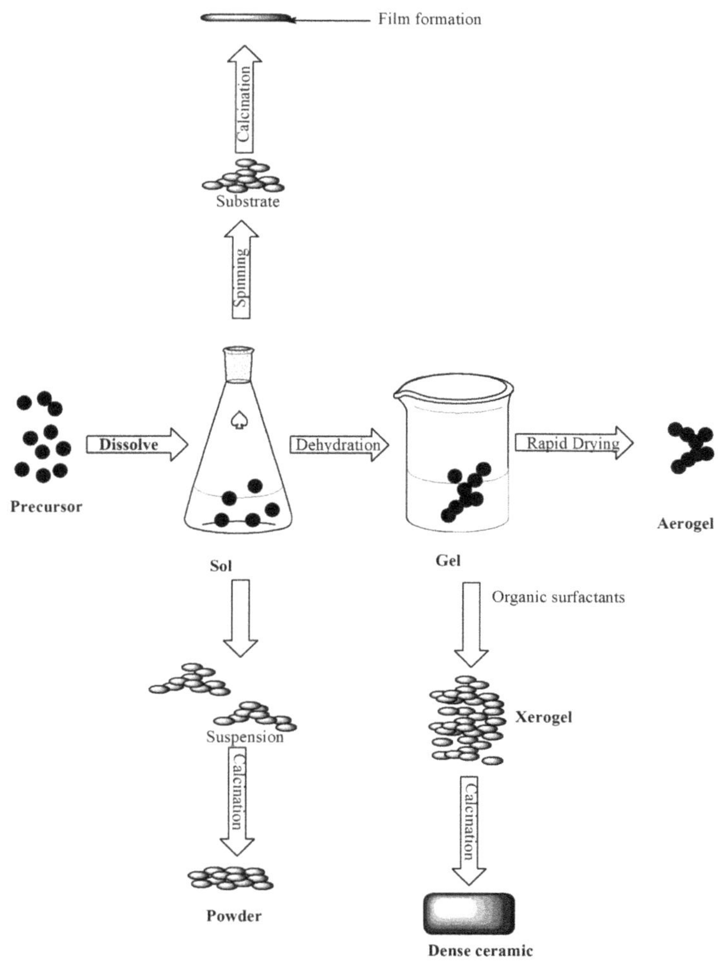


**Figure 2.8.** The sol–gel process and various product manifestations generated by it.

and detoxify heavy metals. They can also reduce metal salts into nanoparticles. The factors which promote the role of micro-organisms as a tool for biosynthesis are the inherent proteins, reducing cofactors, metal resistant genes, enzymes, and organic materials. The various advantages of green biosynthesis which make it suitable for pharmaceutical and biomedical applications are as follows: (i) energy efficiency; (ii) low production expenditure; (iii) less prone to accidents; (iv) safer, less toxic products; and (v) the use of an aqueous medium.

### 2.3.1 Microwave assisted biosynthesis

One of the disadvantages of plant mediated synthesis is the longer reaction times compared to chemical methods. One of the ways to shorten the reaction time is the use of microwave irradiation. Microwave assisted synthesis is a feasible way to perform plant based nanosynthesis as it is both rapid and facile. Other attractive features are the shorter reaction time, minimal energy consumption, and better product yield. Rapid and uniform heating by microwave irradiation provides suitable ambient reaction conditions leading to uniform nucleation and growth [19, 20].

### 2.3.2 The sonochemical approach for biosynthesis

Another green method used in phytonanoparticles is the use of the sonochemical approach [21–23]. The reactions which take place in otherwise ambient conditions are subjected to ultrasound effects arising from acoustic cavitation. Acoustic cavitation implies generation, development, and implosive collapse in a solvent. The diffusion of solvent vapour into the volume of the bubble leads to an increase in the size and the bubble collapses when it reaches its maximum size. 'Hot-spot' theory states that very high temperatures (greater than 5000 K) are obtained on the collapse of this bubble [24, 25]. These extreme conditions give rise to a large number of chemical reactions which eventually lead to the synthesis of nanoparticles. This method has similar advantages to microwave assisted biosynthesis, such as shorter reaction time and better yield. In the next segment we will describe how we employ the above methods to synthesize zero-dimensional nanoparticles.

## 2.4 Chemical synthesis of carbon dots

Carbon-based nanoparticles, such as carbon dots (CDs), are promising zero-dimensional nanoparticles with potential applications ranging from lasing to photovoltaics to biosensing and bioimaging. This is attributed to their strong luminescence accompanied by low toxicity and high chemical stability [26–30]. Laser ablation, arc discharge, electrochemical processes, and plasma assisted synthesis are the commonly used top-down methods for CD synthesis [31]. The bottom-up methods utilized are the pyrolytic method, template assisted synthesis, microwave assisted synthesis, chemical processes, the reverse micelle technique, and the sonochemical method [32–34]. In the following sections some commonly used synthesis methods for CDs will be discussed in detail.

### 2.4.1 Pulsed laser ablation

Simple operation and a shorter reaction time span are the two important aspects which makes pulsed laser ablation (PLA) a preferred method for the nanorange synthesis of CDs [35, 36]. Pulsed laser exposure creates localized spots of high pressure and temperature which in turn are directly able to cut a target into nanoparticles. This high energy and highly accelerated step results in nanoparticles without any residues and prevents any chance of introducing impurities or contamination. The major factors which determine the homogeneity, particle size, and ultimately the quantum yields of the resultant product are as follows [37]: (i) laser energy; (ii) laser pulse width; (iii) ablation time; and (iv) choice of solvents.

Different research groups have worked on the dependence of the properties of the synthesized CDs on the aforementioned parameters. Hu *et al* reported that the nucleation, growth, and size of CDs can be controlled by tuning the laser pulse width [38]. A solid carbon target as a carbon dot source has been utilized to synthesize CDs using the laser ablation method in an acetone solution with laser pulses of 1064, 532, and 355 nm under different irradiation times [39]. They have found that 355 nm laser pulses produced the most efficient CDs. The emission wavelengths of the CDs can be tuned by adjusting the ablation time or altering the laser wavelength. Non-focusing pulsed laser irradiation in toluene and a laser furnace has been applied to synthesize size-controlled CDs [40]. Another target source, such as graphite powders in polyethylene glycol has been used by Nguyen *et al* to synthesize carbon dots by femtosecond laser ablation and have reported smaller carbon dots are produced by increasing the laser irradiation time. One of the drawbacks of this method is the creation of complex functional groups on the surface of the CDs [41]. Recently, Cui *et al* fabricated homogeneous CDs with high quantum yield using a homemade dual-beam PLA system and low-cost carbon cloth, and they reported that the laser ablation time can be shortened by dividing a single laser beam into two laser beams [42].

### 2.4.2 Electrochemical oxidation

Another commonly employed method for carbon dot synthesis is electrochemical oxidation. The major advantages of this process include minimal cost, high purity, good yield, better control over size and morphology, and good reproducibility [43, 44]. Deng *et al* have reported that the applied potentials have a considerable influence on the size of the CDs obtained by electrochemical oxidation of alcohols [44]. Electrochemical oxidation of sodium citrate and urea lead to the synthesis of CDs of size ranging from 1.0–3.5 nm, having good photostability with a quantum yield of 11.4% [45]. Chang *et al* have prepared CDs using *o*-phenylenediamine (OPD) as the carbon source through an electrochemical method with a large Stokes shift [46]. These CDs with large Stokes shift have great potential in sensing systems. This is because they have low background noise, minimal self-quenching and high sensitivity. The electrolytic process leads to polymerization, carbonization, and passivation of OPD at the platinum electrode to form CDs. Hou *et al* not only prepared CDs utilizing an electrochemical method but also used it to synthesize a carbon dot/titanium dioxide ($CD/TiO_2$) nanocomposite. Methylene blue (MB) was degraded using these nanocomposites in the presence of visible light [47]. Recently, copper oxide ($Cu_2O$) was deposited on N-doped carbon quantum dots (N-CQD) using electrochemical methods [48]. Sun *et al* synthesized Fe-N-CQDs by electrolysis of a carbon cloth electrode, which was coated with monoatomic iron-anchored nitrogen-doped carbon (Fe-N-C). The obtained Fe-N-CQDs emitted blue fluorescence and possessed a quantum yield (QY) of 7.5% [49].

### 2.4.3 Sonochemical method

The sonochemical method is a green method which sonicates carbon precursors with known concentration in a suitable solvent for a stipulated time period to produce CDs [50].

Oligomer polyamide resin as a reactant has been used by Chen and co-workers to fabricate CDs through sonication treatment [51]. The resultant CDs are well dispersed, and have low crystallinity and surface rich functional groups. Wang and co-workers sonicated ascorbic acid and ammonia to synthesize fluorescent N-doped CDs via a single step reaction [50]. The synthesis of nitrogen-doped CDs was attempted by Lu *et al* through ultrasonic treatment of dopamine in dimethylformamide (DMF). The research further elaborated that the synthesized CDs possessed bright, temperature dependent fluorescence, and high colloidal and optical stability [52]. Sonochemical treatment was used by Ashritha *et al* to synthesize carbon quantum dots anchored on reduced graphene oxide sheets employing glucose as a precursor. In the course of sonochemical synthesis, ultrasonic radiation created organic radicals and scavenged oxygen functional groups from graphene oxide, forming reduced graphene oxide [53]. Recently, Kumar *et al* formulated doped CD synthesis with metals such as gold, silver, gallium, tin, and lead, through a one-pot sonochemical synthesis process using ultrasound radiation without a catalyst and stabilizer [54, 55].

### 2.4.4 The hydrothermal synthesis method

The synthesis of CDs from organic acids, saccharides, juice, etc, can proceed easily through hydrothermal carbonization. The major advantages of this method are its sustainability and economic viability [56–59]. A single-pot hydrothermal reaction of citric acid with polyethyleneimine yielded blue luminescent CDs with high photoluminescence QY. The resultant CDs possessed great potential as a candidate for optoelectronic devices as they displayed bright photoluminescence, narrow emission spectra, high photostability, and up-converted luminescence [60]. Liu and co-workers synthesized CDs through a 40 min hydrothermal treatment of lignin in the presence of hydrogen peroxide at 180 °C. The hydrothermal treatment was followed by filtration and dialysis with a 3500 Da dialysis membrane for 2 days. The synthesized CDs ranged between 2 and 10 nm [61]. Naik *et al* synthesized nitrogen-doped carbon dots by hydrothermal synthesis for potential biosensing of dopamine (DA) [62]. Water-soluble highly green fluorescent carbon quantum dots were prepared through a simple one step hydrothermal carbonization of waste fish scales by Dhandapani *et al* [63]. Liu and their team also synthesized boron and nitrogen co-doped carbon dots by hydrothermal treatment of branched polyethylenimine and 4-formylphenylboronic acid [64]. Ye *et al* used *p*-phenylenediamine as the carbon source for the preparation of carbon dots via a hydrothermal route. $-NH_2$ functional groups present on CDs exhibit considerable antibacterial activity [65]. Recently Shen *et al* synthesized N-doped CDs by hydrothermal treatment of glucose and m-phenylenediamine without any passivation and reported that graphene-like structured CDs were formed with good photostability, low toxicity, and good water solubility [66]. Cabbage as a carbon source has been used by Bhardwaj *et al* to prepare CDs via a hydrothermal method, carrying oxygen rich –OH and –COOH groups on their surface [67]. A group of scientists synthesized CDs using hydrothermal and microwave assisted methods of citric acid and compares both methods. They found that the hydrothermal method is more significant in the synthesis of CDs with good quantum yield in comparison to microwave assisted synthesis, although the duration of synthesis can be minimized in the case of microwave assisted synthesis [68].

### 2.4.5 The microwave method

The microwave method synthesizes monodispersed carbon dots by legating bonds, using electromagnetic irradiation [69]. Wang *et al* used glycerol with phosphate solution with no surface passivating agents and gave a 14 min microwave irradiation (750 W) to obtain multicolour, photoluminescent CDs [69]. Roshni *et al* obtained N-doped CDs from sesame seeds as a carbon source using microwaves at 800 W [70]. Microwave assisted pyrolysis of citric acid and ethylenediamine generated magneto-fluorescent CDs with an approximate size of 2 nm. The presence of $FeCl_2$ facilitated the reaction [71]. Recently, many natural carbon sources such as guanine, Mexican mint (*Plectranthus amboinicus*) leaves, cotton lint, and *Citrus japonica* have been used by researchers to synthesize CDs via a microwave assisted reflux method [72–75].

### 2.4.6 Thermal decomposition

In the past semiconductor and magnetic nanomaterials have been synthesized widely using thermal decomposition. It is documented that the exposure of organic materials with external heat leads to carbonization and eventual conversion to CDs takes place [76, 77]. Wang and co-workers utilized the thermal decomposition of L-cysteine and 1-butyl 3-methyl imidazolium bromide to generate CDs. AFM images revealed that the thickness of the CD is in the range up to a few layers of graphene [78, 79].

### 2.4.7 The synthesis of carbon dots using molecular fluorophores

A bottom-up chemical synthesis approach was used widely to synthesize CDs via a two-step method, the first step is polymerization and the second step is carbonization of the molecular precursors [80]. Citric acid on its own and in combination with a nitrogen precursor, such as ethylenediamine, urea, or L-cysteine, etc, has been used for convenient synthesis of CDs [81, 82]. It is reported that organic molecules akin to citric acid react with primary amines to form luminescent fluorophoric materials [83–85]. Reckmeier *et al* reported the use of citric acid as a precursor for a molecular fluorophore, which in turn was carbonized to form CDs [86] using ammono thermal synthesis. Citrazinic acid derivative based molecular fluorophores exhibited emission features similar to CDs. These fluorophores absorbed and emitted short wavelengths due to intrinsic core states and long wavelengths of the spectrum attributed to surface states [86]. Gang *et al* reported the synthesis of blue or green fluorescent CDs with the use of acetylacetone, cyclohexanone, 2-hydroxy-2-methylpropiophenone, glutaric dialdehyde, and salicylaldehyde compounds as carbon sources in water or tetrahydrofuran under normal pressure and temperature, 50–95 °C. They found in their study that green photoluminescent CDs can be synthesized using acetylacetone and it exhibited photoluminescence at 520 nm with maximum quantum yield in the temperature range 150–360 K, and these acetylacetone-derived CDs could be utilized in temperature sensing [80].

Recently, some new approaches have been used to synthesize CDs, such as machine learning and the combination of different chemical methods to obtain better results in the field of synthesis processes. The synthesis of highly fluorescent CDs was reported by Han *et al* to monitor $Fe^{3+}$ ions using machine learning assisted synthesis. Machine learning assisted synthesis of CDs was determined by two major factors, namely the mass of precursors and the volume of alkaline catalysts [87]. Sonication followed by a hydrothermal method was utilized to prepare CD and $MoS_2$ nanocomposites [88]. Large-scale synthesis of functionalized CDs with nitrogen (NCDs) and sulphur/nitrogen have both been synthesized using an aldol condensation method under suitable temperature and pressure [89].

### 2.4.8 The biological synthesis of carbon dots

The biocompatibility of CDs from biological sources makes them an interesting candidate for biomedical and bioimaging applications [90–93]. The search for new natural carbon sources is the requirement of today's world, and many scientists have utilized grass, green tea, coffee grounds, plant leaves, algal blooms, garlic, water chestnut, and onion, etc, as carbon sources for the synthesis of CDs [94–98]. Microwave assisted synthesis of photoluminescent CDs using algal blooms as the natural carbon source has been reported. The CDs thus produced exhibited high water solubility, photostability, and luminescence stability in different environments accompanied by the fact that they had low cytotoxicity and excellent cell permeability [99]. Other biological sources, such as oyster mushrooms (*Pleurotus* species) have been utilized for the synthesis of fluorescent blue or green CDs via a hydrothermal carbonization method by Boobalan *et al* and they have demonstrated that these CDs can work not only as colorimetric sensors for heavy metal detection (such as $Pb^{2+}$ ions) but also as fluorescent probes for DNA recognition and they have efficient antibacterial activity. In addition to all these applications, CDs also have anticancer activity against particular breast cancer cells, such as MDA-MB-231 [100]. Gynostemma is a natural biomass used as a precursor for the synthesis of CDs with antioxidant properties *in vitro* and *in vivo* medium using green methods and it was demonstrated by the expression of the mRNA of related antioxidant genes in zebrafish [101]. Highly water-soluble CD was synthesized using curcumin by Pal *et al* in a single step hydrothermal reaction. The synthesized spherical CDs were 4–5 nm in size and exhibited stability in a high colloidal state [102]. Another group of scientists reported that the citrus lemon can also be used as a carbon source to synthesize, via a hydrothermal process, fluorescent nitrogen-doped carbon quantum dots (NCQDs) which are water-soluble, ionically stable, and resistant to photobleaching. High sensitivity, high precision, and good selectivity were observed for $Hg^{2+}$ ions [103]. Fluorescent nitrogen and sulphur co-doped carbon dots (CDs) were synthesized by Zhao *et al* for cellular multicolour imaging with radical scavenging activity using garlic as the carbon source via a hydrothermal method and they reported that the fluorescence of the synthesized CDs is not affected by the presence of metal ions, biomolecules, and high ionic strength environments [104]. *Aloe barbadensis* (*Aloe vera*) extract was employed for the preparation of fluorescent biocompatible CDs. It was reported that the resulting CDs contributed to the blue emission due the presence of –OH groups and these were responsible for the uniform dispersion of the CDs in an aqueous solution. The synthesized CDs worked as photocatalysts for environmental remediation and this was demonstrated by degradation of anionic dye such as eosin yellow; the most salient feature was that it can decompose under visible light irradiation [105].

## 2.5 Chemical synthesis of fullerene

Fullerenes have found excellent applications as building blocks for supramolecular assemblies. The efficacy of these assemblies is greatly increased by a systematic and regular organization of the constituting fullerene molecules in the assembled structures [106]. Metal-porphyrins and carbon nanotube (CNT) nanocomposites with fullerene have various applications in the area of energy conversion, energy storage, electronics, photoelectronics, and biology [107–110]. Islamova *et al* reported new derivatives of fullerenes C60 and C70 containing triazole and deoxythymidine fragments obtained from acetylene methanofullerene and azido thymidine derivatives [111]. Kinzyabaeva *et al* used a sonochemical reaction of C60 with $\alpha$-diols into dichlorobenzene in the presence of solid NaOH to form a fullerene–dioxane adduct [112]. Not only fullerene but also fullerene soot particles with nitroaryl diazonium ions in acetonitrile were prepared using reactions of 2-nitroaniline, 4-nitroaniline, and 2,4-dinitroaniline with isoamyl nitrite by Manafi and co-workers, and the differential thermal analysis (DTA) of functionalized fullerene soot particles by nitroaryl diazonium ions showed the energy releasing property of these compounds [113]. Elessawy *et al* prepared functionalized magnetic fullerene nanocomposites by employing catalytic thermal decomposition of sustainable poly(ethylene terephthalate) (PET) bottle waste. In this reaction ferrocene was used as a catalyst and precursor of magnetite nanoparticles [114]. The synthesis of new hybrid methanofullerenes was reported by Tuktarov *et al* and the cytotoxic activities were studied with respect to different tumour cell lines [115].

### 2.5.1 The biological synthesis of fullerene

The utilization of natural and biological compounds in the synthesis of fullerene have tremendous advantages over chemicals as they provide a high potential for the application of fullerenes in medicine and biology [116]. Fullerenes can form bonds with various radicals, including biologically active ones, due to the presence of free bonds [117]. Fullerene can penetrate cell membranes, modulate ion transport, cross the blood–brain barrier, and has an adhesive and membrane-like potential [118], and these properties can be used for the synthesis of new immunomodulators, adjuvants, and vaccine holders, for example HIV-antigens. For use in biomedical applications, many scientists have synthesized water-soluble fullerene and its derivatives via modifications through carboxylated and polyhydroxylated adducts [119].

In addition, amino acids, peptides, and proteins were also utilized to prepare water-soluble fullerene adducts [120, 121]. Different amino acids such as glycine, paminobenzoic acid, $\omega$-aminocaproic acid, aminobutyric acid, L-proline, L-alanine, methyl-L-alaninate, DL-serine, D-arginine, L-arginine, $\beta$-alanine, valine, cystine, phenylalanine, folacin, carnosine, and glutathione have been used for the synthesis of C60 derivatives [121]. Rasnetsov *et al* present a way of obtaining a new compound group, fullerene-polycarbolic anions, with the general chemical formula $C_{60}H_n$[NH $(CH_2)$ $m$C(O)O–]$_n$, where $m = 1–5$, $n = 2–12$. The synthesized derivatives were characterized by $^1$H and $^{13}$C NMR spectroscopy. They described crystalline hydrates of C60 derivatives with amino-carbonic acids with the general chemical formula and these derivatives were investigated by IR-spectroscopy, and elemental and thermogravimetric analysis [122, 123]. Water-soluble fullerene C60 derivatives with phenylalanine (III), serine (IV), $\beta$-alanine (V), and $\gamma$-phenylbutiryc acid (VI) in the form of salts were synthesized by Hsieh *et al.* They reported that the obtained adducts had five high polar solubilizing groups on one hemisphere of the fullerene cage and the second hemisphere remained non-functionalized and highly hydrophobic, also making them amphiphilic [124]. The Prato reaction in the synthesis of fullerene derivatives with amino acids containing four to five methylene bridges (VIII) has been investigated by many scientists [125]. In this context, Maggini *et al* used 1,3-dipolar cycloaddition of azomethine ylides to C60 (the Prato reaction) to obtained fullerene derivative [126]. C60 derivatives with aspartic and glutamic acids were also synthesized by Zhang *et al* and they also carried out a condensation of N-substituted fulleropyrrolidine with L-aspartic and L-glutamic acids, containing protected $\alpha$-amino- and $\alpha$-carboxyl groups and their characterization was performed using MALDI-TOF, UV–Vis, FT-IR, and $^1$H NMR methods. Fullerene derivatives with peptides were formed in two main ways. In the first method fullerene was functionalized with pre-formed peptide and in the second the peptide was synthesized while the C60-fullerene fragment was incorporated as a synthetic amino acid into the peptide sequence. Romanova *et al* followed the first method to functionalize various dipeptide (L-Ala-L-Ala, D, L-Ala-D, L-Ala, and Gly-L-Val) chains by fullerene C60 via a nucleophilic addition reaction. The structures were identified by 1H NMR and 13C NMR spectroscopy [127]. Synthesis of a methyl glycylglycine-C60-derivative was reported by Wang *et al* via a reaction between an aqueous solution of methyl glycylglycinate and fullerene C60 in alkali medium [128].

## 2.6 Chemical synthesis of nanodiamonds

Nanodiamonds belong to yet another class of zero-dimensional naonomaterials which are biocompatible and chemically robust. They have found potential uses as a fluorescent biomarkers and also as nanosensors. Nanodiamonds have a large network of $sp^3$ bonded carbon lattice, and in this large network some colour centres are formed owing to certain atomic defects. These colour centres absorb in the visible and near-infrared region and also exhibit high photostability [129]. Table 2.2 depicts the dependence of certain physical properties of nanodiamonds on their structure [130].

**Table 2.2.** Structure–property dependence of nanodiamonds.

| S. No. | Physical property | Structural aspect |
|---|---|---|
| 1. | Emission spectrum | Type of colour centre |
| 2 | Brightness of fluorescent nanodiamonds | Number of colour centres per particle |
| 3. | Stability of colour centre | Size and surface morphology of nanodiamonds |

Nanodiamonds can be synthesized using classical techniques such as detonation [131], laser ablation [132], and chemical vapour deposition (CVD) [133]. The major limitation of these methods is that the pressure and temperature cannot be controlled, which in turn affect the control over the properties of the resulting nanoparticles. A recent advance in this aspect is the development of high-pressure-high-temperature (HPHT), wherein a controlled temperature and pressure system can be established. Studies performed under HPHT technique/conditions also revealed that the precursor used in the synthesis have a substantial impact on the final product in terms of size and structure [134].

Researchers have reported that a variety of precursors have been used in the synthesis of nanodiamonds. They include graphite, amorphous carbon, and glassy carbon. It has been observed that extreme conditions are required in the direct conversion of graphite to nanodiamonds (pressure exceeding 10 GPa and temperatures up to 2000 °C–3000 °C), while in the case of amorphous carbon and glassy carbon the conversion to diamond takes place at lower temperatures (1600 °C–1800 °C) [135–137]. The high temperature and high pressure can be reduced to some extent by applying different catalytic media, reported to be from 2000 °C–3000 °C to 1300 °C–1500 °C and 10 GPa to 5–7 GPa, respectively [138, 139]. Recently, diamondoid type hydrocarbons have been used as precursor molecules. The ease of synthesis is attributed to the $sp^3$ hybridized carbon atoms in their structure [140]. Alkahtani *et al* and Spohn *et al* have observed that the extreme conditions (such as temperatures above 1000 °C) employed in the HPHT technique can potentially decompose any organic precursor thereby cutting short any possibilities of engineered nanodiamonds. In order to overcome this they came up with 'moderate HPTP' synthesis where a better control over the synthesis of nanodiamonds can be obtained [141, 142]. In this context Liang and co-workers used germanium iodide ($GeI_4$) for the preparation of nanodiamonds from diamondoid molecules and alkane hydrocarbon under m-HPHT conditions. The conditions of 3.5 GPa pressure and 500 °C temperature are considerably lower than the conditions reported for common HPHT methods for nanodiamond synthesis [143, 144]. Graphene oxide or nitrated polyaromatic hydrocarbons were used as a precursor to synthesize nanodiamonds with average sizes of 2–3 nm using a hydrothermal route at the considerably low temperature of 423 K [145]. Ekimov *et al* synthesized nanodiamonds from hydrocarbons under pressure. The reaction paved way for bulk synthesis of nanodiamonds using germanium iodide or iodine as the mediator and iodinated adamantane, double mixtures of adamantane and naphthalene as substrate. The reaction was further probed for the effect of iodine as a catalyst and was conducted at pressures of 3.5 and 8 GPa [146]. Highly boron-doped nanodiamond was synthesized by the same group at a high static pressure of 8–9 GPa using 9-Borabicyclo[3.3.1]nonane dimer, C16H30B2, as a precursor [147, 148]. The synthesis of nanodiamonds from adamantane at 9.4 GPa was studied in the temperature range of 1500–1600 K [140]. The study reported that at temperatures above 1600 K diamond microcrystals are obtained (formed by the recrystallization of nanodiamonds and graphite), while below 1500 K primarily graphite is obtained. The use of fluorographite and naphthalene as a precursor to produce nanodiamonds resulted in the synthesis of microcrystalline diamonds, however, the characterization of the resultant material revealed the formation of a polycrystalline sample without any adequate control over the crystal size [149]. Again, Ekimov *et al* reported the bulk formation of ultra-small diamonds from brominated and chlorinated adamantanes at a pressure of 8 GPa and demonstrated that brominated adamantane and the temperature of the synthesis can control the size of nanodiamonds [150]. In yet another approach, graphene dispersions were irradiated to synthesize nanodiamonds by a nanosecond Nd YAG laser [151].

### 2.6.1 The biological synthesis of nanodiamonds

Very few studies have been performed on the synthesis of nanodiamond using biological compounds. This is attributed to the fact that the extreme conditions of temperature and pressure, and the expensive and hazardous reaction precursors used in these reactions are too extreme for biological samples [152]. In this context Lin *et al* reported that nanodiamonds have been synthesized from composite material (in the form of paper) synthesized from nanolignin/cellulose nanofibrils. They applied laser writing at a favourable temperature and air pressure and reported that high laser power was the determining factor for the generation of nanodiamonds and graphene nanoribbons. They also obtained carbon nano-onions at a lower laser power as biproducts [153].

## 2.7 Conclusion and perspective

Zero-dimensional carbon nanomaterials have emerged as a rapidly growing and expanding area of research, with a plethora of potential applications. It is important to understand that the selection of the synthesis technique is vital in the development of any material. High purity synthesis is one the most important aspects of materials science, which to a certain extent can be considered as an ever-improving field. Apart from the concerns regarding purity, another vital aspect which cannot be overlooked is the benignity of the synthesis process. The green chemistry approach is steadily taking its rightful place in materials synthesis. The solvent free routes, such as microwave assisted synthesis and ultrasonicated processes, have in fact cemented themselves among the popular synthetic tools. Another emerging field that until recently seemed difficult is the biological synthesis of nanoparticles. The biological synthesis approach is not only environmentally benign but it also has the potential to replicate the natural synthesis process, which is undoubtedly the most efficient and economical! We can definitely say that the synthesis of high purity nanomaterials is one of the most vital aspects of their industrial application and, also, that there is a lot of scope for research in the biological synthesis of these materials.

## Acknowledgments

Dr Tanvi Vats would like to thank Gautam Buddha University, India, for providing a very conducive environment to carry out the research.

## References

[1] Tulinski M and Jurczyk M J M 2017 Nanomaterials synthesis methods *Metrology and Standardization of Nanotechnology: Protocols and Industrial Innovations* (London: Wiley) pp 75–98

[2] Bleeker R A, Troilo L M and Ciminello D P 2004 Patenting nanotechnology *Materials Today* **7** 44–8

[3] Zweck A, Bachmann G, Luther W and Ploetz C 2008 Nanotechnology in Germany: from forecasting to technological assessment to sustainability studies *J. Cleaner Prod.* **16** 977–87

[4] Wang Z L, Liu Y and Zhang Z 2003 *Handbook of Nanophase and Nanostructured Materials Vol. 4: Materials Systems and Applications* (New York: Kluwer Academic/Plenum)

[5] Ingale A G and Chaudhari A N 2013 Biogenic synthesis of nanoparticles and potential applications: an eco-friendly approach *Nanomed. Nanotechnol.* **4** 1–7

[6] Singh S and Khare N 2015 CdS/ZnO core/shell nano-heterostructure coupled with reduced graphene oxide towards enhanced photocatalytic activity and photostability *Chem. Phys. Lett.* **634** 140–5

[7] Yeadon M, Yang J, Ghaly M, Olynick D, Averbach R and Gibson J J I 1997 *Nanophase and Nanocomposite Materials* (Pittsburgh, PA: MRS) pp 179–84

[8] Foord J S, Davies G and Tsang W 1997 *Chemical Beam Epitaxy and Related Techniques* (New York: Wiley)

[9] Kim M, Osone S, Kim T, Higashi H and Takafumi S 2017 Synthesis of nanoparticles by laser ablation: a review *KONA Powd. Part. J.* **34** 80–90

[10] Osseo-Asare K and Arriagada F J 1990 Synthesis of nanosize particles in reverse micro-emulsions Ceram *Trans.* **12** 3–16

[11] Benjamin J S 1976 Mechanical alloying *Sci. Am.* **234** 40–9

[12] Smardz K, Smardz L, Okonska I, Nowak M and Jurczyk M 2008 XPS valence band and segregation effect in nanocrystalline $Mg_2Ni$-type materials *Int. J. Hydrogen Energy* **33** 387–92

[13] Sadrolhosseini A R, Mahdi M A, Alizadeh F and Rashid S A 2019 Laser ablation technique for synthesis of metal nanoparticle in liquid *Laser Technology and its Applications* (Rijeka: Intech)

[14] West B 2009 Characterization of metal nanoparticle interactions with small molecules *MSc Thesis* Queen's University

[15] Ijaz I, Gilani E, Nazir A and Bukhari A 2020 Detail review on chemical, physical and green synthesis, classification, characterizations and applications of nanoparticles *Green Chem. Lett. Rev.* **13** 223–45

[16] Mukherjee P *et al* 2008 Green synthesis of highly stabilized nanocrystalline silver particles by a non-pathogenic and agriculturally important fungus *T. asperellum Nanotechnology* **19** 075103

[17] Prasad K, Jha A K and Kulkarni A R 2007 *Lactobacillus* assisted synthesis of titanium nanoparticles *Nanoscale Res. Lett.* **2** 248–50
[18] Singaravelu G, Arockiamary J, Kumar V G and Govindaraju K 2007 A novel extracellular synthesis of monodisperse gold nanoparticles using marine alga, *Sargassum wightii Greville Colloids Surf.* B **57** 97–101
[19] Huang J *et al* 2007 Biosynthesis of silver and gold nanoparticles by novel sundried *Cinnamomum camphora* leaf *Nanotechnology* **18** 105104
[20] Thirunavoukkarasu M, Balaji U, Behera S, Panda P and Mishra B K 2013 Biosynthesis of silver nanoparticle from leaf extract of *Desmodium gangeticum* (L.) DC. and its biomedical potential *Spectrochim. Acta* A **116** 424–7
[21] Masjedi-Arani M and Salavati-Niasari M 2016 A simple sonochemical approach for synthesis and characterization of $Zn_2SiO_4$ nanostructures *Ultrason. Sonochem.* **29** 226–35
[22] Koda S, Kimura T, Kondo T and Mitome H 2003 A standard method to calibrate sonochemical efficiency of an individual reaction system *Ultrason. Sonochem.* **10** 149–56
[23] Salavati-Niasari M, Javidi J and Davar F 2010 Sonochemical synthesis of $Dy_2(CO_3)_3$ nanoparticles, $Dy(OH)_3$ nanotubes and their conversion to $Dy_2O_3$ nanoparticles *Ultrason. Sonochem.* **17** 870–7
[24] Suslick K S, Choe S-B, Cichowlas A A and Grinstaff M W 1991 Sonochemical synthesis of amorphous iron *Nature* **353** 414–6
[25] Suslick K S and Price G J 1999 Applications of ultrasound to materials chemistry *Ann. Rev. Mater. Sci.* **29** 295–326
[26] Hola K, Zhang Y, Wang Y, Giannelis E P, Zboril R and Rogach A L 2014 Carbon dots—emerging light emitters for bioimaging, cancer therapy and optoelectronics *Nano Today* **9** 590–603
[27] Zhu S *et al* 2013 Highly photoluminescent carbon dots for multicolor patterning, sensors, and bioimaging *Angew. Chem.* **125** 4045–9
[28] Qu S, Wang X, Lu Q, Liu X and Wang L 2012 A biocompatible fluorescent ink based on water-soluble luminescent carbon nanodots *Angew. Chem.* **51** 12215–8
[29] Sau A, Bera K, Mondal P, Satpati B and Basu S 2016 Distance-dependent electron transfer in chemically engineered carbon dots *J. Phys. Chem.* C **120** 26630–6
[30] Sun C *et al* 2015 Combination of carbon dot and polymer dot phosphors for white light-emitting diodes *Nanoscale* **7** 12045–50
[31] Zhang X *et al* 2015 A facile and universal top-down method for preparation of monodisperse transition-metal dichalcogenide nanodots *Angew. Chem.* **54** 5425–8
[32] Kumar R, Singh A and Garg N 2019 Acoustic cavitation-assisted formulation of solid lipid nanoparticles using different stabilizers *ACS Omega* **4** 13360–70
[33] Liu R, Wu D, Feng X and Müllen K 2011 Bottom-up fabrication of photoluminescent graphene quantum dots with uniform morphology *J. Am. Chem. Soc.* **133** 15221–3
[34] Vickers N J 2017 Animal communication: when I'm calling you, will you answer too? *Curr. Biol.* **27** R713–15
[35] Kang S H *et al* 2016 Ultrafast method for selective design of graphene quantum dots with highly efficient blue emission *Sci. Rep.* **6** 38423
[36] Castro H P *et al* 2016 Synthesis and characterisation of fluorescent carbon nanodots produced in ionic liquids by laser ablation *Chem.—Eur. J.* **22** 138–43
[37] Namdari P, Negahdari B and Eatemadi A 2017 Synthesis, properties and biomedical applications of carbon-based quantum dots: an updated review *Biomed. Pharmacother.* **87** 209–22

[38] Hu S, Liu J, Yang J, Wang Y and Cao S 2011 Laser synthesis and size tailor of carbon quantum dots *J. Nanopart. Res.* **13** 7247–52

[39] Reyes D *et al* 2016 Laser ablated carbon nanodots for light emission *Nanoscale Res. Lett.* **11** 1–11

[40] Yu H, Li X, Zeng X and Lu Y 2015 Preparation of carbon dots by non-focusing pulsed laser irradiation in toluene *Chem. Commun.* **52** 819–22

[41] Nguyen V, Yan L, Si J and Hou X 2015 Femtosecond laser-induced size reduction of carbon nanodots in solution: effect of laser fluence, spot size, and irradiation time *J. Appl. Phys.* **117** 084304

[42] Cui L, Ren X, Wang J and Sun M 2020 Synthesis of homogeneous carbon quantum dots by ultrafast dual-beam pulsed laser ablation for bioimaging *Mater. Today Nano* **12** 100091

[43] Liu M, Xu Y, Niu F, Gooding J J and Liu J 2016 Carbon quantum dots directly generated from electrochemical oxidation of graphite electrodes in alkaline alcohols and the applications for specific ferric ion detection and cell imaging *Analyst* **141** 2657–64

[44] Deng J *et al* 2014 Electrochemical synthesis of carbon nanodots directly from alcohols *Chem.—Eur. J.* **20** 4993–9

[45] Hou Y, Lu Q, Deng J, Li H and Zhang Y 2015 One-pot electrochemical synthesis of functionalized fluorescent carbon dots and their selective sensing for mercury ion *Anal. Chim. Acta* **866** 69–74

[46] An Q *et al* 2021 Electrochemical synthesis of carbon dots with a Stokes shift of 309 nm for sensing of $Fe^{3+}$ and ascorbic acid *Dyes Pigments* **185** 108878

[47] Hou Y, Lu Q, Wang H, Li H, Zhang Y and Zhang S 2016 One-pot electrochemical synthesis of carbon dots/$TiO_2$ nanocomposites with excellent visible light photocatalytic activity *Mater. Lett.* **173** 13–7

[48] Muthusankar G *et al* 2018 Electrochemical synthesis of nitrogen-doped carbon quantum dots decorated copper oxide for the sensitive and selective detection of non-steroidal anti-inflammatory drug in berries *J. Colloid Interf. Sci.* **523** 191–200

[49] Sun S *et al* 2021 Electrochemical synthesis of $FeN_x$ doped carbon quantum dots for sensitive detection of $Cu^{2+}$ ion *Green Energy Env.* in press

[50] Wang F, Wang S, Sun Z and Zhu H 2015 Study on ultrasonic single-step synthesis and optical properties of nitrogen-doped carbon fluorescent quantum dots *Fullerenes Nanotubes Carbon Nanostruct.* **23** 769–76

[51] Dang H, Huang L-K, Zhang Y, Wang C-F and Chen S 2016 Large-scale ultrasonic fabrication of white fluorescent carbon dots *Ind. Eng. Chem. Res.* **55** 5335–41

[52] Lu M and Zhou L 2019 One-step sonochemical synthesis of versatile nitrogen-doped carbon quantum dots for sensitive detection of $Fe^{2+}$ ions and temperature *in vitro Mater. Sci. Eng.* C **101** 352–9

[53] Ashritha M *et al* 2021 Experimental and computational studies of sonochemical assisted anchoring of carbon quantum dots on reduced graphene oxide sheets towards the photocatalytic activity *Appl. Surf. Sci.* **545** 148962

[54] Kumar R, Kumar V B and Gedanken A 2020 Sonochemical synthesis of carbon dots, mechanism, effect of parameters, and catalytic, energy, biomedical and tissue engineering applications *Ultrason. Sonochem.* **64** 105009

[55] Kumar V B, Kumar R, Gedanken A and Shefi O 2019 Fluorescent metal-doped carbon dots for neuronal manipulations *Ultrason. Sonochem.* **52** 205–13

[56] Guo Y, Wang Z, Shao H and Jiang X 2013 Hydrothermal synthesis of highly fluorescent carbon nanoparticles from sodium citrate and their use for the detection of mercury ions *Carbon* **52** 583–9

[57] Yang Z-C *et al* 2011 Intrinsically fluorescent carbon dots with tunable emission derived from hydrothermal treatment of glucose in the presence of monopotassium phosphate *Chem. Commun.* **47** 11615–7

[58] Sahu S, Behera B, Maiti T K and Mohapatra S 2012 Simple one-step synthesis of highly luminescent carbon dots from orange juice: application as excellent bio-imaging agents *Chem. Commun.* **48** 8835–7

[59] Qu K, Wang J, Ren J and Qu X 2013 Carbon dots prepared by hydrothermal treatment of dopamine as an effective fluorescent sensing platform for the label-free detection of iron (III) ions and dopamine *Chem.—Eur. J.* **19** 7243–9

[60] Li J-Y *et al* 2017 One-pot hydrothermal synthesis of carbon dots with efficient up-and down-converted photoluminescence for the sensitive detection of morin in a dual-readout assay *Langmuir* **33** 1043–50

[61] Chen W, Hu C, Yang Y, Cui J and Liu Y 2016 Rapid synthesis of carbon dots by hydrothermal treatment of lignin *Materials* **9** 184

[62] Naik V *et al* 2019 Nitrogen-doped carbon dots via hydrothermal synthesis: naked eye fluorescent sensor for dopamine and used for multicolor cell imaging *Appl. Bio Mater.* **2** 2069–77

[63] Dhandapani E, Duraisamy N, Periasamy P J and Vijayakumar P 2021 Highly green fluorescent carbon quantum dots synthesis via hydrothermal method from fish scale *Mater. Today: Proc.* **26** A1–A5

[64] Liu J, Li J, Xu L, Qiao Y and Chen J 2017 Facile synthesis of N, B-doped carbon dots and their application for multisensor and cellular imaging *Ind. Eng. Chem. Res.* **56** 3905–12

[65] Ye Z, Li G, Lei J, Liu M, Jin Y and Li B 2020 One-step and one-precursor hydrothermal synthesis of carbon dots with superior antibacterial activity *Appl. Bio Mater.* **3** 7095–102

[66] Shen T-Y, Jia P-Y, Chen D-S and Wang L 2021 Hydrothermal synthesis of N-doped carbon quantum dots and their application in ion-detection and cell-imaging *Spectrochim. Acta* B **248** 119282

[67] Singh H, Bamrah A, Khatri M and Bhardwaj N 2020 One-pot hydrothermal synthesis and characterization of carbon quantum dots (CQDs) *Mater. Today: Proc.* **28** 1891–4

[68] Ng H M, Lim G and Leo C P 2021 Comparison between hydrothermal and microwave-assisted synthesis of carbon dots from biowaste and chemical for heavy metal detection: a review *Microchem. J.* **165** 106116

[69] Wang Q, Liu X, Zhang L and Lv Y 2012 Microwave-assisted synthesis of carbon nanodots through an eggshell membrane and their fluorescent application *Analyst* **137** 5392–7

[70] Roshni V and Ottoor D 2017 One-step microwave-assisted green synthesis of luminescent N-doped carbon dots from sesame seeds for selective sensing of Fe (III) *Curr. Sci.* **112** 385

[71] Liu R 2021 Facile synthesis of magneto-fluorescent carbon dots by one-step microwave-assisted pyrolysis *J. Alloys Compounds* **855** 157456

[72] Architha N *et al* 2021 Microwave-assisted green synthesis of fluorescent carbon quantum dots from Mexican mint extract for $Fe^{3+}$ detection and bio-imaging applications *Env. Res.* **199** 111263

[73] Li Q *et al* 2021 Rapid microwave-assisted green synthesis of guanine-derived carbon dots for highly selective detection of $Ag^{+}$ in aqueous solution *Spectrochim. Acta* A **248** 119208

[74] Eskalen H, Uruş S, Cömertpay S, Kurt A H and Özgan Ş 2020 Microwave-assisted ultra-fast synthesis of carbon quantum dots from linter: fluorescence cancer imaging and human cell growth inhibition properties *Industr. Crop Prod.* **147** 112209

[75] So R C, Sanggo J E, Jin L, Diaz J M A, Guerrero R A and He J 2017 Gram-scale synthesis and kinetic study of bright carbon dots from citric acid and *Citrus japonica* via a microwave-assisted method *ACS Omega* **2** 5196–208

[76] Wan J-Y, Yang Z, Liu Z-G and Wang H-X 2016 Ionic liquid-assisted thermal decomposition synthesis of carbon dots and graphene-like carbon sheets for optoelectronic application *RSC Adv.* **6** 61292–300

[77] Martindale B C, Hutton G A, Caputo C A and Reisner E 2015 Solar hydrogen production using carbon quantum dots and a molecular nickel catalyst *J. Am. Chem. Soc.* **137** 6018–25

[78] Wang F, Xie Z, Zhang H, Liu C and Zhang Y 2011 Highly luminescent organosilane-functionalized carbon dots *Adv. Funct. Mater.* **21** 1027–31

[79] Brachi P 2020 Synthesis of carbon dots (CDs) through the fluidized bed thermal treatment of residual biomass assisted by $\gamma$-alumina *Appl. Catal.* B **263** 118361

[80] Ye H-G *et al* 2021 Mild bottom-up synthesis of carbon dots with temperature-dependent fluorescence *J. Lumin.* **238** 118311

[81] Qu S *et al* 2016 Toward efficient orange emissive carbon nanodots through conjugated $sp^2$-domain controlling and surface charges engineering *Adv. Mater.* **28** 3516–21

[82] Bera K *et al* 2016 Metamorphosis of ruthenium-doped carbon dots: in search of the origin of photoluminescence and beyond *Chem. Mater.* **28** 7404–13

[83] Reckmeier C J, Wang Y, Zboril R and Rogach A L 2016 Influence of doping and temperature on solvatochromic shifts in optical spectra of carbon dots *J. Phys. Chem.* **120** 10591–604

[84] Schneider J *et al* 2017 Molecular fluorescence in citric acid-based carbon dots *J. Phys. Chem.* C **121** 2014–22

[85] Fang Q *et al* 2017 Luminescence origin of carbon based dots obtained from citric acid and amino group-containing molecules *Carbon* **118** 319–26

[86] Reckmeier C J *et al* 2017 Aggregated molecular fluorophores in the ammonothermal synthesis of carbon dots *Chem. Mater.* **29** 10352–61

[87] Han Y *et al* 2020 Machine-learning-driven synthesis of carbon dots with enhanced quantum yields *ACS Nano* **14** 14761–8

[88] Li N *et al* 2019 Facile synthesis of carbon dots@ 2D $MoS_2$ heterostructure with enhanced photocatalytic properties *Inorg. Chem.* **58** 5746–52

[89] Li L *et al* 2021 Kilogram-scale synthesis and functionalization of carbon dots for superior electrochemical potassium storage *ACS Nano* **15** 6872–85

[90] Sauvage F *et al* 2020 Nanomaterials to avoid and destroy protein aggregates *Nano Today* **31** 100837

[91] Ramanathan S *et al* 2018 Theranostic applications of nanoparticles in neurodegenerative disorders *Int. J. Nanomed.* **13** 5561

[92] Ashrafizadeh M *et al* 2020 Carbon dots as versatile nanoarchitectures for the treatment of neurological disorders and their theranostic applications: a review *Adv. Colloid Interf. Sci.* **278** 102123

[93] Łoczechin A *et al* 2019 Functional carbon quantum dots as medical countermeasures to human coronavirus *ACS Appl. Mater. Interf.* **11** 42964–74

[94] Liu S *et al* 2012 Hydrothermal treatment of grass: a low-cost, green route to nitrogen-doped, carbon-rich, photoluminescent polymer nanodots as an effective fluorescent sensing platform for label-free detection of Cu(II) ions *Adv. Mater.* **24** 2037–41

[95] Zhu L, Yin Y, Wang C-F and Chen S 2013 Plant leaf-derived fluorescent carbon dots for sensing, patterning and coding *J. Mater. Chem.* C **1** 4925–32

[96] Mehta V N, Jha S, Basu H, Singhal R K and Kailasa S K 2015 One-step hydrothermal approach to fabricate carbon dots from apple juice for imaging of mycobacterium and fungal cells *Sens. Actuators* B **213** 434–43

[97] Zhang Z, Sun W and Wu P 2015 Highly photoluminescent carbon dots derived from egg white: facile and green synthesis, photoluminescence properties, and multiple applications *ACS Sustain. Chem. Eng.* **3** 1412–8

[98] Algarra M *et al* 2019 Sustainable production of carbon nanoparticles from olive pit biomass: understanding proton transfer in the excited state on carbon dots *ACS Sustain. Chem. Eng.* **7** 10493–500

[99] Ramanan V *et al* 2016 Outright green synthesis of fluorescent carbon dots from eutrophic algal blooms for *in vitro* imaging *ACS Sustain. Chem. Eng.* **4** 4724–31

[100] Boobalan T *et al* 2020 Mushroom-derived carbon dots for toxic metal ion detection and as antibacterial and anticancer agents *Appl. Nano Mater.* **3** 5910–9

[101] Wei X *et al* 2019 Green synthesis of fluorescent carbon dots from gynostemma for bioimaging and antioxidant in zebrafish *Appl. Mater. Interf.* **11** 9832–40

[102] Pal T, Mohiyuddin S and Packirisamy G 2018 Facile and green synthesis of multicolor fluorescence carbon dots from curcumin: *in vitro* and *in vivo* bioimaging and other applications *ACS Omega* **3** 831–43

[103] Tadesse A, Hagos M, RamaDevi D, Basavaiah K and Belachew N 2020 Fluorescent-nitrogen-doped carbon quantum dots derived from citrus lemon juice: green synthesis, mercury (II) ion sensing, and live cell imaging *ACS Omega* **5** 3889–98

[104] Zhao S *et al* 2015 Green synthesis of bifunctional fluorescent carbon dots from garlic for cellular imaging and free radical scavenging *Appl. Mater. Interf.* **7** 17054–60

[105] Malavika J P *et al* 2021 A sustainable green synthesis of functionalized biocompatible carbon quantum dots from *Aloe barbadensis* miller and its multifunctional applications *Env. Res.* **200** 111414

[106] Benzigar M R *et al* 2018 Highly crystalline mesoporous C60 with ordered pores: a class of nanomaterials for energy applications *Angew. Chem.* **57** 569–73

[107] Tan Z *et al* 2017 Incorporating pyrrolic and pyridinic nitrogen into a porous carbon made from C60 molecules to obtain superior energy storage *Adv. Mater.* **29** 1603414

[108] Rincón-García L *et al* 2016 Molecular design and control of fullerene-based bi-thermo-electric materials *Nature Mater.* **15** 289–93

[109] Liu K *et al* 2019 Spatially confined growth of fullerene to super-long crystalline fibers in supramolecular gels for high-performance photodetector *Adv. Mater.* **31** 1808254

[110] Minami K *et al* 2015 Highly ordered 1D fullerene crystals for concurrent control of macroscopic cellular orientation and differentiation toward large-scale tissue engineering *Adv. Mater.* **27** 4020–6

[111] Islamova L N *et al* 2017 Synthesis of derivatives of fullerenes C60 and C70 containing pharmacophore groups *Mendeleev Commun.* **2** 204–6

[112] Kinzyabaeva Z S and Sharipov G L 2018 A selective synthesis of the fullerene-fused dioxane adduct via heterogeneous reaction of C60 with $\alpha$-diols and NaOH under ultrasonication *Ultrason. Sonochem.* **42** 119–23
[113] Moghadam M M, Zamani M and Pourmousavi S A 2021 Synthesis and characterization of new potential high-energy materials based on fullerene soot nanoparticles and nitroaryl diazonium ions *J. Phys. Chem. Solids* **154** 110101
[114] Elessawy N A, El-Sayed E M, Ali S, Elkady M F, Elnouby M and Hamad H A 2020 One-pot green synthesis of magnetic fullerene nanocomposite for adsorption characteristics *J. Water Process Eng.* **34** 101047
[115] Tuktarov A R *et al* 2020 Hybrid molecules based on fullerene C60 and 5Z, 9Z-dienoic acids: synthesis and cytotoxic activity *Bioorg. Med. Chem. Lett.* **30** 127289
[116] Rašović I 2017 Water-soluble fullerenes for medical applications *Mater. Sci. Technol.* **33** 777–94
[117] Goodarzi S, Da Ros T, Conde J, Sefat F and Mozafari M 2017 Fullerene: biomedical engineers get to revisit an old friend *Mater. Today* **20** 460–80
[118] Zhu X, Xiao S, Zhou D, Sollogoub M and Zhang Y 2018 Design, synthesis and biological evaluation of water-soluble per-O-methylated cyclodextrin-C60 conjugates as anti-influenza virus agents *Eur. J. Med. Chem.* **146** 194–205
[119] Castro E, Garcia A H, Zavala G and Echegoyen L 2017 Fullerenes in biology and medicine *J. Mater. Chem.* B **5** 6523–35
[120] Tabrez S *et al* 2020 Nanoencapsulated dietary polyphenols for cancer prevention and treatment: successes and challenges *Nanomedicine* **15** 1147–62
[121] Li Q, Liu C and Li H 2016 Induction of endogenous reactive oxygen species in mitochondria by fullerene-based photodynamic therapy *J. Nanosci. Nanotechnol.* **16** 5592–7
[122] Pochkaeva E I *et al* 2020 Fullerene derivatives with amino acids, peptides and proteins: from synthesis to biomedical application *Prog. Solid State Chem.* **57** 100255
[123] Rasnetsov L, Shvartsman I Y and Suvorova O 2012 Hydrated N-fullerene-amino acid derivatives, methods for preparing them and based pharmaceutical compositions. *Russian patent* RU2458046C1
[124] Hsieh F-Y *et al* 2017 Water-soluble fullerene derivatives as brain medicine: surface chemistry determines if they are neuroprotective and antitumor *Appl. Mater. Interf.* **9** 11482–92
[125] Watanabe L A, Bhuiyan M P, Jose B, Kato T and Nishino N 2004 Synthesis of novel fullerene amino acids and their multifullerene peptides *Tetrahedron Lett.* **45** 7137–40
[126] Tagmatarchis N and Prato M 2003 The addition of azomethine ylides to [60]fullerene leading to fulleropyrrolidines *Cheminform* **2003** 768–79
[127] Romanova V, Tsyryapkin V and Volpin M E 1994 Addition of amino acids and dipeptides to fullerene C60 giving rise to monoadducts *Russ. Chem. Bull.* **43** 1154–5
[128] Wang N, Li J, Zhu D and Chan T 1995 A C60-derivatized dipeptide *Tetrahedron Lett.* **36** 431–4
[129] Kumar S, Nehra M, Kedia D, Dilbaghi N, Tankeshwar K and Kim K-H 2019 Nanodiamonds: emerging face of future nanotechnology *Carbon* **143** 678–99
[130] Wu Y, Jelezko F, Plenio M B and Weil T 2016 Diamond quantum devices in biology *Angew. Chem. Int. Edn* **55** 6586–98
[131] Alkahtani M H *et al* 2018 Fluorescent nanodiamonds: past, present, and future *Nanophotonics* **7** 1423–53

[132] Guo J *et al* 2016 Modular assembly of superstructures from polyphenol-functionalized building blocks *Nature Nanotechnol.* **11** 1105–11

[133] Wang C, Liu P, Cui H and Yang G W 2005 Nucleation and growth kinetics of nanocrystals formed upon pulsed-laser ablation in liquid *Appl. Phys. Lett.* **87** 201913

[134] Nemanich R J, Carlisle J A, Hirata A and Haenen K 2014 CVD diamond—research, applications, and challenges *MRS Bull.* **39** 490–4

[135] Fang R H, Kroll A V, Gao W and Zhang L 2018 Cell membrane coating nanotechnology *Adv. Mater.* **30** 1706759

[136] Sumiya H 2012 Novel development of high-pressure synthetic diamonds ultra-hard nano-polycrystalline diamonds *Sei Tech. Rev.* **74** 15–23

[137] Liu X *et al* 2011 Crystal growth and characterization of diamond from carbonyl iron catalyst under high pressure and high temperature conditions *Cryst. Growth Design* **11** 3844–9

[138] Sun S *et al* 2014 Synthesis and characterization of hydrogen-doped diamond under high pressure and high temperature *CrystEngComm* **16** 2290–7

[139] Palyanov Y N, Kupriyanov I N, Khokhryakov A F and Borzdov Y M 2017 High-pressure crystallization and properties of diamond from magnesium-based catalysts *CrystEngComm* **19** 4459–75

[140] Ekimov E A, Kudryavtsev O S, Mordvinova N E, Lebedev O I and Vlasov I I 2018 High-pressure synthesis of nanodiamonds from adamantane: myth or reality? *ChemNanoMat* **4** 269–73

[141] Alkahtani M, Lang J, Naydenov B, Jelezko F and Hemmer P 2019 Growth of high-purity low-strain fluorescent nanodiamonds *ACS Photon.* **6** 1266–71

[142] Spohn M *et al* 2018 Poly (1-vinyladamantane) as a template for nanodiamond synthesis *Appl. Nano Mater.* **1** 6073–80

[143] Liang J *et al* 2020 Germanium iodide mediated synthesis of nanodiamonds from adamantane 'seeds' under moderate high-pressure high-temperature conditions *Diamond Rel. Mater.* **108** 108000

[144] Gebbie M A *et al* 2018 Experimental measurement of the diamond nucleation landscape reveals classical and nonclassical features *Proc. Natl Acad. Sci.* **115** 8284–9

[145] Shen Y *et al* 2021 Sub-4 nm nanodiamonds from graphene-oxide and nitrated polycyclic aromatic hydrocarbons at 423 K *ACS Nano* **15** 17392–400

[146] Ekimov E, Kondrina K, Zibrov I, Lyapin S, Lovygin M and Kazanskiy P R 2021 Iodine-mediated high-pressure high-temperature carbonization of hydrocarbons and synthesis of nanodiamonds *Mater. Res. Bull.* **137** 111189

[147] Ekimov E A *et al* 2016 The effect of molecular structure of organic compound on the direct high-pressure synthesis of boron-doped nanodiamond *Phys. Status Solidi* **213** 2582–9

[148] Kondrina K, Kudryavtsev O, Vlasov I, Khmelnitskiy R and Ekimov E A 2018 High-pressure synthesis of microdiamonds from polyethylene terephthalate *Diamond Rel. Mater.* **83** 190–5

[149] Davydov V A, Agafonov V and Khabashesku V N 2017 Pressure-temperature-induced transformations of hydrocarbon-fluorocarbon mixtures into nano-and micron-size diamonds *Eurais. Chem.-Technol. J.* **19** 115–25

[150] Ekimov E, Lyapin S, Grigoriev Y V, Zibrov I and Kondrina K M 2019 Size-controllable synthesis of ultrasmall diamonds from halogenated adamantanes at high static pressure *Carbon* **150** 436–8

[151] Thomas A, Parvathy M and Jinesh K B 2021 Synthesis of nanodiamonds using liquid-phase laser ablation of graphene and its application in resistive random access memory *Carbon Trends* **3** 100023

[152] Zhao J, Lai H, Lu H, Barner-Kowollik C, Stenzel M H and Xiao P 2016 Fructose-coated nanodiamonds: promising platforms for treatment of human breast cancer *Biomacromolecules* **17** 2946–55

[153] Lin Y *et al* 2021 Fabricating nanodiamonds from biomass by direct laser writing under ambient conditions *ACS Sust. Chem. Eng.* **9** 3112–23

# Chapter 3

# The physical and luminescent properties of zero-dimensional carbon nanomaterials

**Rosalin Beura, Mansi Tyagi and Satyabrata Mohapatra**

Materials in the nanoscale (1–100 nm) have received enormous interest due to their peculiar behavior and properties compared to their bulk counterparts. In particular, zero-dimensional (0D) materials have gained increased attention for diverse applications including catalysis, sensing, and hydrogen storage because of their remarkable physicochemical properties. This chapter explores the physical and luminescent properties of the 0D carbon materials exhaustively, focusing on graphene quantum dots (GQDs) and carbon quantum dots (CQDs) and the utilization of these properties for application in various fields.

## 3.1 Introduction

The use of nanoscale carbon allotropes began with the discovery of zero-dimensional fullerenes, followed by 1D carbon nanotubes and then 2D graphene [1]. Zero-dimensional (0D) carbon nanomaterials have attracted enormous interest due to their inherent structural properties, such as a high surface-to-volume ratio and small size, which makes them have more active edges per unit mass. This edge effect and quantum confinement enables them to have pronounced photoluminescent quantum efficiency. Among the various carbon forms the quantum-dot-like entities are the quasi-spherical carbon nanoparticles commonly called carbon dots (CDs), with a diameter below 10 nm. The carbon dot was generated accidentally from electrophoretic purification of multiwalled carbon nanotubes [2]. CDs are gaining increased attention in different research fields for their effective photo-harvesting and photoinduced redox properties, making them an easy rival to fullerenes. Zero-dimensional carbon material, in particular graphene quantum dots (GQDs) and carbon quantum dots (CQDs), have been utilized for various advanced applications starting from sensing, bioimaging, solar photovoltaics, and photocatalysis, to the designing of optoelectrical devices such as solid state light emitting

doi:10.1088/978-0-7503-4048-9ch3  

diodes, because of their excellent properties such as chemical stability, non-toxicity, stable photoluminescence, photo-upconversion, low photobleaching, etc [3]. Various methods have been utilized to obtain carbon dots from a large number of sources that are broadly categorized as top-down methods (bigger to smaller) and bottom-up (smaller to bigger) methods. Top-down methods include cutting down large carbon materials to smaller quantum dots while the bottom-up route involves fabrication from a smaller carbon precursor. A compendious report describing the luminescent properties of carbon nanodots was presented and discussed by Nagaraju and Smirithi [2]. Their review work tried to provide a brief idea of the synthesis of CDs from a wide range of precursor sources and through different synthesis procedures, along with their structural, optical, and electrical properties largely applicable for sensing. Since carbon dots can be prepared from a vast number of source materials, their chemical compositions change accordingly. Hence it is important to analyze their various properties in order to understand their suitability for applications in various fields. In particular, their physical and, more importantly, luminescent properties have been extensively utilized for various applications. It has been observed that variation in the synthesis parameters, such as the method of synthesis, precursors used, operating temperature, duration of reaction, presence of any additional element, etc, leads to noticeable variation in the intrinsic properties of the carbon material and particularly the luminescent properties. The ability to obtain such wide luminescence behavior provides a huge opportunity for applications in various fields including bioimaging, sensing, optoelectonic devices, and photocatalysis. This chapter provides an exhaustive discussion to provide an overall idea of the physical and luminescent properties of 0D carbon materials, in particular CQDs and GQDs, by citing various findings and understanding the concept of their origin of discrepancies reported by different groups. The discussion starts with the physical properties of 0D carbon materials, followed by their luminescent properties, the scope of the material and ends with concluding remarks.

## 3.2 The physical and luminescent properties of zero-dimensional carbon nanomaterials

### 3.2.1 Physical properties

Zero-dimensional carbon material, CDs, consisting of CQDs and GQDs, have been utilized for various advanced applications, from sensing to the design of optoelectrical devices, because of their excellent properties, such as chemical stability, non-toxicity, stable photoluminescence, photo-upconversion, etc. Therefore, it is very important to understand the physical and structural properties of CDs in order to analyze their key features for various applications.

#### *3.2.1.1 Morphology*

As already mentioned, carbon dots are mainly categorized as CQDs and GQDs. Morphology-wise, CQDs are predominantly of quasi-spherical shape and sometimes have a hollow structure, and comprise carbon nanoparticles with a diameter less than 10 nm [4]. On the other hand, GQDs are made up of single or a few layers (<10 layers) of graphene sheets with a diameter less than 10 nm. These small-sized carbon dots can be achieved through various top-down and bottom-up approaches. Uniformly dispersed spherical CQDs with a size distribution of 3–5 nm were synthesized using a hydrothermal method by Kumar *et al* [5]. Qiang *et al* synthesized CQDs with a mean size of about 3–5 nm via an ultrasonic route using potato starch as the precursor [6]. Similarly, Shinde *et al* synthesized GQDs with diameters in the range of 3–4 nm using multiwalled carbon nanotubes through an electrochemical route, where the thickness of the GQDs was estimated to be around 1–2 nm, indicating that each GQD contained 1–3 layers of graphene [7]. Liu *et al* used bottom-up approach to prepare disk-like GQDs with an average diameter of 60 nm and a thickness of 2–3 nm [8]. Single-layer GQDs with a topographic height of about 0.5 nm were prepared from XC-72 carbon black through chemical oxidation [9]. Using different precursor materials and synthetic procedures one can obtain CDs of different sizes.

*3.2.1.2 Composition*

CDs possess a gap of 0.18–0.24 nm within the lattice, which is graphitic in-plane type, and a space of 0.32 nm between the layers. Generally, they are composed of crystalline cores that are similar to amorphous clusters and $sp^2$ hybridized carbon, while certain CDs contain carbon cores similar to diamond, such as $sp^3$ carbon [10]. The major elemental compositions are carbon and hydrogen but due to oxidation of carbon precursors during preparation methods various groups containing oxygen (such as epoxy or ether, hydroxyl, carbonyl, and carboxylic acid groups) are found on the surface of CDs [11]. X-ray photoelectron spectroscopy (XPS) and Fourier transform infrared spectroscopy (FTIR) are useful techniques used to analyze the chemical composition of CDs. Representative XPS and FTIR spectra of CQDs are shown in figure 3.1.

Peng *et al* prepared CQDs by breaking down carbohydrates with nitric acid followed by passivation with amine-terminated compounds [12]. The FTIR spectrum of the as-prepared CQDs revealed the presence of carboxyl groups through a broad peak at 3300 $cm^{-1}$ (vibrational stretching of O–H) and at 1705 $cm^{-1}$ (vibrational stretching of C=O). The ratio of carbon, oxygen, and hydrogen was

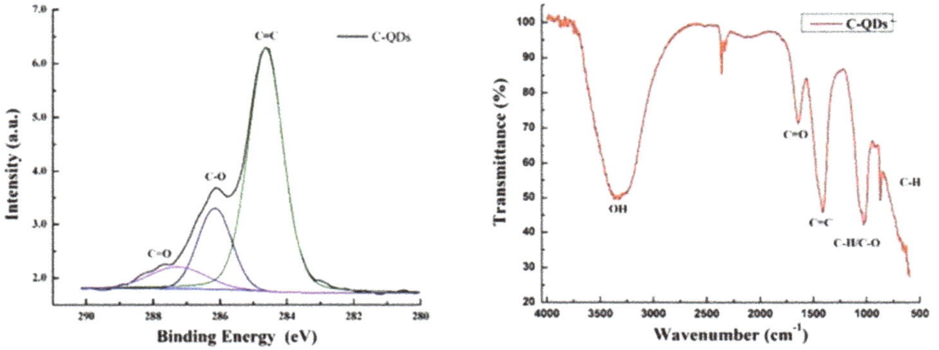


**Figure 3.1.** (a) C 1s XPS spectrum and (b) FTIR spectrum of CQDs. (Adapted from [53].)

estimated to be 2.8:2.3:1 through elemental analysis. The XPS spectrum of CQDs synthesized via a microplasma process showed the O 1s peak around 530 eV and C 1s peak at 284.5 eV [13]. Carbon dots are constituted of 80.31% atomic carbon and 19.68% oxygen. Likewise, the XPS results of GQDs prepared through oxidation and etching of graphite powder [14] showed that GQDs are made up of C and O with a C:O atomic ratio of 1.3. The FTIR spectrum further substantiates the presence of different functional groups (such as alkoxy, hydroxyl, carbonyl, epoxy, and carboxylic acid) on the surface of GQDs that contain a significant amount of oxygen. Peaks were obtained at 1112 $cm^{-1}$ (stretching of C–O from alkoxy groups), 3410 $cm^{-1}$ (O–H), and 1721 $cm^{-1}$ (C=O). The peak for vibration from aromatic C=C was observed at 1620 $cm^{-1}$. The C–O (epoxy) deformation and stretching peaks were detected at 1389 $cm^{-1}$ and 1260 $cm^{-1}$, respectively. The existence of these oxygenated functional groups is said to increase the water solubility of the quantum dots.

#### *3.2.1.3 Crystalline nature*

The presence of carbon bonding has been visualized through observations using Raman spectroscopy, showing two peaks around 1350 and 1600 $cm^{-1}$, referring to the $sp^2$ hybridized disordered carbon and crystalline graphitic carbon, respectively [10]. The Raman spectra of CQDs and GQDs mainly consist of characteristic disorder bands (or D-bands) and graphitic bands (or G-bands). In the case of disordered graphite or glassy carbon, the dangling bonds create vibration that gives rise to a D-band, whereas the vibration of $sp^2$-bonded carbon atom and the $E_{2g}$ mode of graphite in a 2D hexagonal lattice results in a graphitic type band [15]. Purity in the structure of graphitic systems is known from the ratio of intensities ($I_D/I_G$). GQDs have been shown to induce new physical properties compared to their 2D and 1D counterparts due to strong quantum confinement and edge effects. Kumar *et al* prepared CQDs by treating orange juice hydrothermally in ethylenediamine and confirmed their crystalline nature from the intense peaks observed at 1578 $cm^{-1}$ and 1331 $cm^{-1}$ corresponding to the G-and D-bands, respectively. An intensity ratio $I_D/I_G$ value of 0.59 reflected the purity of the CQDs [16]. The presence of structural defects was revealed from the high $I_D/I_G$ ratio of 1.384 obtained from the Raman spectrum of CQDs prepared hydrothermally from single-walled carbon nanotubes [17]. The Raman spectra of GQDs synthesized electrochemically from high purity graphite rods demonstrated a G-band at 1595 $cm^{-1}$ and D-band at 1337 $cm^{-1}$, respectively. The $I_D/I_G$ ratio was obtained as 0.91 which shows the presence of defects in GQDs with disordered structure, coming from the small-sized $sp^2$ clusters [18]. The Raman spectra presented by Li *et al* for electrochemically prepared GQDs revealed a $I_D/I_G$ ratio of 0.5, which demonstrates the high quality of the GQDs [19].

#### *3.2.1.4 Chelating and sorption properties*

CDs are rich in oxygen containing surface functional groups that provide ample active sites for chelation with metal ions. Wang *et al* synthesized liquid fuel derived CDs from crude oil product oxidation [20]. The PL of the CDs quenched on $Cu^{2+}$ ion addition due to the formation of a complex of $Cu^{2+}$ ions with carboxyl, hydroxyl, and N-containing surface functional groups of CDs. The limit of detection (LOD) was estimated to be 0.039 $\mu$M and the PL quenching ability of CDs was utilized to investigate $Cu^{2+}$ ions in rat brain. CDs are considered advantageous as an absorbent material due to their large surface area, small size, and low cytotoxicity [21, 22]. Agarwal *et al* utilized GQDs for the adsorption and removal of oxamyl pesticide from aqueous solution [23]. GQDs were fabricated employing a microwave-assisted hydrothermal approach using citric acid as the raw material. Efficient removal of pesticide was carried out in 25 min of contact time with 0.6 gm of adsorbent dose. Pudza *et al* investigated the adsorption capacity of eco-friendly CDs prepared using tapioca flour as a precursor [24]. CDs successfully removed 80.6% lead ions from solution in a contact time of 260 min, which was attributed to the chemisorption of metal ions on CDs.

*3.2.1.5 Surface modification*

Surface properties of CDs are tailored to enhance their biocompatibility and luminescence properties. Surface modification of CDs can be achieved through various methods including covalent bonding [25, 26], $\pi$–$\pi$ interactions [27], tuning the degree of oxidation [28], and electrostatic interactions [29]. The presence of abundant oxygenated groups on the CD surface assists in covalent bonding with other functional groups [30]. Tam *et al* fabricated amino-functionalized GQDs (a-GQDs) using a hydrothermal technique [31]. The a-GQDs red-shifted to exhibit green fluorescence whereas the pristine GQDs exhibited blue fluorescence. They claimed that the amino groups replaced the existing oxygen groups known to induce non-radiative electron–hole recombination. Thus, the QY of a-GQDs was 32.8% which was higher than pristine GQDs (4.5%). The obtained a-GQDs were utilized for efficient fluorescent sensing of $Cu^{2+}$ ions with a detection limit of 5.6 nM. Ding *et al* reported full color emitting CDs prepared using a hydrothermal route and separated via column chromatography [32]. Although the CDs were similar in size distribution, variation in luminescence from blue to red was attributed to the reduction in band gap due to change in oxidation state of the surface. Li *et al* [33] reduced greenish-yellow GQDs (gGQDs) with $NaBH_4$ and obtained blue GQDs (bGQDs). The QY of the bGQDs increased to 22.9% from the gGQDs which had a QY of 11.7%.

### 3.2.2 Luminescent properties

Optical properties include the observations from ultraviolet (UV) absorption, photoluminescence (PL), PL emission (PLE), and PL quantum yield. Amongst the optical properties, photoluminescence is the dominant property of CDs that has been employed for various applications, such as making them an efficient fluorescent probe in bioimaging as they are nontoxic and chemically inert, utilizing their upconversion property to have wider light absorption range, and for use as a biosensor.

*3.2.2.1 Mechanism of photoluminescence*

The origin of the luminescent behavior of the CDs is due to the electron transfer from $\pi \rightarrow \pi^*$ levels arising from the quantum confinement effect. PL emission changes when the gap between the energy levels ($\pi$ and $\pi^*$) changes, which is caused by the number of fused aromatic C–C rings, the domain size, or the presence of different functional groups on CDs [34]. The bright PL emission due to the electron and hole recombination (radiative) are highly dependent on the excitation wavelength [35]. Various theoretical and experimental findings prove quantum confinement (i.e. decreasing size results in an increase in the energy gap between the valence band (VB) and conduction band (CB), thereby decreasing the emission wavelength) in CDs. The edge effect and surface states in CDs have been shown to induce photoluminescence and also affect the emission profile [28, 36, 37]. The stability of the PL obtained from GQDs has proved to be outstanding compared to fluorescent dyes, as it shows no distinct photobleaching under a UV lamp and the luminescent behavior of GQDs is reported to be strongly dependent on the excitation wavelength ($\lambda_{ex}$).

A handful of works have been published on developing different luminescent carbon materials in a number of ways and the effect of different experimental parameters on the luminescent behavior has also been reported. The following sections provide a brief discussion of the generation of different luminescent carbon materials through different procedures and an understanding of the cause for the differences in the observations.

Zhu *et al* [28] used a solvothermal approach for large production of intense green luminescent GQDs with 11.4% PL quantum yield (using 9,10-bis (phenylethynyl) anthracene). Exciting these GQDs at 320 nm showed a sharp peak at 427 nm with a shoulder at 515 nm. Further, providing an excitation of 375 nm showed one strong peak at 515 nm arising from the surface defects. It was hypothesized that the generation of the PL peak was due to the simultaneous contribution from both the quantum size effect and the surface defects that could have arisen due to the incomplete reduction of GQDs (for example, the epoxy/ether, –OH, C=O and –CO-$NR_2$ groups cause structural defects). It was observed that fluorescent properties depend on the ionic strengths that were tuned by varying the KCl concentration from 0 to 2 M, which showed no variation in the luminescent intensities as well as the peak characteristics. Interestingly, it was observed that the PL behavior was pH dependent. There was a decrease in the PL intensities for a solution having low or high pH and PL remained constant for solutions with pH 4 to pH 8. For pH greater than 12 the peak position blue shifts and the FWHM narrows. The authors have proved that these quantum dots could be used as an excellent biolabeling agent. Gone *et al* [38] investigated the effect of different solvents on the photoluminescence spectrum of GQDs. Few layer [1–4] thick GQDs with a size ranging from 5–8 nm were synthesized using dimethyl sulfoxide (DMSO), dimethylformamide (DMF), and water as the solvent. PL emission was obtained in the blue–green region, which was attributed to the edge defects, size confinement, and functional groups present in GQDs. The edge sites and functional groups of GQDs were modulated by varying the solvents. Armchair edges were found to be higher in GQDs prepared with DMF than the other samples, whereas zigzag sites were found to be dominant in the GQDs synthesized using DMSO. GQDs prepared using DMF showed the highest PL quantum yield of 32%, which is credited to the presence of a higher number of armchair edge sites and additional functional groups compared to the other samples. The QYs with DMSO and water solvent were reported to be 27% and 15%, respectively. This proved that the PL QY of GQDs is dependent on the type of edge defects and functional groups present. The prepared GQDs were utilized to explore the bioimaging application for the detection of cancer cells due to their high PL yield.

Bright blue photoluminescent GQDs were developed by a simple hydrothermal method, where the graphene sheets that were oxidized previously were cut into ultrafine functionalized GQDs [36]. It was observed that due to the hydrothermal conditions the optical and structural properties of GQDs were strongly influenced, as the cutting mechanism completely breaks up the mixed epoxy chains. Findings from the PL analysis showed that the GQDs emitted blue light even in neutral media as opposed to no emission from the graphene sheets even in an alkali medium. An intense peak at 430 nm with a Stokes shift of 110 nm was observed upon exciting with 320 nm. Upon 420 nm excitation the peak position red-shifted along with a reduction in intensity. Recording the PL emission with the strongest luminescence shows that the luminescence was linked directly with transitions occurring at 257 nm and 320 nm in place of the $\pi$–$\pi^*$ transition that is commonly observed. The blue luminescence was thought to originate from a large concentration of free zigzag sites having a carbene-like triplet state $\sigma^1\pi^1$ (i.e. the two orbitals are singly occupied), which has been generated because of the smaller diameter of GQDs (9.6 nm). PL intensity varies inversely on switching between pH 13 and 1, where alkaline conditions resulted in strong luminescence and in acidic conditions the PL intensity was completely quenched. The authors provided a structural model of GQDs in order to understand the observed PL phenomena in different pH conditions.

The size dependent optical properties (PL, PL QY, decay lifetime, EELS) of GQDs (synthesized by a soft-template microwave-assisted hydrothermal route) were studied by Tang *et al* [39] for GQDs with different diameters (1.5 to 3.9 nm ± 0.55 nm) and synthesis times (1 to 9 min). Among the various GQDs prepared, the one with a diameter of 2.4 nm showed the strongest emission for $\lambda_{ex}$ = 350 nm. Further, increasing $\lambda_{ex}$ showed a linear increase in the wavelength of the emission peak ($\lambda_{peak}$) and this phenomenon was supposed to be due to the emissive traps caused by the surface functional group with various energy levels. The values of the peak due to PL excitation and PL spectra red-shifted to 350 and 470 nm, respectively, as compared to the GQD absorption peaks located at 228 nm (the $\pi \rightarrow \pi^*$ transition of C=C) and 282 nm (the $n \rightarrow \pi^*$ transition of C=O). This discrepancy in the position of the peak among different optical properties, such as PLE, PL, and absorption spectra, suggested that GQDs are not regular semiconductors and due to their unique optical properties GQDs are attractive candidates over conventional semiconductor QDs. Relating the QY (the ratio of absorbed photons to emitted photons through fluorescence) of the GQDs to their diameter showed a QY ranging from 5% to 15%, where 2 nm sized GQDs showed the highest QY of 15%. Beyond a diameter of 2 nm, the QY was found to be inversely proportional to the diameter because a larger

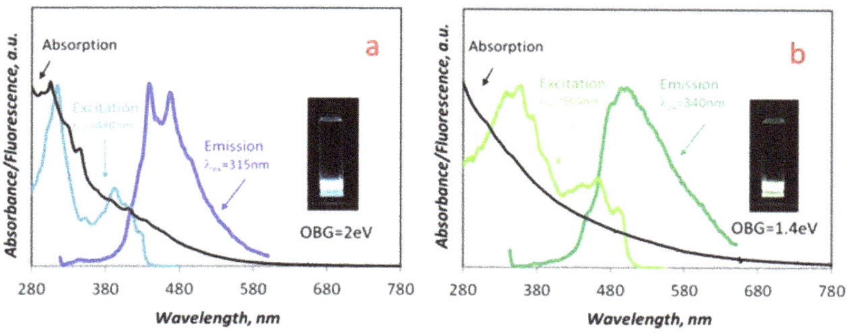


**Figure 3.2.** The optical properties of (a) blue CDs and (b) green CDs, extracted from a heavily sooting flame (HSF) reactor sampled at 6 mm. (Adapted from [40].)

diameter gives a shorter lifetime and hence a lower QY. The smallest size GQDs (1.5 nm) showed a low QY that was due to the low emissive sites on the QD surface.

Green and blue fluorescent carbon nanodots for application in bioimaging sectors [40] were synthesized in controlled gas flame reactors followed by separation from other carbon nanoparticles by extracting with organic solvents such as dichloromethane (DCM) and *N*-methyl-pyrrolidinone (NMP). It was observed that the DCM and NMP extracted CDs showed blue and green fluorescence, respectively, under irradiation of 365 nm UV light (figure 3.2). The fine structured shape absorption spectrum of blue CDs was associated with the peak due to a $\pi$–$\pi^*$ absorption, coming from the absorption due to the complex carbon network of larger poly aromatic hydrocarbon (PAH) ($\geq C_{22}$). The emission spectrum showed two main peaks with a quantum yield of nearly 20%. Contrastingly, green CDs exhibited broad absorption spectra with a tail extending to the visible, which is attributed to the PAH domains of macromolecular carbonaceous material having a lower quantum yield of 2%–6%. Because of the higher quantum yield, blue CDs mask the fluorescence from the green CDs. This versatile feature of CDs provides a challenging path for their application in different bioimaging areas.

A hydrothermal method was used to synthesize double-luminescent-band blue and green CQDs by using 1,3-dihyroxynaphthalene and hydrochloric acid [3]. These CQDs showed two-band luminescence in ethanol and exhibited white fluorescence under illumination of UV light. For understanding the luminous origin the analysis of the UV and PLE spectra of CQDs was very important. A transition from $\pi \rightarrow \pi^*$ for different functional groups was seen as a strong absorption intensity at 237 (C=C–C–O), 282 (C=O), and 338 nm (C–O–C). Similarly, the low intensity absorption bands at 486, 510, and 525 nm were attributed to $n \rightarrow \pi^*$ transitions or overtone $\pi \rightarrow \pi^*$ of the C(=O)O group. In accordance with the emission and excitation spectra variation laws, the green emission band (553 nm) was claimed to arise due to the C(=O)O functional group (486, 510, and 525 nm excitations), blue emission (405 nm) with excitation of 306 nm coming from the surface functional group C(=O)OH, and excitation of 368 nm resulting in both blue and green emissions (420, 433, and 460 nm) arising from the surface functional group C–O–C. It was concluded that the luminescence property originated from surface fluorophores related to carbon–oxygen. Upon fabricating a

CQD based white light emitting diode (WLED), a cold white light source was obtained with a color temperature (CCT) of 6428 K and with Commission Internationale de lEclairage (CIE) coordinates (0.3122, 0.3429). Similarly, a number of researchers have prepared different luminescent CDs through different synthesis methods and functionalization techniques, making them suitable for different applications. Three different kinds of CDs with orange, yellow, and blue luminescence were synthesized using phenylenediamine isomers using a microwave heating method [41]. It was observed that the color as well as the quantum yields (8%–45% ethanol) depend on the extent of oxidation and nitrogen-functional groups (–$NH_2$, C=O/–CONH– and N element) present on the surface. Coating the CDs on starch particles produced three composite phosphors with more than 16% quantum yield that were employed to prepare luminescent blocks and an LED with controlled color temperature. Jiang *et al* also synthesized different luminescent CDs (red, green, and blue) from phenylenediamine precursors, which showed luminescence under excitation of single UV light [42]. The variations in the PL emission were proposed to be due to the differences in size and N content. Direct substitution with nitrogen was proved to drastically change the electronic and chemical properties, making them suitable for a large number of applications. N-doped GQDs (N-GQDs) [43] prepared by microwave synthesis have been shown to possess the widest emission band width (300 nm to 1000 nm), i.e. deep UV to NIR, among any other semiconductor material along with their quantum dot counterparts. The layered structure that created a conjugated system of extensively delocalized $\pi$ electrons accounted for the broad emission properties of the N-GQDs. The deep UV, visible, and NIR emission arose from localized $\pi$ $e^-$ in C=C, partially conjugated $\pi$ $e^-$ in the N-GQDs, and the conjugated $\pi$ $e^-$ present in the layered structure, respectively. This unique emission as well as the photoresponse properties were employed for optoelectronic applications. Sidewise bright blue and bright green fluorescent N-GQDs with quantum yields of 8% and 30% were also synthesized by Ren *et al* and Tam *et al*, respectively, for effective sensing of ferric ion as well as fluorescent ink [44, 45]. Wu *et al* [46] developed highly fluorescent GQDs with strong red, green, and blue luminescence under red, blue, and UV light irradiation and a PL quantum yield of 54.5%, that were used for sensing and imaging (*in vitro* or *in vivo*).

Green synthesis methods from different sources (such as glucose, lemon juice, potato, garlic, kidney beans, etc) have also been adopted for preparing highly luminescent CDs [47–50]. Sun *et al* developed S and N co-doped-carbon dots using a garlic source via a single-step hydrothermal technique [51]. The quantum yield of the prepared quantum dots was increased by adding ethylenediamine (EA) during synthesis. Structural and compositional analysis showed that the content of nitrogen and also the generation of the C–N and C=N were important to increase the PL quantum yield. The prepared CDs were used for metal ion detection, which showed a higher response for selectively detecting $Fe^{3+}$ ions. Figure 3.3 shows the variation in the photoluminescence properties of green synthesized CDs. Similarly, strong green luminescent CDs with quantum yield (14.86% to 24.89%) were hydrothermally synthesized from lemon juice under different temperature, time, aging of precursor, and diluted solvent [49]. Light emission dependent on these varying properties was used for optoelectronics and bioimaging applications.

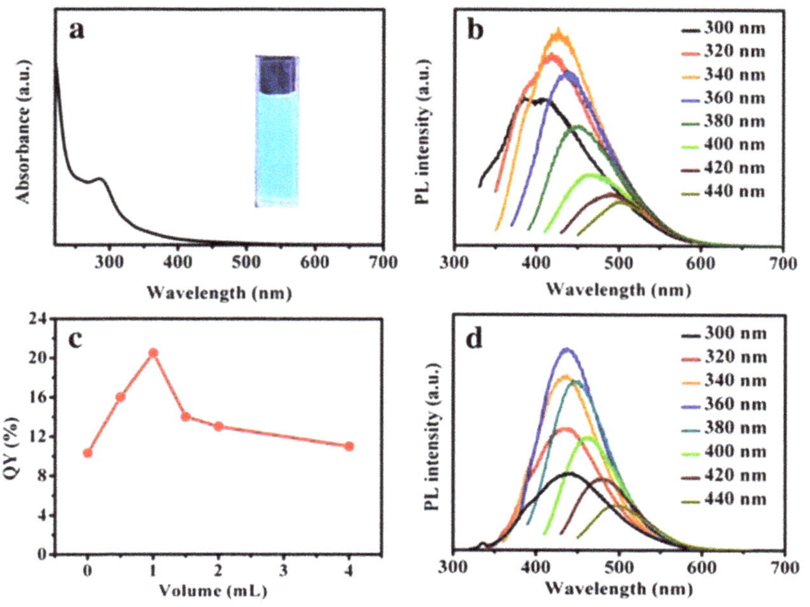


**Figure 3.3.** Absorbance and luminescent properties of CDs: (a) UV–visible absorption, (b) PL emission, (c) PL QY, and (d) emission spectra under different excitation wavelengths. (Adapted from [51].)

Bagheri *et al* [34] adopted a systematic approach to prepare CDs at different pyrolysis temperatures (160 °C–220 °C) and studied the impact of forming more carbonized carbon particles (MCCPs) on the PL behavior, in order to develop a facile smartphone based sensing platform (for sensing metal ions). The interaction of the metal ions with the carboxyl and hydroxyl groups on the CD surface caused a change in the electronic structure as well as exciton distribution in CDs leading to non-radiative exciton recombination via effective transfer of electron or quenching fluorescence emission. This property of PL emission quenching was utilized for sensing of the metal ions present. It was observed that CDs that were prepared at 160 °C for 50 min showed the highest quantum yield (29%) and were used for detecting various metal ions such as $Hg^{2+}$, $Fe^{2+}$, $Mg^{2+}$, $Pb^{2+}$, $Ca^{2+}$, $Zn^{2+}$, and $Cu^{2+}$ ions, which showed the highest sensitivity and selectivity for $Hg^{2+}$ with a 50 nM detection limit (figure 3.4).

In a similar manner, Wang *et al* [52] also synthesized GQDs by thermally decomposing citric acid and studied the effect of structural evolution through varying the synthesis conditions (reaction temperature and duration) on the fluorescence properties. The fraction of uniformly sized small GQDs resulted in blue fluorescence, whereas inhomogeneously distributed large sheets showed low quantum yield fluorescence. Yoo *et al* [37] reported an exclusive review on recent progress on the fluorescent properties of these new classes of carbon nanomaterials tailored through various methods such as surface passivation, doping, and functionalization through various organic and biomolecules, metal nanoparticles, etc.

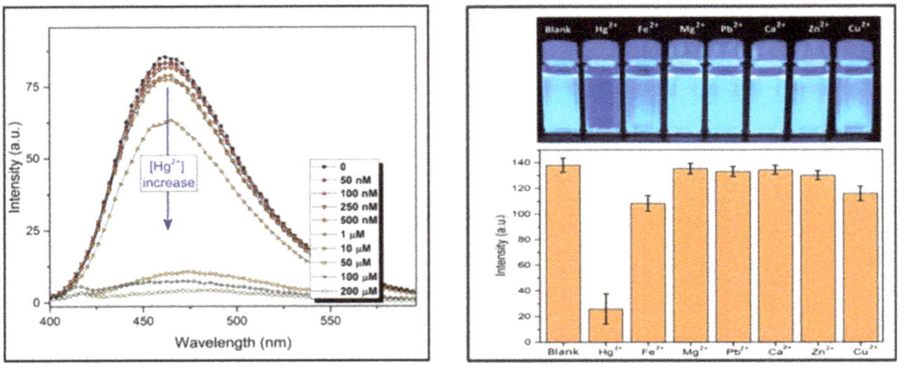


**Figure 3.4.** Quenching of PL emission by CDs (160–50) exposed to different concentration of $Hg^{2+}$ and 100 $\mu$M of various cations under UV irradiation. (Adapted from [34].)

The application of these fluorescent properties of these exciting carbon nanomaterials for heavy metal ion detection is focused on the discussion. Even the anatase and rutile phase of $TiO_2$ can be used to harness the visible light part of the solar spectrum by making a complex photocatalyst with GQDs [10]. Excitation dependent PL upconversion and down-conversion behavior was utilized for the same. Interestingly, the complex photocatalyst rutile system showed a nine times higher catalytic rate compared to the anatase system in degrading methylene blue.

## 3.3 The future scope of these materials

Zero-dimensional carbon materials have been of great utility because of their peculiar properties exhibited due the quantum confinement in all three dimensions. Based on the above discussions, it is clear that both GQDs and CQDs are employable in a broad variety of applications. In particular, the excellent tunable and wavelength dependent luminescent properties of carbon dots have opened up an opportunity for their utilization in the fields of multicolor bioimaging, biosensors, optoelectronic devices, etc. Varying the excitation wavelength can vary the emission wavelength, which can serve the purpose of meeting various application requirements. Hence, there is a long road ahead for exploring strategies to make the best use of carbon dots.

## 3.4 Conclusion and prospects

Carbon dots, in particular CQDs and GQDs, have been used for diverse applications due to their exciting properties. Their physical properties have been observed to vary with different synthesis procedures as well as the precursor material used. Even luminescent properties have been seen to vary with different synthesis procedures. Carbon dots are most importantly known for their luminescent properties. The luminescence of carbon dots is determined to originate from the $\pi$–$\pi$* transition. The difference in the two energy levels influences the nature of the emission directly. Tuning the gap between the energy levels using various methods and by varying the excitation wavelength, different light emitting carbon dots can be generated that are useful for various purposes. This chapter presents an exhaustive discussion to provide an overall idea of the physical as well as the luminescent properties of zero-dimensional carbon materials through results reported by different groups. Hence, it is expected that readers will develop good scientific insight into the physical and luminescent properties of CDs and enhance their understanding of this new class of exciting carbon materials.

## Acknowledgments

The authors acknowledge support from Guru Gobind Singh Indraprastha University for FRGS grant (GGSIPU/DRC/FRGS/2020/1988/12). RB is thankful to CSIR for the Research Associateship (RA) fellowship (09/806(0039)/2020-EMR-I).

## References

[1] Yang F *et al* 2018 Zero-dimensional carbon allotropes—carbon nanoparticles versus fullerenes in functionalization by electronic polymers for different optical and redox properties *ACS Omega* **3** 5685–91

[2] Kottam N and Smrithi S P 2021 Luminescent carbon nanodots: current prospects on synthesis, properties and sensing applications *Methods. Appl. Fluoresc.* **9** 012001

[3] Chen X *et al* 2021 Blue and green double band luminescent carbon quantum dots: synthesis, origin of photoluminescence, and application in white light-emitting devices *Appl. Phys. Lett.* **118** 153102

[4] Baker S N and Baker G A 2010 Luminescent carbon nanodots: emergent nanolights *Angew Chem.—Int. Ed.* **49** 6726–44

[5] Vinod Kumar V, Raman T and Anthony S P 2017 Fluorescent carbon quantum dots chemosensor for selective turn-on sensing of $Zn^{2+}$ and turn-off sensing of $Pb^{2+}$ in aqueous medium and zebrafish eggs *New J. Chem.* **41** 15157–64

[6] Qiang R, Yang S, Hou K and Wang J 2019 Synthesis of carbon quantum dots with green luminescence from potato starch *New J. Chem.* **43** 10826–33

[7] Shinde D B and Pillai V K 2012 Electrochemical preparation of luminescent graphene quantum dots from multiwalled carbon nanotubes *Chem.—A Eur. J.* **18** 12522–8

[8] Liu R, Wu D, Feng X and Müllen K 2011 Bottom-up fabrication of photoluminescent graphene quantum dots with uniform morphology *J. Am. Chem. Soc.* **133** 15221–3

[9] Dong Y *et al* 2012 One-step and high yield simultaneous preparation of single- and multi-layer graphene quantum dots from CX-72 carbon black *J. Mater. Chem.* **22** 8764–6

[10] Zhuo S, Shao M and Lee S 2012 Upconversion and downconversion fluorescent graphene quantum dots: ultrasonic preparation and photocatalysis *ACS Nano* **6** 1059–64

[11] Tajik S *et al* 2020 Carbon and graphene quantum dots: a review on syntheses, characterization, biological and sensing applications for neurotransmitter determination *RSC Adv.* **10** 15406–29

[12] Peng H and Travas-Sejdic J 2009 Simple aqueous solution route to luminescent carbogenic dots from carbohydrates *Chem. Mater.* **21** 5563–5

[13] Ma X, Li S, Hessel V, Lin L, Meskers S and Gallucci F 2019 Synthesis of luminescent carbon quantum dots by microplasma process *Chem. Eng. Process—Process Intensif.* **140** 29–35

[14] Sun Y *et al* 2013 Large scale preparation of graphene quantum dots from graphite with tunable fluorescence properties *Phys. Chem. Chem. Phys.* **15** 9907–13
[15] Gu J, Zhang X, Pang A and Yang J 2016 Facile synthesis and photoluminescence characteristics of blue-emitting nitrogen-doped graphene quantum dots *Nanotechnology* **27** 165704
[16] Kumar D, Singh K, Verma V and Bhatti H S 2014 Synthesis and characterization of carbon quantum dots from orange juice *J. Bionanosci.* **8** 274–9
[17] Ding H, Cheng L W, Ma Y Y, Kong J L and Xiong H M 2013 Luminescent carbon quantum dots and their application in cell imaging *New J. Chem.* **37** 2515–20
[18] Zhang M *et al* 2012 Facile synthesis of water-soluble, highly fluorescent graphene quantum dots as a robust biological label for stem cells *J. Mater. Chem.* **22** 7461–7
[19] Li Y *et al* 2011 An electrochemical avenue to green-luminescent graphene quantum dots as potential electron-acceptors for photovoltaics *Adv. Mater.* **23** 776–80
[20] Wang C, Pan C, Wei X, Yang F, Wu W and Mao L 2020 Emissive carbon dots derived from natural liquid fuels and its biological sensing for copper ions *Talanta* **208** 120375
[21] Long C, Jiang Z, Shangguan J, Qing T, Zhang P and Feng B 2021 Applications of carbon dots in environmental pollution control: a review *Chem. Eng. J.* **406** 126848
[22] Xiao J *et al* 2019 Novel functional fiber loaded with carbon dots for the deep removal of Cr (VI) by adsorption and photocatalytic reduction *J. Environ. Sci. (China)* **83** 195–204
[23] Agarwal S, Sadeghi N, Tyagi I, Gupta V K and Fakhri A 2016 Adsorption of toxic carbamate pesticide oxamyl from liquid phase by newly synthesized and characterized graphene quantum dots nanomaterials *J. Colloid Interface Sci.* **478** 430–8
[24] Pudza M Y, Abidin Z Z, Rashid S A, Yasin F M, Noor A S M and Issa M A 2020 Eco-friendly sustainable fluorescent carbon dots for the adsorption of heavy metal ions in aqueous environment *Nanomaterials.* **10** 1–19
[25] Guo Y, Zhang J, Zhang W and Hu D 2019 Green fluorescent carbon quantum dots functionalized with polyethyleneimine, and their application to aptamer-based determination of thrombin and ATP *Microchim. Acta* **186** 717
[26] Huang Y *et al* 2016 A dual-channel fluorescent chemosensor for discriminative detection of glutathione based on functionalized carbon quantum dots *Biosens. Bioelectron.* **86** 748–55
[27] Li H, Zhang Y, Wang L, Tian J and Sun X 2011 Nucleic acid detection using carbon nanoparticles as a fluorescent sensing platform *Chem. Commun.* **47** 961–3
[28] Zhu S *et al* 2011 Strongly green-photoluminescent graphene quantum dots for bioimaging applications *Chem. Commun.* **47** 6858–60
[29] Jin H, Gui R, Wang Y and Sun J 2017 Carrot-derived carbon dots modified with polyethyleneimine and Nile blue for ratiometric two-photon fluorescence turn-on sensing of sulfide anion in biological fluids *Talanta* **169** 141–8
[30] Yan F, Jiang Y, Sun X, Bai Z, Zhang Y and Zhou X 2018 Surface modification and chemical functionalization of carbon dots : a review *Microchim. Acta* **185** 424
[31] Van T T and Choi W M 2018 One-pot synthesis of highly fluorescent amino-functionalized graphene quantum dots for effective detection of copper ions *Curr. Appl Phys.* **18** 1255–60
[32] Ding H, Yu S, Wei J and Xiong H 2016 Full-color light-emitting carbon dots with a surface-state-controlled luminescence mechanism *ACS Nano* **10** 484–91
[33] Li L L *et al* 2012 A facile microwave avenue to electrochemiluminescent two-color graphene quantum dots *Adv. Funct. Mater.* **22** 2971–9
[34] Bagheri Z *et al* 2017 New insight into the concept of carbonization degree in synthesis of carbon dots to achieve facile smartphone based sensing platform *Sci. Rep.* **7** 1–11

[35] Liang W, Bunker C E and Sun Y P 2020 Carbon dots: zero-dimensional carbon allotrope with unique photoinduced redox characteristics *ACS Omega* **5** 965–71
[36] Pan D, Zhang J, Li Z and Wu M 2010 Hydrothermal route for cutting graphene sheets into blue-luminescent graphene quantum dots *Adv. Mater.* **22** 734–8
[37] Yoo D, Park Y, Cheon B and Park M H 2019 Carbon dots as an effective fluorescent sensing platform for metal ion detection *Nanoscale Res. Lett.* **14** 272
[38] Rajender G, Goswami U and Giri P K 2019 Solvent dependent synthesis of edge-controlled graphene quantum dots with high photoluminescence quantum yield and their application in confocal imaging of cancer cells *J. Colloid Interface Sci.* **541** 387–98
[39] Tang L, Ji R, Li X, Teng K S and Lau S P 2013 Size-dependent structural and optical characteristics of glucose-derived graphene quantum dots *Part. Part. Syst. Charact.* **30** 523–31
[40] Russo C, Apicella B and Ciajolo A 2019 Blue and green luminescent carbon nanodots from controllable fuel-rich flame reactors *Sci. Rep.* **9** 1–8
[41] Liu C *et al* 2018 Orange, yellow and blue luminescent carbon dots controlled by surface state for multicolor cellular imaging, light emission and illumination *Microchim. Acta* **185** 539
[42] Jiang K *et al* 2015 Red, green, and blue luminescence by carbon dots: full-color emission tuning and multicolor cellular imaging *Angew Chem.—Int. Ed.* **54** 5360–3
[43] Tang L *et al* 2014 Deep ultraviolet to near-infrared emission and photoresponse in layered n-doped graphene quantum dots *ACS Nano* **8** 6312–20
[44] Van T T, Trung N B, Kim H R, Chung J S and Choi W M 2014 One-pot synthesis of N-doped graphene quantum dots as a fluorescent sensing platform for $Fe^{3+}$ ions detection *Sensors Actuators* B **202** 568–73
[45] Ren Q, Ga L and Ai J 2019 Rapid synthesis of highly fluorescent nitrogen-doped graphene quantum dots for effective detection of ferric ions and as fluorescent ink *ACS Omega* **4** 15842–8
[46] Wu X, Tian F, Wang W, Chen J, Wu M and Zhao J X 2013 Fabrication of highly fluorescent graphene quantum dots using L-glutamic acid for *in vitro*/*in vivo* imaging and sensing *J. Mater. Chem.* C **1** 4676–84
[47] Lu W *et al* 2012 Economical, green synthesis of fluorescent carbon nanoparticles and their use as probes for sensitive and selective detection of mercury(II) ions *Anal. Chem.* **84** 5351–7
[48] Lu W, Qin X, Asiri A M, Al-Youbi A O and Sun X 2013 Green synthesis of carbon nanodots as an effective fluorescent probe for sensitive and selective detection of mercury(II) ions *J. Nanopart. Res.* **15** 1343
[49] Hoan B T, Tam P D and Pham V H 2019 Green synthesis of highly luminescent carbon quantum dots from lemon juice *J. Nanotechnol.* **2019** 1–9
[50] Tripathi K M, Tran T S, Tung T T, Losic D and Kim T 2017 Water soluble fluorescent carbon nanodots from biosource for cells imaging *J. Nanomater.* **2017** 7029731
[51] Sun C *et al* 2016 Synthesis of nitrogen and sulfur co-doped carbon dots from garlic for selective detection of $Fe^{3+}$ *Nanoscale Res. Lett.* **11** 1–9
[52] Wang S, Chen Z G, Cole I and Li Q 2015 Structural evolution of graphene quantum dots during thermal decomposition of citric acid and the corresponding photoluminescence *Carbon* **82** 304–13
[53] Dager A, Uchida T, Maekawa T and Tachibana M 2019 Synthesis and characterization of mono-disperse carbon quantum dots from fennel seeds: photoluminescence analysis using machine learning *Sci. Rep.* **9** 1–10

# Chapter 4

# Toxicological overview of zero-dimensional carbon nanomaterials

**V Akhil, A Arathi, X Joseph and P V Mohanan**

Over the past few years there has been a surge in the development of zero-dimensional carbon nanomaterials due to their remarkable physiochemical properties. The structural diversity and the excellent optical and thermal properties of these materials enable them to find potential roles in medicine, agriculture, and energy-related fields. However, the biomedical applications of these materials are still under investigation as they requires extensive toxicity studies and safety assessments. The present chapter addresses the significance of toxicity evaluation and further covers the cytotoxic responses of zero-dimensional carbon nanomaterials such as carbon quantum dots, graphene quantum dots, and fullerenes.

## 4.1 Introduction

Nanotechnology has seen a significant expansion in almost all fields of research in recent decades. Nanoparticle mediated applications have become a day-to-day affair. The catalogue of nanomaterials is expanding worldwide. Carbon-based nanomaterials are finding their unique place owing to their structural dimensions and excellent physicochemical characteristics [1]. Based on their structural dimensions, carbon nanomaterials are classified mainly into zero-dimensional (dimensionless), one-dimensional, two-dimensional, and three-dimensional nanomaterials. The unrivalled features of carbon nanomaterials have paved the way for the emerging platform of theranostics [2].

Zero-dimensional carbon nanomaterials are among the widely explored carbon-based nanomaterials. In contrast to the other carbon nanomaterials, zero-dimensional carbon nanomaterials possess several advantages: ease of production, excellent physicochemical properties, and low-cost production. Carbon quantum dots (CQDs), graphene quantum dots (GQDs), and fullerenes are widely explored materials among

doi:10.1088/978-0-7503-4048-9ch4  

the zero-dimensional carbon nanomaterials. They have established applications in many important research areas, such as nanomedicine [3], disease diagnosis and drug delivery [4], cancer therapy [5], sensing [6], and electrocatalysis [7]. The characteristic high surface-to-volume ratio is responsible for the increased catalytic activities of nanomaterials. Their size, in the nanometre-scale, allows nanomaterials to present quantum effects, thereby producing various characteristic fluorescent, magnetic, and electric properties compared to bulk materials [8].

The expanding application of nanotechnology in various industrial and biological arenas has raised universal interest regarding the consequences of nanoparticle exposure in biological systems [9]. Biological safety is an essential aspect of all therapeutics and medical devices intended to have a medical application. In an attempt to resolve the queries concerning the safety of nanomaterials, toxicity assessment is being conducted extensively in *in vitro* and *in vivo* systems [10]. The toxicity analysis of newly synthesized nanomaterials in various cell types and *in vivo* models has aided researchers in unravelling the cellular pathways and toxicity mechanisms associated with it. The cell–nanoparticle interaction is strictly monitored by multiple methods of microscopy and biochemical assays. The toxicity of nanomaterials is elucidated by observing the viability, morphology, enzyme production, histopathology, and ROS and RNS production that helps distinguish between the standard and affected systems. The *in vitro* and *in vivo* effects of nanoparticles are being studied worldwide with plenty of available literature [11]. These reports provide a better insight into how nanomaterials cause toxicity. The existing toxicity assays were developed for assessing the chemical toxicity. However, newer methods of toxicity analysis are being established by revamping current techniques and making them specific for determining the toxicity of nanomaterials.

Numerous nanomaterials have been synthesized and their applications have been shown in many aspects of agriculture, such as plant growth, plant health, disease management, and crop protection from pests. Unlike other semiconductor-based nanomaterials, carbon nanomaterials provide the advantage of low toxicity and increased biocompatibility. Hence various studies have highlighted the applications of carbon nanomaterials in agriculture [12]. With increasing production of nanomaterials for industrial applications and extensive research being carried out on the industrial applications of CQDs, the environmental release of these nanomaterials is a significant concern. These nanomaterials would finally end up in bodies of water, and hence their effects on aquatic systems need to be assessed in detail.

The most discussed characteristics of zero-dimensional nanomaterials include their biocompatibility, due to their wide potential for various biological applications. Hence zero-dimensional nanomaterials are being studied extensively to elucidate their toxicity in different *in vitro* and *in vivo* conditions. This chapter provides an overview of the significance of toxicity evaluation of zero-dimensional carbon nanomaterials, mainly CQDs, GQDs, and fullerenes. Moreover, this chapter features the important factors affecting the toxicity of zero-dimensional carbon nanomaterials.

## 4.2 Approaches/methods for the assessment of toxicity

The increased application and production of nanomaterials have resulted in increased distribution and accumulation of these materials in the environment, potentially posing serious public health issues. Hence, to minimize the potential risks and predict the side effects associated with the application of nanomaterials in various organisms, a combination of physicochemical characterization techniques and different types of *in vitro*, *in vivo*, and *in silico* assays are required (figure 4.1). The biological responses and toxicity of nanomaterials generally depend on their charge, size, coating, morphology, and chemical composition. The investigation of nanoparticle induced toxicity and the cellular mechanism of cell death associated with nanoparticle exposure can be carried out by various techniques, including spectroscopic measurements, biomarker detection, and oxidative stress detection in both *in vitro* and *in vitro*.

### 4.2.1 *In vitro* analysis

*In vitro* analysis is one of the primary types of studies conducted to identify the toxic response of nanomaterials or chemical substances on cultured mammalian cell lines. The treatment of nanomaterials on various cell lines indicates the effect of the materials in terms of cell viability. Moreover, *in vitro* studies should give a detailed understanding of the toxicity mechanisms and specific pathways regulating the toxicity. *In vitro* studies include various assays that can determine the toxicity profile of nanomaterials by quantification of various factors affecting cell death.

Cell viability studies using a 3-(4,5-dimethylthiazol-2-yl)-2,5-diphenyl tetrazolium bromide (MTT) assay are considered the most common and standard methods.

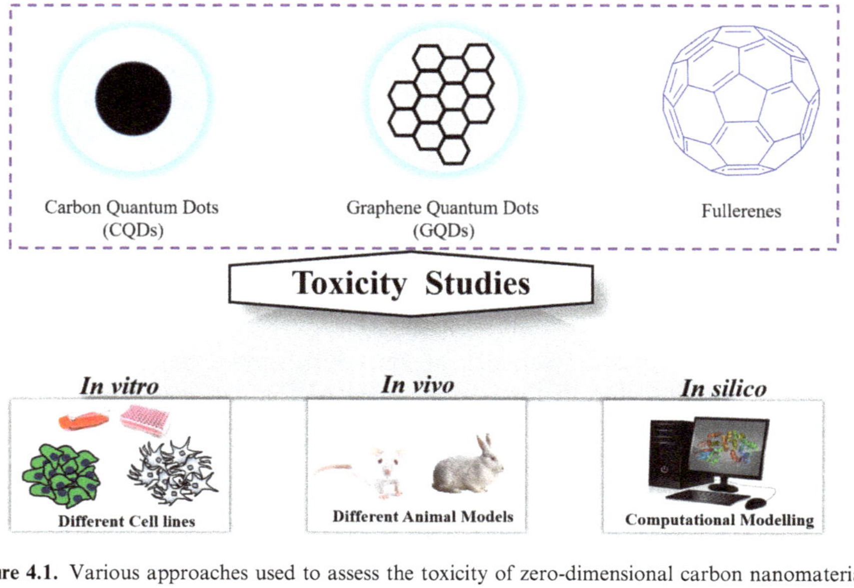


**Figure 4.1.** Various approaches used to assess the toxicity of zero-dimensional carbon nanomaterials.

The NAD(P)H-dependent oxidoreductase enzymes present in the active mitochondria of viable cells reduce the yellow MTT reagent and produce purple insoluble formazan crystals [13]. The insoluble formazan crystals can be solubilized by adding suitable solubilization solutions such as dimethyl sulphoxide (DMSO) and can be quantified spectrophotometrically by detecting absorbance at 500 to 600 nm using a multi-well plate reader. The amount of formazan crystal will be proportional to the number of viable cells present in the population [14]. The MTT is positively charged and can penetrate the cell. However, the modified tetrazolium salts, such as 3-(4,5-dimethylthiazol-2-yl)-5-(3-carboxymethoxyphenyl)-2-(4-sulfophenyl)-2H-tetrazolium (MTS), 2,3-bis-(2-methoxy-4-nitro-5-sulfophenyl)-2H-tetrazolium-5-carboxanilide (XTT), and water-soluble tetrazolium salts (WST-1, -3, -4, -5, -8 etc), used for cell viability assays, are negatively charged and are used with a combination of intermediate electron acceptor reagents [15]. They do not require any solubilizing solution addition step to solubilize the formazan crystals. Intermediate electron acceptors, such as phenazine ethyl sulphate or phenazine methyl sulphate, used along with MTS can penetrate the cell and be reduced either in the cytoplasm or on the cell surface itself and exit the cells, further converting the tetrazolium salt to the coloured soluble formazan crystal. A sodium 3′-(1-(phenylaminocarbonyl)-3,4-tetrazolium)-bis (4-methoxy6-nitro) benzene sulphonic acid hydrate assay (XTT) also quantifies the viable cells based on the ability of active cells to convert the yellow tetrazolium salt to orange formazan dye, which can be measure directly by using a multiwall plate reader [16].

Also, the cell viability and cytotoxicity of cells can be measured using resazurin, the active component present in Alamar blue. Metabolically active cells can reduce the non-fluorescent blue dye, resazurin, to a fluorescent pink compound, resorufin. The formation of resorufin can be measured by either fluorimetry or colorimetry [17, 18]. Unlike the MTT assay, the resazurin based cell viability assay does not require any solubilizing agent addition stage.

Neutral red or 3-amino-7-dimethylamino-2-methylphenazine hydrochloride is a weakly cationic dye that can penetrate the cell membrane through non-ionic passive diffusion. It concentrates and retains in the lysosome upon cell entry, where it becomes charged [19]. Hence, the neutral red uptake assay is based on the viable cells' ability to maintain the pH gradient by producing ATP. Dead cells cannot maintain the pH gradient; hence, the neutral red dye would not be retained in the lysosome. The retained chromophore can be extracted using acidified ethanol solution. The released dye can be quantified by measuring the absorbance at 540 nm using a multi-plate reader, and it will be proportional to the cell number present in the population [20].

Cell death or cell membrane damage can lead to the release of lactate dehydrogenase (LDH), a soluble cytoplasmic oxidoreductase enzyme that catalyses pyruvate to lactate interconversion to the surrounding environment [21]. The released LDH can be quantitatively measured by spectrophotometrically using a tetrazolium salt, iodonitrotetrazolium (INT). The reduced nicotinamide adenine dinucleotide (NADH) produced by LDH during oxidation of lactate to pyruvate enables the conversion of yellow tetrazolium salt to a red formazan compound in the presence of an electron receptor [22]. The formation of formazan product will be directly proportional to the LDH released and hence the cell damage or death in the population [23]. LDH release can also be measured using fluorometric analysis with the aid of resazurin. The NADH produced during the oxidation of lactate to pyruvate converts the non-fluorescent resazurin to highly fluorescent resorufin.

Dichloro-dihydro-fluorescein diacetate (DCFH-DA) assay is another important assay that determines cytotoxicity through the amount of reactive oxygen species (ROS) generated inside the cells [24]. Quantification of ROS is highly significant since it is directly linked to the viability of cells. Increased ROS generation can lead to cellular apoptosis and autophagy.

The non-destructive and inexpensive sulforhodamine B (SRB) assay measures the cellular protein content, and it can be correlated with the total number of cells present in the population [25]. The aminoxanthane dye, sulforhodamine B, binds to the protein content present in the cell through the electrostatic interaction with primary amino groups under a mildly acidic environment. Hence, a fixation step using trichloroacetic acid is required. The development of colour can be measured by adding a solubilizing agent (10 mM Tris base solution) and spectrophotometrically using a multi-plate reader at 510 nm [26].

### 4.2.2 *In vivo* analysis

*In vivo* methods of toxicity analysis are performed in animals, in particular rats, mice, and rabbits. The *in vivo* characterization gives an overall evaluation of toxicity in the whole organism. The kinetics of a nanomaterial that enters a living system must be evaluated thoroughly. The methods to assess the absorption, distribution, and elimination of nanomaterials are primarily characterized to evaluate the toxicity. Histopathology, haematology, and serum chemistry analysis are performed in experimental animals to assess the toxicity [10]. However, although *in vivo* animal tests can provide physiologically relevant data towards the study of nanomaterial toxicity, they are also associates with ethical issues. Animal study experiments should be carried out using the 3Rs principle—replace, reduce, and refine. The main goal of implementation of the 3Rs principle is to reduce the animals required for experiments. Therefore, various agencies, such as Organization for Economic Cooperation and Development (OECD), have implemented guidelines for acute toxicity experiments, such as eye irritation, dermal toxicity, lethal dose 50, oral toxicity, etc. The acute oral toxicity test involves the oral administration of nanomaterials in animals. The toxicological response can be observed by monitoring the potential skin symptoms or toxic signs, including losing body weight and behavioural changes, or changes in food or water consumption. Various biochemical parameter analyses, such as haematological and histopathological analyses, also enable the evaluation of the material's potential toxicity. Acute eye irritation involves applying nanomaterials in the conjunctival sac of one eye of the animal, and the other eye serves as a control. The eye reaction needs to be monitored for evaluating the toxicity of the material. For dermal exposure, the material is applied to the skin. The monitoring of skin symptoms or toxic signs and histopathological analysis of skin biopsies will give the toxicity profile of the materials [27].

### 4.2.3 *In silico* analysis

The combination of *in vitro* experiments with computer-based modelling enhances the understanding of cellular mechanisms at the atomic levels. With the motive of producing biologically safer nanomaterials, *in silico* modelling is a novel method that could replace animal experiments. The practical impossibility of screening all the nanomaterials synthesized for their toxicity has paved the way for bioinformatics based computational applications for determining the toxicity. Several databases, including ModNanoTox, NanoHub, NanoDatabank and NanoMILE, and OCHEM, allow nanoparticle toxicity detection. In nanotoxicology studies, molecular dynamics (MD), molecular docking, and quantitative structure–activity relationship (QSAR) studies are the bioinformatics tools used to predict nanomaterial toxicity [28].

The interaction of small molecules with larger molecules through computational studies is performed in molecular docking studies. All the possible orientations and conformations between larger and smaller molecule are elucidated. Docking methods are employed to analyze the toxicity by understanding the binding affinity and orientation between the molecules. Another exciting and advantageous method is the QSAR assay. It is a computer-based assay that calculates the toxic effects as a function of various molecular descriptors. Machine learning and mathematical statistics are used to interpret the linkage between chemical structure and biological activity. The QSAR assay is one of the potential assays to predict the toxicity profile of newly discovered nanomaterials [29]. For example, the cytotoxicity of fullerenes to *Salmonella typhimurium* was detected using quasi-SMILES descriptors obtained with the Monte Carlo method [30]. The physicochemical properties of nanomaterials are investigated using molecular dynamics. The ethical consideration in the conventional toxicity assessment using animal models can be completely ignored in molecular dynamics studies. Moreover, the combination of all three approaches can significantly contribute to the detailed understanding of the safety profiles and future applications of zero-dimensional carbon-based nanomaterials.

## 4.3 Physicochemical factors affecting the toxicity

Like other nanomaterials, several factors contribute to the toxicity of CQDs, GQDs, and fullerenes [31]. The toxicity is mainly dependent on the physicochemical parameters of the material. Starting from the synthesis route of zero-dimensional nanomaterials, the size of the material, morphology of the material, surface functionalization, charge over the surface, and doping with another atom can contribute towards the toxicity as mentioned earlier (figure 4.2). These factors can change the mode of interaction of the nanomaterials with cell organelles leading to toxic responses.

### 4.3.1 Synthesis route

The synthesis route is one of the significant issues related to the toxicity of nanomaterials. Carbon-based zero-dimensional nanomaterials such as carbon and

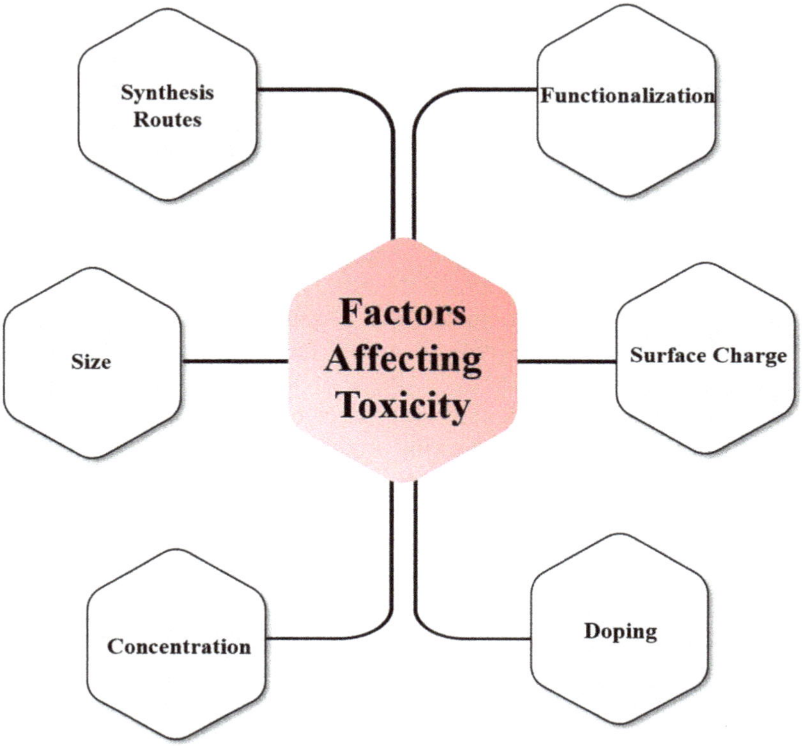


**Figure 4.2.** Illustration showing the different factors that affect the toxicity of zero-dimensional carbon nanomaterials.

graphene quantum dots are synthesized mainly through either bottom-up or top-down approaches. The top-down approach involves the synthesis of the desired nanomaterial from a bulk material by crumbling it down into smaller materials. Bottom-up refers to the synthesis of the nanomaterial from the smallest precursor using chemical or physical forces to assemble the molecules. Even though there are various reports on the toxicity of zero-dimensional carbon nanomaterials, no conclusive studies or reports are available for detailed toxicity. The starting materials required for the synthesis vary from simple to concentrated acids according to the synthesis approach. Typically, the top-down approach requires highly toxic reagents and concentrated acids or alkalis in the reaction schemes. Moreover, the cutting down of bulk material in the top-down approach results in sharp edges of the nanomaterials, affecting cell viability.

The different synthesis routes of CQDs may contribute to the toxicity of these particles. The importance of various synthesis routes in the toxicity profiles of carbon quantum dots is unexplored. CQDs produced by microwave synthesis from glucose were analysed for their toxicity. 16HBE cells treated with different concentrations of CQDs (1 to 200 $\mu g\ mL^{-1}$) were exposed to the cells to determine the toxic effects. The CQDs reduced cell viability and induced oxidative stress in human bronchial epithelial cells. Although the cells showed signs of cytotoxicity, further studies are needed to confirm whether these effects arise due to the cell lines preferred or the characteristic differences in the CQDs used in the study [32]. Citric acid pyrolysis is a common technique used in the production of CQDs that find its application in drug delivery and bioimaging. The toxicity of carbon quantum dots produced through citric acid pyrolysis was studied on breast cancer cell lines [33]. Even after the exposure of CQDs for 24 h, the cells showed less toxicity. The study pointed out that the toxic nature of CQDs arises only at higher concentrations and after a prolonged period of irradiation; that is, the toxicity is dependent on the concentration and exposure time. The optimization of synthesis protocols can thereby reduce the cytotoxicity of CQDs.

### 4.3.2 Size

A material is defined to be in its nano form when its size is estimated to be below 100 nm. The toxicity of nanomaterials depends greatly on their size. The entry and exit of nanomaterials in and out of a biological system depend on their size. The resident times of these nanomaterials in the systems determine the toxicity of the materials. The behaviour of nano-sized materials in biological systems can be very complex and could even alter the cell's normal behaviour, leading to toxic effects. The distribution of nanomaterials in *in vivo* systems can be attributed to the size of these materials. The particle size is inversely proportional to the biological and chemical activity [34].

Carbon nanomaterials are reported to exhibit size dependent toxicity in various cell lines. The improved biocompatibility of GQDs when compared with GO is primarily due to their size difference. Micrometre sized graphene oxide induced significant cytotoxicity when compared to GQDs. The cellular internalization of nanometre sized GQDs was greater compared to GO. The intracellular ROS, loss of mitochondrial potential, and cell cycle aberrations were significantly lower for GQDs [35]. The toxicity of CQDs, GQDs, and GO was assessed in green gram sprouts. The CQDs and GQDs were found to have lower cytotoxicity when compared to GO and CNTs [36]. The absence of synthesis methods to produce controlled size fullerenes restricts the size-based toxicity analysis of fullerenes. The size dependent cytotoxicity of C60 fractions were assessed in A549 cell lines. The study revealed that fullerenes with smaller size have higher toxicity compared to the fractions with higher size [37].

### 4.3.3 Concentration (dose)

Concentration or dosage is also essential when it comes to the toxicity of zero-dimensional carbon nanomaterials. Most studies reported the concentration-dependent toxicity of nanomaterials. The toxic concentration of each nanomaterial can vary depending upon the surface charge, surface functionalisation, and the type of cell used. Most studies report that the zero-dimensional GQDs were non-toxic in low concentrations and toxic in high concentrations [38]. In all cases, the carbon-based zero-dimensional materials were non-toxic up to a limit of 100 g $mL^{-1}$.

Fullerenes induced cell death and protected the cells from oxidative stress in a concentration-dependent mechanism. They exhibited these dual effects on hippocampal neurons. The neuron viability was increased in lower concentrations, and higher concentrations exhibited neuronal cytotoxicity on MTT analysis [39].

### 4.3.4 Surface functionalization

Surface functionalization of nanomaterials is primarily done to tune the nanomaterial's physical and chemical properties and thereby enhance their biocompatibility. Surface chemistry has proven to be an essential factor determining the toxicity of nanoparticles. The functionalization of carbon nanomaterials with various biologically relevant molecules and facilitating their applications in drug delivery and bioimaging are being researched extensively. The chemical nature and charge of the functional groups greatly determine the toxicity of surface-functionalized carbon nanomaterials. Surface functionalization can also alter the cellular uptake pathway and intracellular traffic, which can directly or indirectly affect the toxicity of these nanomaterials. Various amino acids, biopolymers, and biomolecules are being used for coating the surfaces of nanomaterials. Depending on the chemical nature of the coating materials, the surface functionalization can render charge over the surface of zero-dimensional carbon nanomaterials. The effect of surface charge on the toxicity of zero-dimensional carbon nanomaterials is discussed in the next section.

The emission wavelength and efficacy are determined greatly by the surface chemistry of GQDs. The surface chemistry also determines the toxicity of these nanoparticles. The cell toxicity of GQDs surface-functionalized with ethylenediamine in HeLa cells showed very low toxicity, and the cell viability was above 80% up to a concentration of 125 $\mu$g mL$^{-1}$. The organic molecule functionalization is shown to have little effect on *in vitro* toxicity [40]. The GQDs were surface-functionalized by carboxylation, and the resultant GQDs exhibited increased solubilization. The *in vitro* cytotoxicity analysis of carboxylated GQDs showed no evident acute toxicity or any morphological changes. The intravenous exposure of carboxylated GQDs for the assessment of *in vivo* toxicity indicated its accumulation in the spleen, kidney, liver, and tumours. Haematological analysis revealed the non-toxic nature of carboxylated GQDs in rats at dosages of 5 and 10 mg kg$^{-1}$ [41].

Surface modification enhanced the effects of fullerenes *in vitro* and *in vivo*. The surface modification improved the solubility of fullerenes in different solvents advantageous to biological systems. The modification of fullerenes from hydrophobic to hydrophilic could greatly impact the cellular interactions, thereby reducing the toxicity and enhancing biocompatibility. The hydroxyl group functionalization of fullerenes increased the cellular interactions of the material as they increased the solubility of fullerenes. While elucidating the toxicity of such hydroxylated fullerenes there were contradictory results. The cytotoxic effects of hydroxylated fullerenes were studied in L929, rat C6 glioma, and human glioma cell lines. The hydroxylated fullerenes were shown to delay cell death compared to the pure fullerenes [42]. Hydroxylated fullerenes evidently showed cytotoxicity in

Chinese hamster ovarian (CHO) and lung (CHL) cells but had no effects on L929 fibroblasts. The $LC_{50}$ was found to be L929 < CHO < CHL [43]. The cytotoxicity of HEK cells with fullerenes having a different degree of hydroxylation was assessed, and the results show that the highest concentration of the highest degree of hydroxylated fullerenes was toxic to HEK cell lines [44].

### 4.3.5 Surface charge

Surface functionalization with positively or negatively charged molecules can impart charge to CQDs, GQDs, and fullerenes. The charge of the QDs can be of greater interest due to the fact that the charge contributes to the uptake as well as the toxicity of the material. The surface charge can be negative, positive, or neutral. The covalent or non-covalent functionalization with amine moieties obtains a positively charged surface, whereas the negatively charged surface contains carboxylic acid functionalization. Surface charge can be substantial in the retention and clearance of nanomaterials in the body.

To assess the role of surface charge in toxicity, carbon QDs were functionalized with various molecules bearing charged moieties [45]. The synthesized CQDs themselves were negatively charged whereas polyethylenimine (PEI) and polyethylene glycol (PEG) functionalization imparted positive and negative charge, respectively. Positively charged CQDs exhibited the highest toxicity when compared to the negatively charged and neutral CQDs. The pristine CQDs with a negative charge over the surface did not induce any toxicity towards the cells. The authors stressed the importance of a neutral charge over the CQD surface for various biomedical applications.

The effect of different surface charges of CQDs was studied in human umbilical cord mesenchymal stem cells (hUCMSCs). The charged CQDs showed more uptake in the cell compared with uncharged CQDs. The cell uptake differentiation and proliferation of the mesenchymal cells suggested that higher positive charges induced higher cytotoxicity. Even so, the functionalization of a weakly positive charged molecule could possess higher biocompatibility and improved labelling efficiency in hUCMSCs [46].

In the case of graphene quantum dots, the treatment of anionic GQDs synthesized using acid functionalization at higher concentrations showed excellent biocompatibility when treated with human umbilical vein endothelial cells (HUVEC). There was no change in the cell morphology even at a concentration greater than 300 $\mu$g $mL^{-1}$ [47]. In contrast, the neutral GQDs exhibited toxicity when compared to the carboxylic acid functionalized GQDs. Similarly, another study showed the influence of charge over the GQDs in the uptake and clearance in melanoma cells. The synthesized GQDS showed similar size but with two different surface charges (positive and negative). Both the GQDs exhibited excellent biocompatibility but there was a difference in the clearance. Amine functionalized (positive charged) GQDs showed a rapid clearance when compared to the slow clearance of the acid functionalized (negatively charged) GQDs [48].

### 4.3.6 Doping

Doping of nanomaterials improves the fluorescence intensity or fluorescence efficiency, which is of great interest in bioimaging. Doping of zero-dimensional carbon nanomaterials normally uses the incorporation of heteroatoms such as boron, nitrogen, fluorene, sulphur, and phosphorous into the general core of the nanomaterials.

Carbon QDs were doped with nitrogen and lanthanides to improve the imaging applications [49]. The cytotoxicity analysis of CQDs co-doped with N, Gd, and Yb on HeLa cells, NIH-3T3 cells, and Chinese hamster ovary cells showed that the doped CQDs did not significantly affect the cell viability or proliferation. The carbon quantum dots exhibited high stability in biological media and enhanced the multi-model bioimaging. Coating CQDs with amino acids aids in tuning the physicochemical properties of CQDs. Amino acid functionalized CQDs did not show cellular toxicity on HeLa cells even at a concentration of 100 $\mu$g $mL^{-1}$. The biocompatibility of such functionalized carbon QDs was reported to be excellent and could be used for various biological applications [50].

The surface chemistry of graphene quantum dots also governs their toxicity. Doping GQDs with various materials has shown an increase in cellular viability. Reports suggest that doping raw GQDs with foreign atoms such as nitrogen and boron reduces the toxic effects. The boron-doped graphene QDs exhibited increased cell viability even at a concentration of 4.0 mg $mL^{-1}$ [51]. Cytotoxicity analysis of nitrogen-doped GQDs against HeLa cells also showed less toxicity and the cells were able to tolerate a concentration of up to 14 $\mu$g $mL^{-1}$. This study concluded that the GQDs were biocompatible in nature and suitable for application in bioimaging [52]. Since N doped GQDs have a potential application in cellular imaging, their *in vivo* toxicity and transgenerational effects were studied in *Caenorhabditis elegans*. The nematodes did not show any toxic behaviour when exposed to N-GQDs [53].

## 4.4 The toxicity of zero-dimensional carbon nanomaterials

### 4.4.1 Carbon quantum dots (CQDs)

The purification of single walled carbon nanomaterials through electrophoresis [54] accidentally led to the identification of carbon quantum dots, a novel kind of zero-dimensional carbon nanomaterial. CQDs caught the eye of the research field with their excellent physicochemical and biocompatibility characteristics. The conventional semiconductor quantum dots gained a potential competitor with the emergence of CQDs owing to their advantages such as chemical stability, environmental friendliness, stable photoluminescence, antibacterial properties, and lower cytotoxicity. The one step and simple synthesis approaches that can be used for the production of carbon dots also made them an excellent replacement for heavy metals, which had higher toxicity when incorporating into biological systems. Laser ablations [55, 56], pyrolysis [57], electrochemical synthesis [58], microwave assisted synthesis [59], and a number of green synthesis protocols [60, 61] are being used for the synthesis of carbon quantum dots. The applications of CQDs range from chemical and biosensing, bioimaging, electro- and photocatalysis, bioimaging, and nanomedicine [62]. This vast arena of applications also encourages researchers to analyze the potential toxicity of CQDs, in particular as they find multiple applications in the biological arena, including nanomedicine, and can affect the natural environment.

Carbon dots prepared from an improved nitric acid oxidation method showed low toxic effects on mice and rats under a specific dosage of carbon dots. No clinical toxic effects were observed in acute toxicity analysis in a comparison of the control (0.9% NaCl) and treated (51 mg carbon dots $kg^{-1}$ body weight) groups of animals. The major organs analysis did not reveal any abnormality and there was no evidence of genotoxicity either [63]. The carbon dots prepared from a cheap and natural source, orange juice, were also analysed for their photoluminescent properties and biocompatibility [64]. The CDs were efficiently taken up by human osteosarcoma (MG-63) cells and showed no cytotoxicity even at a high dose (200 $\mu g\ mL^{-1}$) and long incubation time (72 h).

The cytotoxicity and *in vivo* toxicity of fluorescent carbon dots surface passivated by oligomeric PEG molecules intended for bioimaging applications were analysed. The fluorescent dots were non-toxic to human breast cancer MCF-7 and human colorectal adenocarcinoma HT-29 cells. The study also revealed that even after the prolonged exposure of CQDs (28 days), the treated mice did not show any significant toxicity responses at dosages above the specific values commonly used for imaging studies [65]. The toxicity of carbon dots was explored using a $^1H$ NMR-based metabolic approach supplemented with biochemical analysis and histopathology. There was no obvious evidence of any tissue impairments in the histopathology analysis. Except for the slight change in the albumin levels in serum and AChE activity in the liver and kidneys, there were no significant changes in the biochemical analysis either. Nevertheless, the study also reported that the CQDs affected the cell membrane, immune system, and normal liver clearance on analysing the NMR metabolomic profiles [66]. The *in vivo* systemic and repeated dose toxicity of citric acid-based carbon dots were evaluated in ICR mice [67]. The cellular toxicity effects in HepG2, HL-7702, MCF-7, and HUVEC cell lines showed no significant cytotoxicity. The biochemical and histopathological analysis did not reveal any abnormalities in the male and female mice compared to the control animals. The study concluded that the citric acid-based carbon QDs are biocompatible and hence are potential molecules for biolabeling and molecular imaging.

In contradiction to the previous reports, the cytotoxic evaluation of CQDs with positive, negative, and neutral charges conducted by Havrdova *et al* showed a toxic nature when treated with mice fibroblast cells NIH/3T3 (figure 4.3). The viability studies, ROS generation studies, and flow cytometric analysis suggested a significant change in the toxicity of CQDs with a difference in the surface charge [45]. The cell cycle analysis showed the nuclear entry of the PEI modified positively charged CQDs and the subsequent inhibition of the G0/G1 and G2/M phases. The pristine CQDs with a negative charge over the surface did not induce any toxicity towards the cells. Similarly, the neutral CQDs prepared through the modification using PEG also did not induce any toxicity, as evidenced by the ROS generation studies and cell cycle analysis.

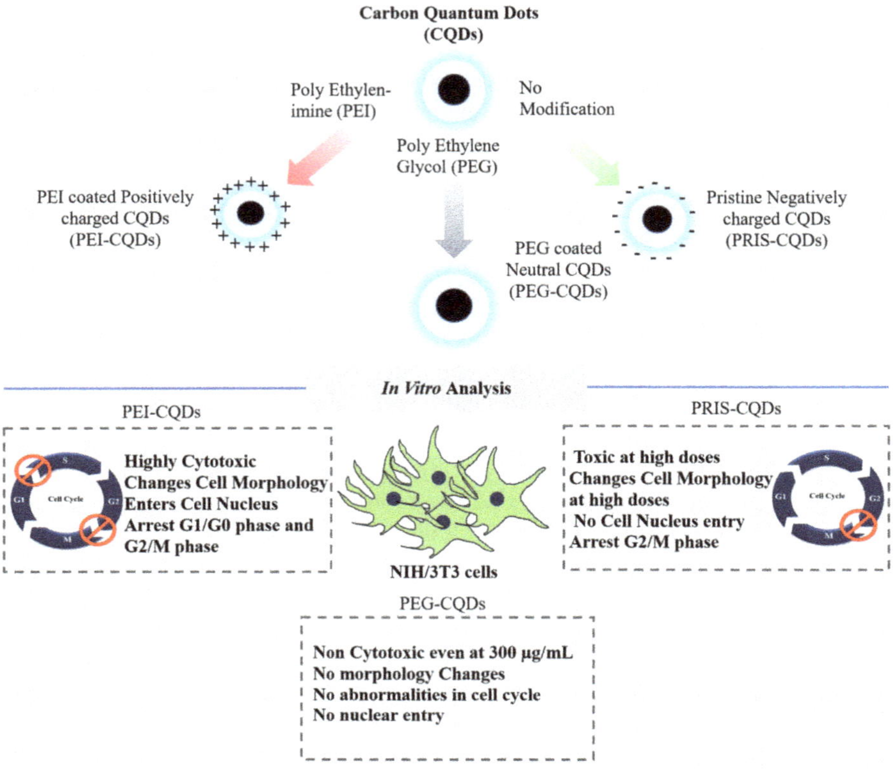


**Figure 4.3.** Toxic responses of CQDs.

CQDs are reported to enhance the biocompatibility and photocatalytic activities of various molecules. A CQD/$ZnFe_2O_4$ composite was synthesized and the *in vitro* toxicity of the composite was evaluated in A549 cells. Concentration-dependent cell viability was reported in the CQD/$ZnFe_2O_4$ composite and $ZnFe_2O_4$. The cell viability of CQD/$ZnFe_2O_4$ was higher compared to $ZnFe_2O_4$ which caused more lethal effects. Hence the CQD/$ZnFe_2O_4$ composite was marked biocompatible and enhanced photocatalysis [68].

Even though the biocompatibilities of CQDs are well studied, the ecological aspects of their toxicity have been explored less. Early studies confirmed that CQDs were non-toxic and safe for the environment. CQD probes were used in zebra fish for fluorescent imaging [69]. Using zebra fish as a model, the toxicity of CQDs was tested and it was reported that CQDs did not intervene with the development of the embryonic stages of zebra fish. However, the analysis of toxicity in algal systems proved an adverse result. The potential environmental hazards of CQDs were assessed by studying *Chlorella pyrenoidosa*, a freshwater green algae as a model organism [70]. The toxicity of CQDs at different trophic levels was analysed in zebra fish, zooplanktons, and phytoplankton [71]. The zebra fish showed negligible toxicity when exposed to a 200 $\mu$g mL$^{-1}$ concentration of CQDs. In contrast, the CQDs induced immobility and mortality in zooplanktons and inhibited nutrition absorption and photosynthesis in phytoplankton, all in a dose dependent and time dependent manner.

The structural and functional changes occurring in different embryos due to the exposure of various materials can be termed as developmental toxicity. Male and female embryos/larvae of rare minnows (*Gobiocyprisrarus*) were exposed to CQDs to investigate the toxic effects on different developmental stages. The study concluded that the rare minnow embryos exhibited developmental toxicity characterized by decreased spontaneous movements and body length, increased heart rate, and other developmental defects when exposed to higher concentrations of CQDs. The generation of oxidative stress, the inefficiency of DNA repair, and altered gene expression were the mechanisms that related to the developmental toxicity of the CQD treated embryos [72]. The fluorescence performance and the toxicity concerns of conventional semiconductor QDs have brought CQDs into the spotlight for *in vivo* imaging applications. Their excellent biocompatibility and photostability crown CQDs as next-generation molecules for *in vitro* and *in vivo* bioimaging.

CQDs, being the newest member of the group and possessing excellent biocompatibility, have been studied extensively for their effects on plant and plant growth promoting factors. Water-soluble luminescent carbon dots intended to be used as a biocompatible fluorescent ink exhibited low or zero toxicity to plants and animals. Bean sprouts grown in carbon dots containing solution showed no signs of toxicity and did not intervene in plant growth. The highly biocompatible nature of the carbon dots synthesized indicates their potential use in various applications of fluorescent ink technology [73]. The carbon dots were non-toxic and biocompatible in green gram sprouts compared to the CdTe QDs and Au NPs. The aqueous solution of 25 $\mu$g mL$^{-1}$ of CDs showed no signs of growth retardation or any other toxic behaviour even after 60 h of growth [74]. However, the toxic nature of CDs in plant physiology was also reported in addition to their biocompatible nature. The phytotoxicity assessment of CDs in *Arabidopsis thaliana* through a physiological, metabolomic, and transcriptomic analysis revealed that the CDs could cause a concentration-dependent phytotoxicity from a range of 62.5, 125, 250, 500, and 1000 mg L$^{-1}$. The study provided insight into the potential risk of CQDs and understand the biomarkers contributing to the toxicity [75]. Although studies are being carried out on the potential use of CQDs in agriculture, there is a need for more substantial research to elucidate the applications of CQDs in agriculture completely. So far, the studies in the literature show that CQDs are of no or low toxicity on different cell lines. The cytotoxicity of carbon dots is valid only at higher concentrations. These higher concentrations are surpassing those intended for their applications in a living system. The carbon quantum dots exhibit negligible toxicity *in vitro*, *in vivo*, and in environmental toxicity analysis. Even though the CQDs caused dose dependent toxicity they displayed a minimal reduction in cell viability at the concentrations required for bio-labelling.

### 4.4.2 Graphene quantum dots (GQDs)

As an essential member of the carbon family, GQDs have gained wide attention among the scientific community due to their multifaced applications. Owing to their unique features, such as a large surface area and excellent mechanical properties, GQDs have shown their immense versatility in various applications including sensing, energy conversion, bioimaging, nanomedicine, single electron transistors, spintronics, memory devices, and so on. It is necessary to discuss the synthesis of GQDs when it comes to toxicity. GQDs are widely synthesized using the top-down approach by cutting down large reduced graphene oxide, graphene oxide, carbon nanotubes, and carbon black into smaller particles of size less than 100 nm. Hence the synthesis approach can be a parameter leading to toxicity in the case of GQDs. Along with raw graphene quantum dots, other classes of doped quantum dots, such as nitrogen-doped GQDs and boron-doped GQDs, are also reported in the literature.

The safety of GQDs for biomedical applications is paramount due to their similar chemical nature compared to GO. *In vitro* and *in vivo* toxicity analysis of GO revealed its inherent toxicity. A detailed toxicological analysis will give information on the efficient distribution and clearance of GQDs from biological systems and identify cellular interactions that can be used as a starting point for future biomedical applications. The majority of the toxicity evaluation of GQDs revealed its non-toxic effects. Nurunnabi and co-workers explored the cytotoxicity and *in vivo* distribution of GQDs by synthesizing a new carboxylated green fluorescent GQD with improved water solubility using citric acid. The authors utilized KB, MDA-MB231, A549 cancer cells, and MDCK normal cell lines for identifying the toxicity of GQDs. *In vitro* studies at lower concentrations of carboxylated GQDs did not reveal any abnormalities in the cell lines. At higher (500 $\mu$g mL$^{-1}$) concentrations the human epidermal cancer cells were slightly sensitive to the GQDs with a viability percent less than 60. Overall, the *in vitro* studies using MTT and LDH assays showed the non-toxic nature of synthesized carboxylated GQDs. *In vivo* studies through intravenous injection of carboxylated GQDs in nude mice revealed the accumulation of GQDs in the liver, spleen tumour, and kidney after 24 h. The accumulation pattern suggested the distribution of GQDs on the whole body through blood circulation. Long term exposure of the GQDs for 22 days with different doses (5 and 10 mg kg $^{-1}$) also did not induce any toxic effects [41].

The studies of GQDs with human gastric cancer MGC-803 and breast cancer MCF-7 cells highlighted the cellular uptake and cytotoxicity of the GQDs. GQDs synthesized by a photo-Fenton reaction showed effective uptake through caveolae-mediated endocytosis. Shorter incubation with the cells ended up in the accumulation of GQDs in endoplasmic reticulum whereas more prolonged incubation showed nuclear accumulation. Both the cells showed excellent cell viability even after the treatment of 400 $\mu$g mL$^{-1}$ GQDs up to 24 h. Moreover, lower ROS generation and no mitochondrial damage by the GQDs towards the cancer cell lines showed the non-toxic effect of GQDs on these cancer cell lines [35]. Zhang and co-workers also reported the non-toxic nature of the synthesized GQD and

GQD-PEG through *in vitro* and *in vivo* analysis. The results indicated the fast metabolizing nature of GQDs in kidneys and tumours, retaining this ability irrespective of exposure route [76].

To elucidate the effects and mechanistic aspects of the interaction of GQDs with the immune system, Zhou and co-workers exposed different concentrations of GQDs to THP-1 macrophages. The investigation concluded the drastic dose dependent effects of GQDs towards macrophages. At lower concentrations, the GQDs exhibited negligible toxicity, whereas at higher concentrations they generated high levels of ROS. This, in turn, ended up in increased levels of Bax, Bad, and activated caspase 3, caspase 9, and beclin 1, etc, leading to apoptotic and autophagic cell death. Cytokine production was lower in high concentrations compared to low concentrations of GQDs. GQDs also induced an inflammatory response via the p38MAPK and NF-kB mediated signalling pathways in THP-1 activated macrophages, as shown in figure 4.4 [77].

Although *in vitro* studies exhibited the non-toxic nature of synthesized GQDs towards human cell lines, there is still uncertainty regarding the safety of GQDS when it comes to practical biomedical applications. The hidden mechanistic aspects of the cellular interaction of GQDs and their less explored side effects of the GQDs are the major hurdles hindering its broad biomedical applicability.

Genotoxicity evaluation of GQDs on NIH-3T3 cells showed the toxic nature of GQDs synthesized using the conventional Hummers' method. The treatment of GQDs with an average lateral size of 40 nm revealed elevated levels of gene expression of DNA damage related proteins such as p53, Rad 51, and OGG1. These results were surprising due to the non-toxic behaviour of GDS at the cellular level.

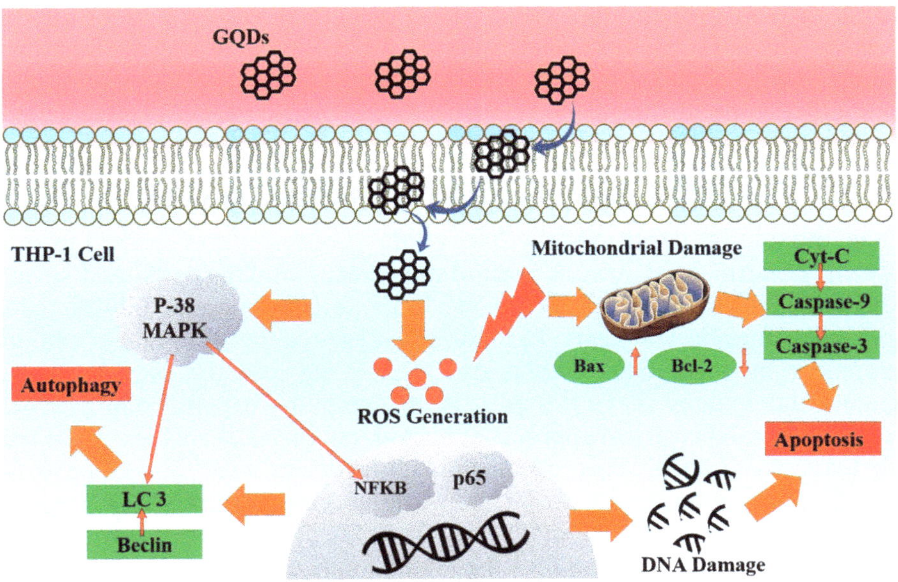


**Figure 4.4.** Toxic response and mechanism of GQDs.

The ROS generation of the GQDs was pointed out to be the primary reason for the DNA damage. The DCFH-DA assay to determine intracellular ROS generation confirmed the increased generation of ROS after treatment for 24 h [78].

Functionalized GQDs were also not so far ahead in the race when it comes to toxicity. Interactions between RBC membranes and nitrogen-doped GQDs of smaller size reported the effect of size in altering the morphology and haemolysis of red blood cells. Surface-enhanced infrared absorption spectroscopy (SEIRAS) studies showed that the hydrophilic nature of the N-GQDs resulted in reduced haemolysis compared to GO. The hydrophilic N-GQDs interacted with the hydrophilic inner membrane of the RBC, thereby disturbing the conformation and structure of the membrane layer resulting in the formation of echinocytes. These slight morphological changes did not affect the membrane integrity. SEIRAS studies highlighted the importance of considering the size and chemical nature of the GQDs intended for biomedical applications [79].

Wang and co-workers evaluated the plausible transgenerational issues and their mechanisms related to the exposure of GQDs on *Caenorhabditis elegans* as an animal model. The results suggested the non-toxic behaviour of synthesized N-GQDs towards wild type as well as mutant nematodes. N-GQDs tend to accumulate in the intestines of both wild type and mutant models. Moreover, there were no signs of N-GQD accumulation in the embryos and F1 progeny of treated nematodes. This indicates the efficient removal of N-GQDs from the nematode body through excretion without crossing the intestinal barriers and getting into the secondary organs [53]. *In vivo* studies on the exposure of zebra fish to N-GQDs revealed the substantial disturbance of the endogenous redox enzyme system through stopping anti-oxidant enzyme activities. Furthermore, the N-GQDs triggered the enzyme cytochrome P450 mono-oxygenase in the embryos of zebra fish. Altogether, the N-GQDs completely affected the anti-oxidant enzyme functions by inhibiting various electron transfer processes [80].

In another study, the exposure of normal GQDs caused harmful effects to zebra fish embryos with an increased mortality rate and a decrease in heart rate in a dose dependent manner. A larval behavioural assay showed a drastic decrease in the swimming distance and speed of the larvae while exposing them to high concentrations. In addition, exposure to a 200 $\mu g\ mL^{-1}$ concentration resulted in various embryonic abnormalities such as precardial oedema, vitalline cysts, and spine and tail bending. Fluorescence images showed the accumulation of GQDs predominantly in the intestine and heart. These results suggested the inhibition of embryonic development by the GQDs with a size of 2–5 nm and the prevalence of the developmental toxicity of GQDs [81]. Along with the experimental approaches there were several theoretical approaches to studying the mechanistic aspects related to the toxicity of GQDs. Theoretical simulations determined the relation between the orientation and interaction of GQDs with the cell membrane. GQDs can enter into the cell membrane quickly and disrupt them when they interact through a sharp corner. The sharp corner, as well as the closed angle configuration, can effectively minimize the interactive free energy. Most of the computational studies revealed the atomic details of the penetrating nature and interactions of GQDs into the membrane [82].

Overall, the *in vitro, in vivo,* and computational studies do not give conclusive information regarding the toxicity of GQDs. In most cases the synthesized GQDs were found to be non-toxic at a lower concentration. Mainly, GQDs showed dose dependent cytotoxic behaviour in all the reports. A higher concentration of GQDs irrespective of the surface modifications showed toxicity in all the approaches. However, there remains a critical gap in the understanding of the underlying mechanism of the toxicity [38]

### 4.4.3 Fullerenes

For three decades of research, fullerenes have engaged the interest of scientific communities in exploring its multidisciplinary applications. Fullerenes are much exploited for their anti-oxidant properties for applications where cell damage by oxidation is preferred and in applications where oxidative damage is advantageous, in particular in non-biological applications. The low solubility of fullerenes in water and polar solvents was considered a disadvantage. However, conjugation with molecules imparting hydrophilic surfaces increased their solubility. The functionalization of fullerenes was exploited by researchers around the globe to improve their biological applications, in particular in the field of drug delivery and nanomedicine. The alteration of the surface properties of fullerenes produced several fullerene derivatives with diverse properties [83].

The toxicity of different water-soluble aggregates of fullerenes was analysed in Chinese hamster lung fibroblasts cells. The cytotoxicity analysis and apoptosis assay did not reveal any morphological differences, suggesting that the cells were viable in the culture medium with different aggregates of fullerenes. The maximum dose of the fullerene aggregates did not increase the apoptosis rate. The different aggregate sizes of fullerenes did not impart any toxicity to cells [84].

Since fullerenes are widely used in many consumer products, including cosmetic formulations and therapeutics, toxicity should be of significant concern. This could also result in the release of fullerenes into ecological systems and cause severe environmental burdens. Toxicity of fullerenes may arise due to the solvents used for their preparation, or the by-products. The generation of ROS, activation of caspase 3/7, and up-regulation of gene expression of immune system processes that could severely alter the health of cells are the proposed mechanisms inducing toxicity (figure 4.5). The ROS mediated cytotoxicity of fullerenes was analysed with the help of experimental and mathematical modelling in different suspensions of fullerenes [85]. The study was conducted in various cell lines and concluded that fullerenes' toxicity depends on the residual solvents.

The fullerenes exhibited dose dependent and cell type dependent toxicity in *in vitro* and *in vivo* analysis. There was no relevant toxicity of fullerenes in HEK 293 cells up to 18 $\mu g\ mL^{-1}$. Above this concentration there was an expected toxic response. The intraperitoneal administration of fullerenes in mice also showed dose dependent toxicity. There was no evidence of toxicity in the histological analysis in a

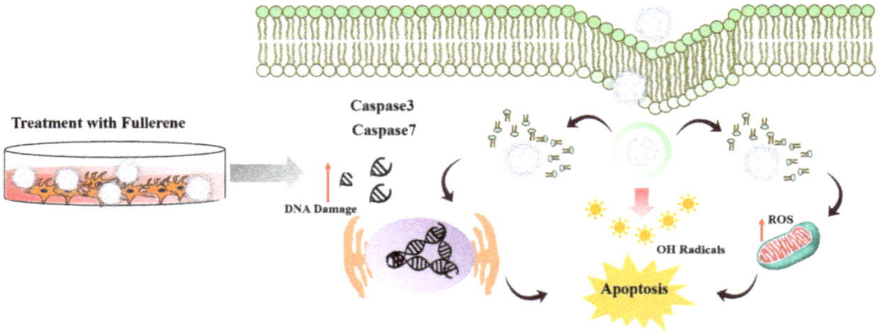


**Figure 4.5.** Toxic responses of fullerenes.

concentration between 75 and 150 mg $kg^{-1}$. $IC_{50}$ and $LD_{50}$ confirmed the low toxicity of fullerenes C60 [86]. Various studies conducted around the globe concluded that fullerenes do not impose any significant toxicity. No acute or sub-acute toxicity was observed with pristine C60. The functionalisation of fullerenes with biocompatible compounds was also shown to behave with less toxicity.

The solubility of pristine fullerenes limits their biological application. The functionalization of fullerenes can increase the aqueous solubility of fullerenes with various functional groups. Among the functionalization carried out on fullerenes, polyhydroxy fullerenes do not demonstrate any chronic or acute toxicological effects on freshwater organisms; however, they increased the culture density and simulated the reproduction of the green algae *Pseudokirchneriella subcapitata*, increased the lifespan of the invertebrate *Ceriodaphniadubia*, enhanced the growth of *Aspergillus niger*, and enabled longer hypocotyls in *Arabidopsis thaliana* without altering any complex physiological processes [87]. However, polyhydroxy fullerenes on isolated mitochondria demonstrated a dose dependent swelling and mitochondrial permeability transition which further reduced mitochondrial membrane potential [88]. Underivatized fullerenes are scarcely soluble in water, and form water stable aggregates. The hydroxylation of fullerenes significantly decreases the cytotoxicity associated with pristine fullerenes. Functionalisation with compounds such as dextran was performed to elucidate the toxic response of fullerenes. L929 fibroblasts showed cell viability above 80% when treated with dextran coated fullerenes of concentrations up to 80 $\mu g\ mL^{-1}$. The coated fullerene exhibited dose dependent ROS generation. L929 fibroblasts did not show any difference in mitochondrial potential or lysosome integrity. The nucleus was intact, and the actin filaments were stable [89]. However, in C6 glial cells the dextran coated fullerenes exhibited a deteriorating cellular mechanisms with increasing concentration of particles. The cellular viability, mitochondrial membrane potential, and lysosomal integrity were affected with increasing concentration of dextran coated fullerenes [90].

Fullerenes are good organic contaminant absorbers. Hence the presence of fullerene can modify the overall toxic effects of organic micro-contaminants. Fullerenes can enter into river and other aquatic systems and interact with various contaminants present in the ecosystem. A study conducted by Freixa and co-workers demonstrated the potential consequences of the exposure of river biofilm to fullerene and fullerene with organic micro-contaminants (OMC) . The exposure to fullerene did not cause any toxic effect on river biofilm. However, the exposure of OMCs along with fullerenes modified the OMCs' toxicity. The exposure of diuron with fullerene reduced its toxicity, but the exposure of triclosan with fullerenes increased its toxicity. This may be attributed to the structural difference in the OMCs, the proportion of OMCs and fullerenes in aquatic systems, and the competition of fullerene contaminants blocking the biological cell membrane receptors [91]. Various therapeutic bioactivities of fullerenes have been successfully studied by multiple research groups. Although the unique physicochemical characteristics of fullerene can contribute to a wide variety of biomedical applications, its toxicity needs to be studied thoroughly to determine the proper dose for optimum medical applications. Therefore, fullerenes have been studied for potential toxicity in various cell lines and organisms.

The increased application of fullerenes in different fields is expected to cause various types of exposure toxicity in humans. Fullerenes could enter the body through oral, nasal, and dermal routes and as a consequence might elicit toxic responses in the skin, lungs, and gastrointestinal tract. Therefore, understanding the toxicokinetics of fullerenes is important. The natural defence mechanisms of the human body have several barriers to prevent the entry of any foreign particles. This can prevent the entry, block the absorption, and inhibit the distribution of particles. If the fullerenes can breach these preventive mechanisms, there are possibilities for fullerene induced toxicity in the body [92].

The DNA damage caused by fullerenes and their derivatives was studied *in vitro* and *in vivo*. The exposure of 100 $\mu$g $mL^{-1}$ C60 did not induce DNA damage in mouse lung epithelial cells. The C60 did not induce cell death within 24 h or in a long-time exposure of 576 h. Strand breaks were not increased with C60 treatment, and the results indicated the low genotoxicity potential of C60 [93]. The genotoxicity of ethanol and water suspension of fullerenes were assessed for their genotoxic effects in human lymphocytes. This was claimed to be the first genotoxicity analysis for colloidal fullerenes. A comet assay revealed genotoxic effects in ethanol to water and aqueous suspensions of fullerenes; 2.2 $\mu$g $L^{-1}$ for the aqueous suspension and 4.2 $\mu$g $L^{-1}$ for the ethanol to water suspension induced genotoxicity [94]. The oxidative damage of DNA by the oral administration of fullerenes was tested in rats. 8-oxo-2′-deoxyguanosine accumulation within the cells is an indication of oxidative stress. The elevated level of 8-oxo-2-deoxyguanosine in the liver and lungs, along with the mRNA expression of 8-oxoguanine DNA glycosylase, indicated oxidative stress. This suggests the possibility of DNA damage [95].

Even though the dose is a major factor determining the toxic responses to fullerenes, several other factors need to be taken into account when mentioning toxicity. As mentioned in this chapter, several factors such as size, charge, and surface modifications can contribute to toxicity. To sum up, the toxicity of fullerenes must be thoroughly evaluated before use. A generalized overview cannot determine the individual toxicity of different fullerene derivatives. The collection of data regarding the *in vivo* toxic behaviour must be ensured before validation for medicinal applications.

## 4.5 Conclusion and future challenges

The zero-dimensional carbon nanomaterials, comprising CQDs, GQDs, and fullerenes have attracted researchers as they possess better biocompatibility than inorganic nanomaterials. Their small size and high surface-to-volume ratio provide them ascendancy in biological applications. Nevertheless, all novel materials intended for biological applications must undergo rigorous toxicity evaluation to analyse the possible hazards that could arise with the usage of the materials. Also, the exposure of these nanoparticles intentionally or unintentionally is expected. Therefore, understanding the toxicological overview is critical for zero-dimensional carbon nanomaterials due to their multifaceted applications in living systems. The toxicity of zero-dimensional carbon nanomaterials depends on various physicochemical characteristics such as size, concentration, surface chemistry, and synthesis routes. The interaction of these nanomaterials with *in vitro* and *in vivo* systems could also contribute to the toxicity of the materials.

A detailed toxicity analysis incorporating all these aspects would aid in predicting the effects of nanomaterials in living systems. In general, unlike the nanomaterials or quantum dots prepared from heavy metals, the biocompatibility of carbon and graphene quantum dots is comparatively high. Most of the current toxicity evaluations rely on the measurement of cell viability studies. However, the results from these studies can sometimes be unreliable. The toxicity studies and evaluation of these materials are still in their infancy due to the lack of conclusive evidence and mechanisms for the generated toxicity. Even though numerous reports are available on the toxicity of zero-dimensional nanomaterials, the detailed cellular interactions and mechanisms remain unclear.

## Acknowledgements

The authors wish to express their thanks to the Director and Head, Biomedical Technology Wing, Sree Chitra Tirunal Institute for Medical Sciences and Technology (Govt. of India), Trivandrum, Kerala, India, for their support and for providing the infrastructure to carry out this work. PVM, AVG, AA, and JX thank the Department of Science and Technology, Govt. of India, New Delhi for financial support (DST/TDT/DDP-04/2018(G)).

## References

[1] Scida K, Stege P W, Haby G, Messina G A and García C D 2011 Recent applications of carbon-based nanomaterials in analytical chemistry: critical review *Anal. Chim. Acta* **691** 6–17

[2] Patel K D, Singh R K and Kim H W 2019 Carbon-based nanomaterials as an emerging platform for theranostics *Mater. Horiz.* **6** 434–69

[3] d'Amora M and Giordani S 2018 Smart nanoparticles for biomedicine *Carbon Nanomaterials for Nanomedicine* ed G Ciofani 1st edn (New York: Elsevier) ch 7 pp 103–13

[4] Zhang R and Olin H 2012 Carbon nanomaterials as drug carriers: real time drug release investigation *Mater. Sci. Eng.* C **32** 1247–52

[5] Liu Z, Robinson J T, Tabakman S M, Yang K and Dai H 2011 Carbon materials for drug delivery and cancer therapy *Mater. Today* **14** 316–23

[6] Wang Z, Hu T, Liang R and Wei M 2020 Application of zero-dimensional nanomaterials in biosensing *Front. Chem.* **8** 320

[7] Dave K and Gomes V G 2019 Carbon quantum dot-based composites for energy storage and electrocatalysis: mechanism, applications and future prospects *Nano Energy* **66** 104093

[8] Mostofizadeh A, Li Y, Song B and Huang Y 2011 Synthesis, properties, and applications of low-dimensional carbon-related nanomaterials *J. Nanomater.* **2011** 685081

[9] Gupta R and Xie H 2018 Nanoparticles in daily life: applications, toxicity and regulations *J. Environ. Pathol. Toxicol. Oncol.* **37** 209–30

[10] Kumar V, Sharma N and Maitra S S 2017 *In vitro* and *in vivo* toxicity assessment of nanoparticles *Int. Nano Lett.* **7** 243–56

[11] Cha K E and Myung H J 2007 Cytotoxic effects of nanoparticles assessed *in vitro* and *in vivo* *J. Microbiol. Biotechnol.* **17** 1573–8

[12] Mukherjee A, Majumdar S, Servin A D, Pagano L, Dhankher O P and White J C 2016 Carbon nanomaterials in agriculture: a critical review *Front. Plant Sci.* **7** 172

[13] Mosmann T 1983 Rapid colorimetric assay for cellular growth and survival: application to proliferation and cytotoxicity assays *J. Immunol. Methods* **65** 55–63

[14] Hansen M B, Nielsen S E and Berg K 1989 Re-examination and further development of a precise and rapid dye method for measuring cell growth/cell kill *J. Immunol. Methods* **119** 203–10

[15] Berridge M V, Herst P M and Tan A S 2005 Tetrazolium dyes as tools in cell biology: new insights into their cellular reduction *Biotechnol. Annu. Rev.* **11** 127–52

[16] Roehm N W, Rodgers G H, Hatfield S M and Glasebrook A L 1991 An improved colorimetric assay for cell proliferation and viability utilizing the tetrazolium salt XTT *J. Immunol. Methods* **142** 257–65

[17] Page B, Page M and Noel C 1993 A new fluorometric assay for cytotoxicity measurements *in-vitro* *Int. J. Oncol.* **3** 473–6

[18] O'brien J, Wilson I, Orton T and Pognan F 2000 Investigation of the Alamar Blue (resazurin) fluorescent dye for the assessment of mammalian cell cytotoxicity *Eur. J. Biochem.* **267** 5421–6

[19] Repetto G, Del Peso A and Zurita J L 2008 Neutral red uptake assay for the estimation of cell viability/cytotoxicity *Nat. Protoc.* **3** 1125–31

[20] Ates G, Vanhaecke T, Rogiers V and Rodrigues R M 2017 Assaying cellular viability using the neutral red uptake assay *Cell Viability Assays* (New York: Humana Press) pp 19–26

[21] Burd J F and Usategui-Gomez M 1973 A colorimetric assay for serum lactate dehydrogenase *Clin. Chim. Acta* **46** 223–7

[22] Korzeniewski C and Callewaert D M 1983 An enzyme-release assay for natural cytotoxicity *J. Immunol. Methods.* **64** 313–20

[23] Chan F K, Moriwaki K and De Rosa M J 2013 Detection of necrosis by release of lactate dehydrogenase activity *Immune Homeostasis* (Totowa, NJ: Humana Press) pp 65–70

[24] Aranda A, Sequedo L, Tolosa L, Quintas G, Burello E, Castell J V and Gombau L 2013 Dichloro-dihydro-fluorescein diacetate (DCFH-DA) assay: a quantitative method for oxidative stress assessment of nanoparticle-treated cells *Toxicol. in Vitro* **27** 954–63

[25] Vichai V and Kirtikara K 2006 Sulforhodamine B colorimetric assay for cytotoxicity screening *Nat. Protoc.* **1** 1112–6

[26] Orellana E A and Kasinski A L 2016 Sulforhodamine B (SRB) assay in cell culture to investigate cell proliferation *Bio-protocol* **6** e1984

[27] Clichici S and Filip A 2015 *In vivo* assessment of nanomaterials toxicity *Nanomaterials: Toxicity and Risk Assessment* (London: IntechOpen) pp 93–121

[28] Huang H J, Lee Y H, Hsu Y H, Liao C T, Lin Y F and Chiu H W 2021 Current strategies in assessment of nanotoxicity: alternatives to *in vivo* animal testing *Int. J. Mol. Sci.* **22** 4216

[29] Puzyn T, Rasulev B, Gajewicz A, Hu X, Dasari T P, Michalkova A, Hwang H M, Toropov A, Leszczynska D and Leszczynski J 2011 Using nano-QSAR to predict the cytotoxicity of metal oxide nanoparticles *Nat. Nanotechnol.* **6** 175–8

[30] Shinohara N, Matsumoto K, Endoh S, Maru J and Nakanishi J 2009 *In vitro* and *in vivo* genotoxicity tests on fullerene C60 nanoparticles *Toxicol. Lett.* **191** 289–96

[31] Gupta A, Kumar S and Kumar V 2019 Challenges for assessing toxicity of nanomaterials *Biochemical Toxicology: Heavy Metals and Nanomaterials* (London: IntechOpen)

[32] Zhang X, He X, Li Y, Zhang Z, Ma Y, Li F and Liu J 2013 A cytotoxicity study of fluorescent carbon nanodots using human bronchial epithelial cells *J. Nanosci. Nanotechnol.* **13** 5254–9

[33] Esfandiari N, Bagheri Z, Ehtesabi H, Fatahi Z, Tavana H and Latifi H 2019 Effect of carbonization degree of carbon dots on cytotoxicity and photo-induced toxicity to cells *Heliyon* **5** e02940

[34] Ershov B G, Abkhalimov E V, Solovov R D and Roldughin V I 2016 Gold nanoparticles in aqueous solutions: influence of size and pH on hydrogen dissociative adsorption and Au (III) ion reduction *Phys. Chem. Chem. Phys.* **18** 13459–66

[35] Wu C, Wang C, Han T, Zhou X, Guo S and Zhang J 2013 Insight into the cellular internalization and cytotoxicity of graphene quantum dots *Adv. Healthcare Mater.* **2** 1613–9

[36] Li X, Zhou Z, Lu D, Dong X, Xu M, Wei L and Zhang Y 2014 The effect of pristine carbon-based nanomaterial on the growth of green gram sprouts and pH of water *Nanoscale Res. Lett.* **9** 1–6

[37] Song M, Yuan S, Yin J, Wang X, Meng Z, Wang H and Jiang G 2012 Size-dependent toxicity of nano-C60 aggregates: more sensitive indication by apoptosis-related Bax translocation in cultured human cells *Environ. Sci. Technol.* **46** 3457–64

[38] Wang S, Cole I S and Li Q 2016 The toxicity of graphene quantum dots *RSC Adv.* **6** 89867–78

[39] Zha Y Y, Yang B, Tang M L, Guo Q C, Chen J T, Wen L P and Wang M 2012 Concentration-dependent effects of fullerenol on cultured hippocampal neuron viability *Int. J. Nanomed.* **7** 3099

[40] Qian Z, Ma J, Shan X, Shao L, Zhou J, Chen J and Feng H 2013 Surface functionalization of graphene quantum dots with small organic molecules from photoluminescence modulation to bioimaging applications: an experimental and theoretical investigation *RSC Adv.* **3** 14571–9

[41] Nurunnabi M, Khatun Z, Huh K M, Park S Y, Lee D Y, Cho K J and Lee Y K 2013 *In vivo* biodistribution and toxicology of carboxylated graphene quantum dots *ACS Nano* **7** 6858–67

[42] Isakovic A *et al* 2006 Distinct cytotoxic mechanisms of pristine versus hydroxylated fullerene *Toxicol. Sci.* **91** 173–83

[43] Su Y, Xu J Y, Shen P, Li J, Wang L, Li Q, Li W, Xu G T, Fan C and Huang Q 2010 Cellular uptake and cytotoxic evaluation of fullerenol in different cell lines *Toxicology* **269** 155–9

[44] Saathoff J G, Inman A O, Xia X R, Riviere J E and Monteiro-Riviere N A 2011 *In vitro* toxicity assessment of three hydroxylated fullerenes in human skin cells *Toxicol. in Vitro* **25** 2105–12

[45] Havrdova M, Hola K, Skopalik J, Tomankova K, Petr M, Cepe K, Polakova K, Tucek J, Bourlinos A B and Zboril R 2016 Toxicity of carbon dots—effect of surface functionalization on the cell viability, reactive oxygen species generation and cell cycle *Carbon* **99** 238–48
[46] Yan J *et al* 2018 The effect of surface charge on the cytotoxicity and uptake of carbon quantum dots in human umbilical cord derived mesenchymal stem cells *Colloids Surf.* B **171** 241–9
[47] Yan C, Hu X, Guan P, Hou T, Chen P, Wan D, Zhang X, Wang J and Wang C 2020 Highly biocompatible graphene quantum dots: green synthesis, toxicity comparison and fluorescence imaging *J. Mater. Sci.* **55** 1198–215
[48] Ramana L N, Dinh L N and Agarwal V 2021 Influence of surface charge of graphene quantum dots on their uptake and clearance in melanoma cells *Nanoscale Adv.* **3** 3513–21
[49] Bouzas-Ramos D, Cigales Canga J, Mayo J C, Sainz R M, Ruiz Encinar J and Costa-Fernandez J M 2019 Carbon quantum dots codoped with nitrogen and lanthanides for multimodal imaging *Adv. Funct. Mater.* **29** 1903884
[50] Pandit S, Behera P, Sahoo J and De M 2019 *In situ* synthesis of amino acid functionalized carbon dots with tunable properties and their biological applications *ACS Appl. Bio Mater.* **2** 3393–403
[51] Hai X, Mao Q X, Wang W J, Wang X F, Chen X W and Wang J H 2015 An acid-free microwave approach to prepare highly luminescent boron-doped graphene quantum dots for cell imaging *J. Mater. Chem.* B **3** 9109–14
[52] Li R, Liu Y, Li Z, Shen J, Yang Y, Cui X and Yang G 2016 Bottom-up fabrication of single-layered nitrogen-doped graphene quantum dots through intermolecular carbonization arrayed in a 2D plane *Chem.—A Eur. J.* **22** 272–8
[53] Zhao Y, Liu Q, Shakoor S, Gong J R and Wang D 2015 Transgenerational safety of nitrogen-doped graphene quantum dots and the underlying cellular mechanism in *Caenorhabditis elegans Toxicol. Res.* **4** 270–80
[54] Xu X, Ray R, Gu Y, Ploehn H J, Gearheart L, Raker K and Scrivens W A 2004 Electrophoretic analysis and purification of fluorescent single-walled carbon nanotube fragments *J. Am. Chem. Soc.* **126** 12736–7
[55] Cui L, Ren X, Wang J and Sun M 2020 Synthesis of homogeneous carbon quantum dots by ultrafast dual-beam pulsed laser ablation for bioimaging *Mater. Today Nano.* **12** 100091
[56] Doñate-Buendía C, Fernández-Alonso M, Lancis J and Mínguez-Vega G 2020 Pulsed laser ablation in liquids for the production of gold nanoparticles and carbon quantum dots: from plasmonic to fluorescence and cell labelling *J. Phys.: Conf. Ser.* **1537** 012013
[57] Shang W, Cai T, Zhang Y, Liu D and Liu S 2018 Facile one pot pyrolysis synthesis of carbon quantum dots and graphene oxide nanomaterials: all carbon hybrids as eco-environmental lubricants for low friction and remarkable wear-resistance *Tribol. Int.* **118** 373–80
[58] Niu F, Ying Y L, Hua X, Niu Y, Xu Y and Long Y T 2018 Electrochemically generated green-fluorescent N-doped carbon quantum dots for facile monitoring alkaline phosphatase activity based on the $Fe^{3+}$-mediating ON–OFF–ON–OFF fluorescence principle *Carbon* **127** 340–8
[59] Eskalen H, Uruş S, Cömertpay S, Kurt A H and Özgan Ş 2020 Microwave-assisted ultra-fast synthesis of carbon quantum dots from linter: fluorescence cancer imaging and human cell growth inhibition properties *Ind. Crops Prod.* **147** 112209
[60] Sabet M and Mahdavi K 2019 Green synthesis of high photoluminescence nitrogen-doped carbon quantum dots from grass via a simple hydrothermal method for removing organic and inorganic water pollutions *Appl. Surf. Sci.* **463** 283–91

[61] Ramezani Z, Qorbanpour M and Rahbar N 2018 Green synthesis of carbon quantum dots using quince fruit (*Cydonia oblonga*) powder as carbon precursor: application in cell imaging and $As^{3+}$ determination *Colloids Surf.* A **549** 58–66

[62] Lim S Y, Shen W and Gao Z 2015 Carbon quantum dots and their applications *Chem. Soc. Rev.* **44** 362–81

[63] Wang K, Gao Z, Gao G, Wo Y, Wang Y, Shen G and Cui D 2013 Systematic safety evaluation on photoluminescent carbon dots *Nanoscale Res. Lett.* **8** 1–9

[64] Sahu S, Behera B, Maiti T K and Mohapatra S 2012 Simple one-step synthesis of highly luminescent carbon dots from orange juice: application as excellent bio-imaging agents *Chem. Commun.* **48** 8835–7

[65] Yang S T *et al* 2009 Carbon dots as nontoxic and high-performance fluorescence imaging agents *J. Phys. Chem.* C **113** 18110–4

[66] Hong W, Liu Y, Li M H, Xing Y X, Chen T, Fu Y H, Jiang L, Zhao H, Jia A Q and Wang J S 2018 *In vivo* toxicology of carbon dots by $^{1}$H NMR-based metabolomics *Toxicol. Res.* **7** 834–47

[67] Zheng X, Shao D, Li J, Song Y, Chen Y, Pan Y, Zhu S, Yang B and Chen L 2015 Single and repeated dose toxicity of citric acid-based carbon dots and a derivative in mice *RSC Adv.* **5** 91398–406

[68] Huang Y, Liang Y, Rao Y, Zhu D, Cao J J, Shen Z, Ho W and Lee S C 2017 Environment-friendly carbon quantum dots/$ZnFe_2O_4$ photocatalysts: characterization, biocompatibility, and mechanisms for NO removal *Environ. Sci. Technol.* **51** 2924–33

[69] Kang Y F, Li Y H, Fang Y W, Xu Y, Wei X M and Yin X B 2015 Carbon quantum dots for zebrafish fluorescence imaging *Sci. Rep.* **5** 1–2

[70] Xiao A, Wang C, Chen J, Guo R, Yan Z and Chen J 2016 Carbon and metal quantum dots toxicity on the microalgae *Chlorella pyrenoidosa Ecotoxicol. Environ. Safety* **133** 211–7

[71] Yao K, Lv X, Zheng G, Chen Z, Jiang Y, Zhu X, Wang Z and Cai Z 2018 Effects of carbon quantum dots on aquatic environments: comparison of toxicity to organisms at different trophic levels *Environ. Sci. Technol.* **52** 14445–51

[72] Xiao Y Y, Liu L, Chen Y, Zeng Y L, Liu M Z and Jin L 2016 Developmental toxicity of carbon quantum dots to the embryos/larvae of rare minnow (*Gobiocypris rarus*) *BioMed Res. Int.* **2016** 4016402

[73] Qu S, Wang X, Lu Q, Liu X and Wang L 2012 A biocompatible fluorescent ink based on water-soluble luminescent carbon nanodots *Angew. Chem. Int. Ed.* **51** 12215–8

[74] Song Y, Feng D, Shi W, Li X and Ma H 2013 Parallel comparative studies on the toxic effects of unmodified CdTe quantum dots, gold nanoparticles, and carbon nanodots on live cells as well as green gram sprouts *Talanta* **116** 237–44

[75] Chen J, Liu B, Yang Z, Qu J, Xun H, Dou R, Gao X and Wang L 2018 Phenotypic, transcriptional, physiological and metabolic responses to carbon nanodot exposure in *Arabidopsis thaliana* (L.) *Environ. Sci.: Nano.* **5** 2672–85

[76] Chong Y, Ma Y, Shen H, Tu X, Zhou X, Xu J, Dai J, Fan S and Zhang Z 2014 The *in vitro* and *in vivo* toxicity of graphene quantum dots *Biomaterials* **35** 5041–8

[77] Qin Y, Zhou Z W, Pan S T, He Z X, Zhang X, Qiu J X, Duan W, Yang T and Zhou S F 2015 Graphene quantum dots induce apoptosis, autophagy, and inflammatory response via p38 mitogen-activated protein kinase and nuclear factor-κB mediated signaling pathways in activated THP-1 macrophages *Toxicology* **327** 62–76

[78] Wang D, Zhu L, Chen J F and Dai L 2015 Can graphene quantum dots cause DNA damage in cells? *Nanoscale* **7** 9894–901

[79] Wang T, Zhu S and Jiang X 2015 Toxicity mechanism of graphene oxide and nitrogen-doped graphene quantum dots in RBCs revealed by surface-enhanced infrared absorption spectroscopy *Toxicol. Res.* **4** 885–94

[80] Deng S, Fu A, Junaid M, Wang Y, Yin Q, Fu C, Liu L, Su D S, Bian W P and Pei D S 2019 Nitrogen-doped graphene quantum dots (N-GQDs) perturb redox-sensitive system via the selective inhibition of antioxidant enzyme activities in zebrafish *Biomaterials* **206** 61–72

[81] Wang Z G *et al* 2015 Toxicity of graphene quantum dots in zebrafish embryo *Biomed. Environ. Sci.* **28** 341–51

[82] Liang L, Peng X, Sun F, Kong Z and Shen J W 2021 A review on the cytotoxicity of graphene quantum dots: from experiment to simulation *Nanoscale Adv.* **3** 904–17

[83] Johnston H J, Hutchison G R, Christensen F M, Aschberger K and Stone V 2010 The biological mechanisms and physicochemical characteristics responsible for driving fullerene toxicity *Toxicol. Sci.* **114** 162–82

[84] Kyzyma O A *et al* 2019 State of aggregation and toxicity of aqueous fullerene solutions *Appl. Surf. Sci.* **483** 69–75

[85] Markovic Z *et al* 2007 The mechanism of cell-damaging reactive oxygen generation by colloidal fullerenes *Biomaterials* **28** 5437–48

[86] Prylutska S V *et al* 2019 *In vitro* and *in vivo* toxicity of pristine C60 fullerene aqueous colloid solution *Fuller. Nanotub. Carbon Nanostructures* **27** 715–28

[87] Gao J, Wang Y, Folta K M, Krishna V, Bai W, Indeglia P, Georgieva A, Nakamura H, Koopman B and Moudgil B 2011 Polyhydroxy fullerenes (fullerols or fullerenols): beneficial effects on growth and lifespan in diverse biological models *PLoS One* **6** e19976

[88] Yang L Y, Gao J L, Gao T, Dong P, Ma L, Jiang F L and Liu Y 2016 Toxicity of polyhydroxylated fullerene to mitochondria *J. Hazard. Mater.* **301** 119–26

[89] Ashtami J, Anju S and Mohanan P V 2019 Conformity of dextran-coated fullerene C70 with L929 fibroblast cells *Colloids Surf.* B **184** 110530

[90] Biby T E, Prajitha N, Ashtami J, Sakthikumar D, Maekawa T and Mohanan P V 2020 Toxicity of dextran stabilized fullerene C60 against C6 Glial cells *Brain Res. Bull.* **155** 191–201

[91] Freixa A, Acuña V, Gutierrez M, Sanchís J, Santos L H, Rodriguez-Mozaz S, Farré M, Barceló D and Sabater S 2018 Fullerenes influence the toxicity of organic micro-contaminants to river biofilms *Front. Microbiol.* **9** 1426

[92] Aschberger K, Johnston H J, Stone V, Aitken R J, Tran C L, Hankin S M, Peters S A and Christensen F M 2010 Review of fullerene toxicity and exposure—appraisal of a human health risk assessment, based on open literature *Regul. Toxicol. Pharm.* **58** 455–73

[93] Jacobsen N R, Pojana G, White P, Møller P, Cohn C A, Smith Korsholm K, Vogel U, Marcomini A and Loft S; Wallin H 2008 Genotoxicity, cytotoxicity, and reactive oxygen species induced by single-walled carbon nanotubes and C60 fullerenes in the FE1-Muta™ Mouse lung epithelial cells *Environ. Mol. Mutagen.* **49** 476–87

[94] Dhawan A, Taurozzi J S, Pandey A K, Shan W, Miller S M, Hashsham S A and Tarabara V V 2006 Stable colloidal dispersions of C60 fullerenes in water: evidence for genotoxicity *Environ. Sci. Technol.* **40** 7394–401

[95] Folkmann J K, Risom L, Jacobsen N R, Wallin H, Loft S and Møller P 2009 Oxidatively damaged DNA in rats exposed by oral gavage to C60 fullerenes and single-walled carbon nanotubes *Environ. Health Perspect.* **117** 703–8

# Chapter 5

# The application of zero-dimensional carbon nanomaterials as sensors for environmental pollutant detection

**Garima Rathee, Gaurav Bartwal, Sumit Kumar, Jyotsna Rathee and Pratima R Solanki**

The properties of zero-dimensional (0D) carbon nanomaterials, including graphene and carbon quantum dots (GQDs and CQDs), have produced great interest in their use in sensing applications. Their marvelous fluorescent, electronic, chemiluminescent, photoluminescent, and electrochemiluminescent characteristics provide them with plausible sensing. This chapter reviews the elementary information on GQDs and CQDs before concentrating on their broader sensor applicability, followed by the discussion of the future scope for GQDs and CQDs nanomaterials in sensor applications.

## 5.1 Introduction

With the rapid growth in population and escalation of industrial and agricultural activities, a tremendous quantity of pollutants has been released in the environment in a short period of time. Hence, in the current century, much effort has been focused on maintaining the environment sustainably while keeping the pace with scientific advancements. The types of waste discarded into the environment are often carcinogenic, harmful, or highly lethal, for instance, pharmaceuticals/personal care products (PPCPs), heavy metals, poisonous inorganic gases, volatile organic compounds (VOCs), dyes, and endocrine-disrupting chemicals (EDCs). Such pollutants aggressively disrupt the balance of nature, increase their harmfulness at an alarming rate, and present a terrible threat to the environment. To cope with the challenges associated with water and air remediation, numerous physical, chemical, biological, and physical-chemical techniques have been introduced by researchers from every corner of the world [1]. Although these techniques are effective, they suffer from certain drawbacks due to which such techniques are not employed for

doi:10.1088/978-0-7503-4048-9ch5  

practical applications. Moreover, biomedical approaches are incapable of the complete elimination of highly carcinogenic and bio-refractory pollutants [2]. The traditional methods, such coagulation, sedimentation, filtration, and centrifugal separation, have also produced carcinogenic secondary contaminants and increased the burden on ecosystems [3]. The unsafe and continuous occurrence of organic pollutants, along with the co-occurrence of innumerable inorganic contaminants in the last few decades, have escalated the focus on environmental remediation to achieve total mineralization, thus concentrate towards the lower energy consumption to spark the awareness of sustainable environment.

One of the fast-evolving fields in the last few decades has been nanotechnology. Numerous nanomaterials (NMs) with minor or more complicated changes to their physicochemical and physical characteristics have been fabricated by playing with the atomic arrangement, chemical combination, and dimensions of NMs [4]. The term 'nano' was invented in 1974 by Norio Taniguchi. Among the various NMs, the zero-dimensional (0D) NMs have been considered as the flag-bearer of nanotechnology. 0D NMs have more active edge sites because of their characteristic structural features, for example, large surface-to-volume ratios (S:V) and ultrasmall sizes. Owing to their important properties, such as quantum confinement and edge effects, 0D NMs manifest unique enhanced characteristics of chemiluminescence and excellent photoluminescence quantum efficiency [5]. To date, various 0D NMs have been discovered, such as fullerenes, carbon quantum dots (CQDs), graphene quantum dots (GQDs), inorganic quantum dots (IQDs), polymer dots, noble metal nanoparticles (NPs), magnetic NPs, and up-conversion NPs [6–8]. These materials have proven their broad applicability in environmental remediation, energy, sensing, nanodevices, material sciences, photovoltaic science, catalysis, and biomedicine [9–12]. An illustrative representation of zero-dimensional (0D) NMs is given in figure 5.1.

By altering the nanomaterial size and changing it to a 0D structure, the newly produced structure could be provided with unique features that might vary from those of higher dimensional NMs in a more advantageous way. 0D NMs have proved to be typically quasi-spherical or spherical NPs with a diameter $\leqslant$ 100 nm [13]. Due to their unique characteristics such as biocompatibility, chemical inertness, photoluminescence (wavelength dependent), optical stability, and cellular permeability, 0D NMs provide outstanding adaptability for biomedical applications, for example, biochips, biosensors, cosmetics, nanomedicines, and bioelectronics. Among all the 0D NMs, the carbon-based zero-dimensional NMs have gained significant attention from researchers from distinct fields.

Lately, continuous efforts have been made to fabricate affordable and facile approaches to solve environmental issues by exploiting functional and eco-friendly novel materials. Such novel environment cleansing materials include a large number of carbon NMs. Carbon NMs, comprising graphene, carbon dots (CDs), mesoporous carbon (MC), and carbon nanotubes (CNTs), have been widely evaluated and have emerged as an admirable and robust sensing platform for environmental applicability,

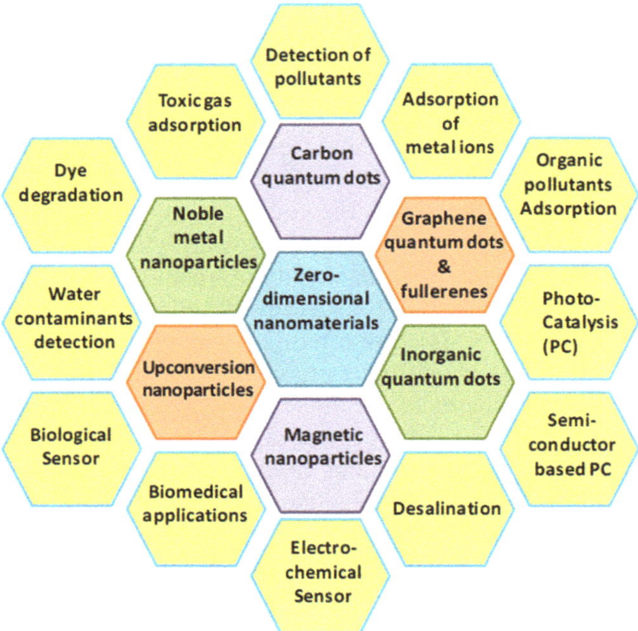


**Figure 5.1.** Zero-dimensional (0D) nanomaterials and applications.

for example, the testing and elimination of chemical contaminants [14]. In this chapter we attempt to assemble all the extraordinary research on the applicability of NMs for reducing environmental pollutants.

## 5.2 Zero-dimensional carbon-based nanomaterials

In nanotechnology, carbon-based NMs are considered as extremely valuable materials because of their low-cost fabrication, multifunctional surface modifications, and lower intrinsic toxicity. Due to these outstanding properties, carbon NMs have emerged as potential candidates for numerous biological applications such as bioimaging, biosensing, and drug delivery. Among all the existing carbon NMs, 0D carbon NMs exhibited unmatched optical characteristics and electrical properties, higher quantum yield, and lower toxicity, and hence can be employed in the production of sensors with micro-dimensions, allowing enhanced efficiency and judicial power utilization.

### 5.2.1 Graphene quantum dots (GQDs)

Among the 0D graphene-based NMs, GQDs are characterized by very thin graphitized planes matching the atomic dimensions (generally 1–2 layers with a thickness $\leqslant 2$ nm) with smaller transverse dimensions (typically $\leqslant 10$ nm) [15]. Photoluminescence (PL) is one of the valuable characteristics exhibited by GQDs. Numerous studies have established that fine-tuning the morphology, dopants, or dimensions can help to alter the PL emission wavelength and excitation of GQDs. These tunable properties of GQDs

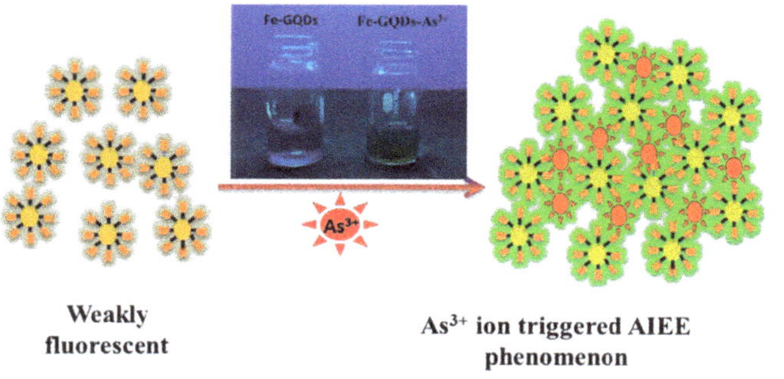


**Figure 5.2.** Proposed mechanism behind the $As^{3+}$ triggered aggregation-induced enhanced emission phenomenon via complex formation. (Reproduced with permission from [18]. Copyright 2019 Royal Society of Chemistry.)

are the main reason for their applicability as sensors in detecting environmental pollutants and their use in biosensing and bioimaging. Additionally, due to numerous oxygen-rich functional groups at the GQDs' edge, they possess biocompatibility and excellent water stability [16]. Moreover, because of benefits such as good conductivity, anti-bleaching, and luminescence stability, the applicability of GQDs has been extended in biosensing [17]. GQDs sensors are extensively employed in identifying multiple biomarkers and ions along with the diagnosis of major diseases. Due to their extraordinary PL properties, these NMs have been used in fabricating numerous optical biosensors to identify numerous metal ions [15].

For example, Pathan *et al* fabricated a novel aggregation-induced boosted PL sensor using magnetic GQDs (Fe-GQDs) to test the presence of arsenic ions in polluted water with a high degree of sensitivity and selectivity [18]. Figure 5.2 illustrates the mechanism behind the detection ability of magnetic GQDs (Fe-GQDs). Furthermore, Qian *et al* fabricated a fluorescent sensor using graphene oxide and a GQDs–aptamer probe to identify lead. As the designed sensor exhibits extraordinary biocompatibility for both GO and GQDs, this result is expected to be helpful for the identification of lead *in vitro* as well as *in vivo* [19]. Additionally, GQD based sensors also play a role in identifying biomarkers, for example, dopamine, ascorbic acid, amino acid, and DNA. Recently, a rapid and sensitive fluorescence turn-on sensor using GQDs (orange emission) was designed to detect ascorbic acid. Such a fabrication approach delivers a sensor with the benefits of selectivity and universality. Also, there was no involvement of any heavy metals, resulting in no heavy metal pollution.

### 5.2.2 Carbon quantum dots (CQDs)

CQDs are also denoted as carbon dots (CDs). They are also quasi-spherical particles with significant fluorescent properties and a diameter less than 10 nm. Compared to the previously discussed GQDs, these particles have lower crystallinity because of increased surface defects and less $sp^2$ carbon [5]. CQDs exhibit outstanding optical

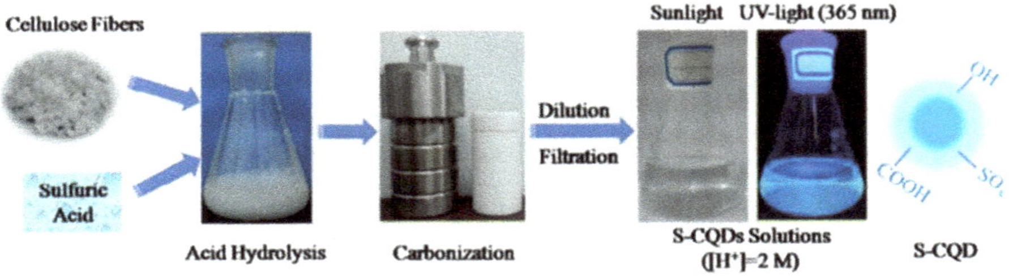


**Figure 5.3.** Proposed fabrication approach of S-doped carbon quantum dots. (Reproduced with permission from [22]. Copyright 2016 Royal Society of Chemistry.)

characteristics in chemiluminescence (CL), electrochemiluminescence (ECL), and fluorescence (FL), due to which they are widely exploited in drug delivery biosensing and bioimaging. Analogous to GQDs, these can be functionalized and fabricated conveniently and quickly. Their surface chemical characteristics, optical characteristics, biocompatibility, and surface reaction performance can be enhanced by surface modifications or doping [20]. The application of CQDs in the sensing of metal ions is developing swiftly, and numerous CQD-dependent fluorescent and electrochemical sensors have been introduced with excellent performance. For example, a functionalized CQD-altered gating electrode has been developed by Fan *et al* for the early detection of copper ions [21].

Moreover, CQDs have also shown applications in the quantitative and ultrafast identification of heavy metal ions present in contaminated water. Moreover, innovative sulfur-doped CQDs (S-CQDs) have been engineered with admirable sensitivity and selective identification of Fe (III) ions in pH = 0 solutions [22] (figure 5.3). These NMs displayed a higher quantum yield, strong acidophilia, and higher luminescence in a highly acidic medium.

### 5.2.3 Fullerenes

Fullerene is a molecular allotrope of carbon which was discovered in 1985 by Kroto *et al.* $C_{60}$ is one of the common fullerenes and comprises five to six $sp^2$ hybridized carbon rings, making a condensed icosahedron. Fullerenes provide structural stability, larger surface volume ratios, and higher electron affinity [23]. Due to these properties fullerenes are employed in multiple applications, comprising medicine, electronics, and biology. Additionally, they also provide excellent biocompatibility and affinity towards multiple organic molecules. Therefore, they could be employed for the construction of numerous biosensors.

Fullerenes offer a larger electroactive surface area, due to which they are used for constructing electrochemical sensors for detecting amino acids. For example, a electrochemical sensor based on $C_{60}$ has been fabricated to differentiate quantitatively between D- and L-serine [24]. Such a biosensor could be considered as a promising device for the early detection of schizophrenia in patients. Furthermore, a biosensor which is fullerene based and has photoelectrochemical (PEC) properties was also developed and fabricated to detect DNA. A smart PEC $Co_3O_4$ fullerene based biosensor was designed to achieve the ultrasensitive testing of DNA [25].

This section has discussed the three categories of 0D carbon NMs that can be used to synthesize different sensors, such as PL, electrochemical, electronic, and electrochemiluminescence. These fabricated sensors can be used widely to detect ions, pollutants, disease markers, and biomolecules. Although zero-dimensional carbon NMs have received significant attention from many researchers, the research is still considered to be in the initial stages. Also, the currently used fabrication techniques fail to control the structure of the NMs at the atomic level due to the inadequacy of the characterization techniques which can monitor the step-by-step progress. Additionally, the mechanisms behind the various characteristics, such as fluorescence quenching, which is frequently presented by zero-dimensional carbon NMs, is not well explained. In the upcoming decades, accuracy in the synthesis methods is expected to improve remarkably and, thus, the characteristics of 0D carbon NMs might be studied in detail to design biosensors and sensors for environmental remediation and clinical applications.

## 5.3 The biocompatibility, nanotoxicology, and cytotoxicity of 0D carbon nanomaterials

One of the most critical factors for an ideal sensor is its biocompatibility. A outstanding bio-benign nature is essential for numerous biological uses of 0D carbon NMs, particularly in bioimaging. A few studies related to the lower toxicity of GQDs in mammalian systems have been reported [26]. *In vitro* studies using HeLa cells to test GQDs were performed by Chong *et al* [26]. The results showed the lower cytotoxicity of GQDs due to their ultrasmall size and higher oxygen content. Moreover, the rapid removal of GQDs from kidneys with no accretion of GQDs was observed in an *in vivo* biodistribution assay of a mouse model. In contrast, graphene oxide (GO) and GQD injections revealed that GQDs presented no pronounced toxic effect on the mouse model, whereas GO showed severe toxicity, even resulting in the mouse mortality due to the accumulation of GO.

The cytotoxicity was reduced with distinct functionalization of GQDs with different groups (comprising COOH, $NH_2$, and CO-$N(CH_3)_2$ and the results were determined by employing a mixture of thiazolyl blue colorimetric (MTT), flow cytometry analysis, and trypan blue assay in human neural glioma C6 and human A549 lung carcinoma cells [27]. All three types of GQDs exhibited good biocompatibility. In short, functional group modifications on GQDs have exhibited a smaller impact on biocompatibility and cytotoxicity. Moreover, the impact of functional groups on GQD cytotoxicity was also evaluated using the production ability of reactive oxygen species (ROS), which was applicable for all the GQD derivatives. By schematic deactivation of distinct groups on GQDs (OH, COOH, or –C=O), the authors discovered that the –C=O groups on GQDs exhibited a noticeable impact on ROS production [28]. Most prominently, elimination of the oxygen-containing functional groups on GQDs might boost their photostability and reduce their cytotoxicity. These findings have provided significant support for establishing an improved strategy for determining GQD biocompatibility. Based on a radiotherapy study by the Ge group [29], the excellent biocompatibility and lower cytotoxicity of oxygen group functionalized GQDs have allowed their successful use as active nano-radiosensitizers. The results demonstrate that the GQDs boosted the sensitivity of colorectal carcinoma cells to ionizing irradiation. Through their study, the authors designed a safe, economical, and efficient nano-radiosensitizer for tumor radiotherapy, thus expanding the possible uses of CQDs *in vivo*.

## 5.4 Environmental pollutants

Pollutants discharged into ecosystems (such as the air, water, and soil) can be categorized as synthetic organic (endocrine disruptors, forensic or pharmaceutical substances, herbicides, pesticides, VOCs) and inorganic (nitrate, heavy metals, phosphates, toxic inorganic gases) [30]. Table 5.1 summarizes a few characteristic pollutants affecting both air and water. The concentration of these pollutants is irrelevant to their toxicity as all the pollutants have many damaging effects on human physiology at tiny concentrations. Furthermore, two types of toxic gases, VOCs and inorganic gases, have been the root cause of trouble in air pollution. Toxic inorganic gases comprise $H_2S$, $NO_x$, $SO_x$, and S-containing gases. $NO_x$ and $SO_x$ are formed through repetitive reactions from the nitrogen and sulfur impurities present in fossil fuels. From general point of view, $SO_2$ and $NO_2$ are two among many characteristic air pollutants with a vast historical impact that has left sharp and horrible trails. VOCs, on the other hand, are released from oil refining, transportation, storage, oil gas evaporation, industry emission, automobile exhaust emission, waste material dumping, etc. These can evaporate easily ambient pressure and temperature because of their lower boiling point and high vapor pressure. VOCs are typically well defined as 'compounds of carbon'. However, carbon dioxide, ammonium carbonate, carbonic acid, carbon monoxide, metallic carbides/carbonates, and carbon dioxide are excluded. The emission of VOCs into the air has resulted in many severe environmental issues, for example, mutagenic, pathogenic, and even fatal illnesses.

In environmental water contamination, multiple heavy metal and inorganic ions occur at the surface and in groundwater, along with various other organic moieties, for example, pesticides, polycyclic aromatic hydrocarbons, pharmaceuticals, PPCPs, herbicides, EDCs, dyes, etc. Furthermore, heavy metals ions ($Zn^{2+}$, $Cu^{2+}$, $Co^{2+}$, $Cd^{2+}$, and $Pb^{2+}$) are highly carcinogenic for aquatic life, human beings, plants, and animals [31]. For the reasonable limits of heavy metals in contaminated water, fixed permitted exposure parameters have been set and are evaluated by the World Health Organization (WHO) and US Environmental Protection Agency (EPA). The permitted limits for $Zn^{2+}$, $Fe^{2+}$, and $Cu^{2+}$ lie within 1–10 ppm, and those for $Hg^{2+}$, $Cd^{2+}$, and $Pb^{2+}$ are 1–10 ppb [32]. EDCs are capable of disrupting hormonal activities by interfering with the reproductive systems of humans and wildlife. Hence, contact with minute concentrations of EDCs could potentially lead to many diseases (such as diabetes, cancer, and neurobehavioral issues) [33].

**Table 5.1.** Examples of common environmental pollutants.

| Chemical | Environment | Applications |
|---|---|---|
| | | **VOC** |
| Benzene | Indoor air | Industrial chemical, a component of gasoline, solvent |
| Cyclohexane | Indoor air | Industrial chemical and solvent |
| Acetone | Water, atmosphere | Industrial chemical and solvent |
| Toluene | Indoor air | Industrial chemical and solvent |
| Trichloroethylene | Indoor air, water | Industrial chemical and solvent |
| | | **Inorganic toxic gases** |
| Nitrogen oxide | Atmosphere | Industry exhaust, metal smelting, fossil fuel burning |
| Sulfur dioxide | Atmosphere | Industry exhaust, combustion of S-containing minerals |
| Hydrogen sulfide | Atmosphere | Waste from the petrochemical industry |
| Carbon monoxide | Atmosphere | Automobile exhaust, inadequate combustion of carbonaceous substances |
| Ammonia | Atmosphere | Gas, chemical, and medical industry |
| | | **Metal ions** |
| Cobalt (II)(IV) | Water, soil, atmosphere | Nuclear power plants, industrial waste, mineral mining |
| Cadmium (II) | Water, soil, atmosphere | Mineral mining, industrial waste |
| Nickel (II) | Water | Mineral mining, industrial waste |
| Lead (II) | Water, soil, atmosphere | Automobile exhaust, mineral mining, industrial waste |
| Chromium (VI) | Water, atmosphere | Fossil fuel, industrial waste |
| Mercury (II) | Water, soil, atmosphere | Aquaculture, industrial waste |
| | | **EDCs** |
| Bisphenol A | Water | Industrial chemical |
| 17 $\alpha$-ethinyl estradiol | Water | Pharmaceutical |
| Perfluorooctanoic acid | Water | Emulsifier, surfactant |
| Phenol | Water | Medicine, solvent, industrial chemical |
| 2,4-dichlorophenol | Water, soil | Pesticide, solvent, pharmaceutical intermediate |
| Atrazine | Water | Pesticide |
| 17$\beta$-estradiol (E2) | Water | Pharmaceutical |
| Phthalate | Water | Plasticizer |

| | Dyes | |
|---|---|---|
| Rhodamine 6G | Water | Chemical analysis and detection, dyeing |
| Rhodamine B | Water | Cosmetics industry, dyeing, chemical analysis and detection |
| Methyl orange | Water | Chemical analysis and detection, dyeing |
| Methylene blue | Water | Chemical and medical analysis and identification, dyeing |

PPCPs are designed minute organic moieties which are persistent and abundant, and thus have an endocrine disruption tendency in aquatic organisms [34]. Artificial dyes are widely used in the cosmetic, food, paper, textile, and printing industries, resulting in harmful effects on the environment when discharged into water [35]. Pharmaceutical products are another class of pollutants that are regularly found in water bodies, for example, antibiotics, hormonal drugs, muscle relaxants, and pain killers. These show distinct influences on the environment and living beings, and long-term ingestion of mixtures, even in minute quantities, is hazardous [36].

## 5.5 Electrochemical techniques for environmental applications

For the elimination of pollutants from water and air, numerous traditional methods have been introduced. Dust removal, oxidation, condensation, combustion, adsorption, and catalysis are typically practical approaches to monitor and control gaseous contaminants. For the removal of water contaminants, there are four significant types of approaches, comprising physical-chemical (membrane filtration, adsorption, extraction, and ion exchange), physical (sedimentation, filtration, and centrifugal separation), biological (anaerobic and aerobic processes), and chemical (precipitation, coagulation/flocculation, and oxidation–reduction) [1, 37–40].

The advanced oxidation process (APO) is a practical and greener approach for the degradation of organic pollutants, as numerous potent oxidizing radicals, for example, singlet oxygen ($^1O_2$), hydroxyl (•OH), and sulfate ($•SO_4^-$), are produced during AOPs. Further, the pollutants, mainly organic contaminants, are oxidized into less carcinogenic moieties (carbon dioxide and water) by the radicals mentioned above. The designed AOPs comprise Fenton oxidation, photoelectrocatalysis, photocatalysis, photochemical oxidation, electrochemical oxidation, and ozone oxidation. Electrochemical methods are more advantageous than traditional water treatment methods and have undergone swift progress in the last few decades as they generally use electrons as 'cleansing agents'.

Additionally, carcinogenic contaminants with highly corrosive properties could be treated by using electrochemical techniques because of their excellent anticorrosion characteristics. Moreover, the selectivity and the reaction rate of electrochemical approaches in eliminating environmental pollutants are very high. Also, electrochemical methods can use multipurpose paths under moderate conditions and the complete function can be computerized easily. Hence, these methods have been employed for the oxidation of organic pollutants, for the elimination of heavy metals with significant toxicity, for diminishing harmful oxyanions, and most importantly to sterilize water. Due to the involvement of electron enhancement, electrochemical methods are effective in the preservation of environmental interfacial procedures with compelling higher rate by systematically and precisely monitoring the electron transfer methods [41]. These methods can be categorized into electrochemical reduction, electro-osmosis, electrosorption, electrochemical oxidation, electrocoagulation, and photoelectrochemical oxidation. The electric coagulation technique, which is popularly known as the electric floating technique, separates the pollutants with the condensation of cations on a common pollution particle. In the electro-osmosis approach, the detection of concentration and separation of ionized contaminants is achieved by evaluating the movement of liquid by using an advanced ion-selective membrane. This approach is mainly based on the intrinsic characteristics of the chosen membrane and electric field applied externally [42]. Electrosorption is a form of absorption catalyzed by generating a potential on a charged electron surface or a flow of charged ions in the direction of oppositely charged electrodes after being placed in an electric field [43].

The electrochemical oxidation process is further subdivided into direct and indirect oxidation. Contaminants are converted (oxidized) directly into harmless moieties at the anode electrode through direct oxidation. In contrast, the contaminants are oxidized with the help of highly reactive intermediates, which are generated *in situ* from cathodic oxygen reduction with indirect oxidation. Such a method is typically employed for dehydrogenation of chlorinated hydrocarbons and the retrieval of these metals. Finally, the photoelectrocatalytic oxidation process, using photo-induced charge carriers, has proved to be an efficient route for working on a semiconductor electrode surface. Therefore, this method provides the combined benefits of the photo- and electrochemical methods, and hence it has been widely valued in the treatment of environmental pollution.

## 5.6 Applications of 0D carbon nanomaterial-based sensors for the detection of environmental pollutants

The outstanding characteristics of 0D carbon NMs make them potential candidates for sensing. CQDs have been adopted preferentially, anticipating their great potential as fluorescent probes to detect multiple analytes because of their fluorescent characteristics. Generally, the sensors fabricated with GQDs and CQDs can be divided into electrochemical, optical, luminescence, and photoelectrochemical sensors.

### 5.6.1 Photoluminescence chemosensors

0D carbon NMs, in particular CQDs and GQDs, have both been exploited extensively as fluorescent probes owing to their ability to detecting heavy metal ions and pesticides. As a fact of statistical research, metal ions with environment fluctuation probability have proved as the most broadly investigated analytes via

CQDs [44]. Therefore, sensors for detecting different metal ions ($Ag^{+}$, $Fe^{3+}$, $Hg^{2+}$, $Cd^{2+}$, $Cu^{2+}$, $Zn^{2+}$, $Au^{3+}$, $Co^{2+}$, $Pb^{2+}$, $Pd^{2+}$, $Cr^{6+}$, $Ni^{2+}$, $Mn^{2+}$, $Al^{3+}$, $Sn^{2+}$, $Be^{2+}$, $Bi^{3+}$, and $K^{+}$) have been introduced [45]. In most studies, the data show that CQD luminescence quenching is a common phenomenon in the presence of metal ions, and is suitable for their detection and exploration. It was also observed that variety arises from the substance from which the CQDs are obtained or the procedure used to generate the CQDs. Characteristically, the estimations must not include any surface alterations, resulting in questionnaires' gate opening related to the selectivity. An example is the study by Zhang *et al*, who established the quenching of N-doped CQD fluorescence by $Hg^{2+}$, which was attributed to modification of the N-CQD surface states affected by the $Hg^{2+}$. The data also presented selectivity for $Hg^{2+}$ in the presence of other metal ions at high concentrations.

Correspondingly, a recent study using 4.0 nm CQDs discovered the influence of a range of metal ions, comprising $Al^{3+}$, $Mn^{2+}$, $Mg^{2+}$, $Hg^{2+}$, $Zn^{2+}$, $Pb^{2+}$, $Cu^{2+}$, $Cd^{2+}$, $Ag^{+}$, $Ni^{2+}$, $Co^{2+}$, $Fe^{2+}$, and $Fe^{3+}$, on the CQDs' photoluminescence intensity [46]. It was estimated that the fluorescence intensities of the CQDs were quenched substantially in the presence of $Fe^{3+}$ solution. Therefore, the CQDs were employed to analyze $Fe^{3+}$ in tap water having relevant data with an unidentified sample compared to what estimated by atomic absorption spectroscopy. However, serious questions related to selectivity remain unsolved—what is so special about N-CQDs which makes them highly selective and preferred for $Hg^{+}$, whereas CQDs have special advantages for $Fe^{3+}$? To complicate the situation, another report was published that demonstrated the fluorescence of CQDs fabricated using camphor as the starting reagent was successfully quenched by Fe (II) but not by Fe (III) [47].

These aspects highlight the need to recognize the transduction mechanism and how the different fabrication techniques may provide selectivity for only one metal instead of another, and suggest that there is significant potential in employing GDQ and PET for organic species sensing. However, there is a need to improve the selectivity, in contrast to use unmodified CQDs with inaccurate surface states. A report related to GO and GQDs as Pb (II) sensors monitored the fluorescence turn-on process dependent on PET. The designed sensor had the potential to detect lead with a wide linear range, not exceeding a concentration 400 nM, and ultra-fast feedback times for the lower detection limit of the lead samples with a concentration 0.6 nM [19].

Moreover, this sensor could also differentiate other metals from lead metal lines, maintaining higher sensitivity and excellent reproducibility. Fluctuations of CQD fluorescence data have proved to be advantageous in identifying numerous inorganic molecules [48]. The detection of nitrogen dioxide ($NO_2$) gas in one study involved QD modified aerogels as the sensors [49]. In the study, the fabricated porous CQD modified aerogel in a solid state exhibited strong fluorescein performance. In this study, significant fluorescence quenching by $NO_2$ gas was observed, however, no such quenching was noticed when the modified aerogel was exposed directly to an atmosphere of pure nitrogen.

Furthermore, in a separate study, thiol surface-functionalized CQDs in an aqueous medium were fabricated and used to detect arsenite in water sources [48].

The study demonstrated an enhancement in the modified-CQDs' fluorescence intensity with arsenite addition. Based on these results, it could be suggested that the fabricated CQD probe exhibited higher selectivity for the detection of arsenite. Finally, organic moieties, such as aniline, organic thiols, diniconazole, ascorbic acid, and glucosamine have been tested using CQDs [50]. The sensitivity of CQD-based sensing probes for detecting organic moieties based on the fluorescence quenching could be enhanced with higher quantum yields of CQDs. An ultrasonic technique was used to fabricate an innovative S and N-doped CQD (N,S-CQDs) employing ionic liquids as the sole precursor. The obtained sensor was found to be effective for the detection of pesticides with longer photoluminescence lifetimes [51].

### 5.6.2 Chemiluminescence chemosensors

Chemiluminescence sensors, with their benefits of easy instrumentation, wider range, high sensitivity, and no surrounding interference have gained interest in sensing. In a study using CQDs enriched and doped with nitrogen and sulfur (N,S-CQDs), $KMnO_4$ produced strong chemiluminescence compared to other oxidants. Moreover, the impact of metal ions on the $KMnO_4$–N,S-CQDs chemiluminescent arrangement was also determined. The results demonstrated that the chemiluminescence intensity weakened on contact with copper (II) ions [52]. The sensor was used for identifying $Cu^{2+}$ ions in dilute water and human plasma trials. Chemiluminescence in association with CQDs has been employed for organic moiety identification. An illustration is the application of hydrothermally designed S,N-CQDs for controlling indomethacin [52]. The study illustrated the fabrication of a chemiluminescent system for the identification of indomethacin by combining the designed S,N-CQDs with numerous oxidants consisting of $NaIO_4$, $H_2O_2$, $KMnO_4$, and $K_3Fe(CN)_6$. The data demonstrated that the $KMnO_4$–S,N-CQD reaction showed the maximum sensitivity for identification of indomethacin with 65 $\mu g\ L^{-1}$.

### 5.6.3 Electrochemiluminescence chemosensors

Due to good its controllability, high sensitivity, and lower background, electrochemiluminescence (ECL) has progressively emerged as a powerful and significant analytical tool in various fields, for example, immunoassays, pharmaceutical investigations, and pollutant identification [53]. By combining CQDs with various materials, CQD-dependent composites have been introduced to identify chlorinated phenols, microRNA, sophoridine, chlorpromazine, and pentachlorophenol employing ECL [16]. For example, a CQD-based sensing platform was fabricated by involving a modifier of ruthenium metal with polyvinyl alcohol and composite film. The designed platform was used for early ECL detection of sophoridine [54] with a $5.0 \times 10^{-11}$ M detection limit. The presence of CQDs enhanced the conductivity, increased the transmission rate of electrons, and behaved like an electron subway permitting abundance amount of $Ru(bpy)_3^{2+}$ in the system. This work was used to design a ruthenium and gelatin based CQD composite film sensor for identifying chlorpromazine (detection limit = $8.0 \times 10^{-11}$ M) [54].

### 5.6.4 Electrochemical chemosensors

The role of GQDs and CQDs has also been exploited in electrochemical sensors. For example, Pd NP-functionalized N-GQDs@N-doped carbon hollow nanospheres have been designed for the sensing of hydrogen peroxide [55]. In this combination, the Pd NPs and N-GQDs were used for signal amplification as active electrocatalysts in electrochemical sensing. These sensors provided a very irrelevant response towards electroactive moieties comprising cathodic petrification from $O_2$ and anodic interference from uric acid, ascorbic acid, and dopamine, dodged interference signals, and displayed satisfying activity with lower detection limit, high sensitivity, and shorter response duration for actual and rapid identification of $H_2O_2$ excreted by cancer cells. In table 5.2 we have summarized a few 0D carbon nanostructures sensors for environmental pollutants.

**Table 5.2.** Some 0D carbon nanostructures sensors for environmental pollutant sensing.

| 0D based sensor | Transduction mechanism | Analyte | Surface functionalization | Ref. |
|---|---|---|---|---|
| CQDs (source: egg white) | Fluorescence quenching | $Fe^{3+}$ | | [56] |
| CQDs (source: blueberry) | Fluorescence quenching | $Fe^{3+}$ | | [57] |
| CQDs (source: $\beta$-cyclodextrin) | Colorimetry | $Fe^{3+}$ | | [58] |
| CQDs (source: BSA) | Fluorescence quenching | $Fe^{3+}$ | Thiol groups | [59] |
| B-GQD | Fluorescence quenching | $Fe^{3+}$ | | [60] |
| Ionic liquid-derived CQDs | Fluorescence quenching | $Cu^{2+}$ | | [61] |
| TPEA-CQDs | Fluorescence quenching | $Cu^{2+}$ | Tris(pyridine-2-ylmethyl) ethane-1,2- diamine) | [62] |
| CdTe-CQDs | Fluorescence quenching | $Cu^{2+}$ | Colorimetry carboxyl-modified CdTe QDs | [63] |
| Valine-derived CQDs | Fluorescence quenching | $Hg^{2+}$ | | [64] |
| DNA-CQDs/GO | Phosphorescence recovery | $Hg^{2+}$ | Thymine | [65] |
| CQDs (source: camphor) | Fluorescence quenching | $Cd^{2+}$ | | [47] |
| Carbon dots (source: sago waste) | Fluorescence quenching | $Cr^{2+}$ | | [66] |
| N,S-CQDs | Inner filter effect (IFE) quenching | $Cr_2O_7^{2-}$ | | [67] |

(*Continued*)

**Table 5.2.** (*Continued*)

| 0D based sensor | Transduction mechanism | Analyte | Surface functionalization | Ref. |
|---|---|---|---|---|
| CQDs rooted agarose hydrogel | Colorimetry | $Cr^{6+}$ | | [68] |
| Carbon dots (source: glucose) | Fluorescence quenching | $K^{+}$ | | [69] |
| CQDs from honey | Fluorescence quenching | $Au^{3+}$ | | [70] |
| CQDs (source: citric acid monohydrate) | Fluorescence enhancement | $Zn^{2+}$ | Quinoline derivatives | [71] |
| CQDs (source: sago waste) | Fluorescence quenching | $Zn^{2+}$ | | [66] |
| CQDs (source: camphor) | Fluorescence quenching | $Zn^{2+}$ | | [47] |
| CQDs | Fluorescence enhancement | $Co^{2+}$ | Cetyltrimethylamm onium bromide | [72] |
| B-GQDs | Fluorescence enhancement | $Al^{3+}$ | | [73] |
| CQDs (source: sago waste) | Fluorescence quenching | $Ni^{2+}$ | | [66] |
| CQDs (source: sago waste) | Fluorescence quenching | $Co^{2+}$ | | [66] |
| Dual emission CQDs | Fluorescence quenching | $Bi^{3+}$ | | [74] |
| Cyclam-functionalized CQDs | Fluorescence recovery | $S^{2-}$ | 1,4,8,11-tetraazacyclotetradecane (Cyclam) | [75] |
| CQDs (source: ammonium citrate) | Fluorescence quenching and recovery | $SO_3^{2-}$ | | [76] |
| CQDs (source: sago waste) | Fluorescence recovery | $Sn^{2+}$ | | [66] |
| CQDs (source: sago starch) | Fluorescence recovery | $Sn^{2+}$ | | [77] |
| CQDs (source: sago waste) | Fluorescence quenching | $Ca^{2+}$ | | [66] |
| CQDs (source: citric acid) | Fluorescence quenching | $Pd^{2+}$ | | [78] |
| N-CQDs-Pd@Au hexoctahedrons | Electrochemical quenching | $Pd^{2+}$ | | [79] |
| CQDs rooted agarose hydrogel | Colorimetry | $Pb^{2+}$ | | [68] |

| | | | | |
|---|---|---|---|---|
| GQDs-graphene oxide | Fluorescence recovery | $Pb^{2+}$ | | [19] |
| CQDs rooted agarose hydrogel | Colorimetry | $Mn^{2+}$ | | [68] |
| CQDs-nitrogen-rich supramolecular | Fluorescence quenching | $SeO_3^{2-}$ | Supramolecular ligand | [80] |
| CQDs (source: whey) | Fluorescence quenching | $Se^{4+}$ | Amine rich ligand | [81] |
| $Fe_3O_4$@$SiO_2$@CQDs | Fluorescence recovery | $F^-$ | | [82] |
| N-CODs | Fluorescence recovery | $I^-$ | | [83] |
| N-CQD-$Hg^{2+}$ | Fluorescence recovery | $I^-$ | | [84] |
| Ditheritheritol-CQDs | Fluorescence enhancement | Arsenite | Ditheritheritol | [48] |
| N-GQD@N-carbon@Pd | Electrochemical enhancement | $H_2O_2$ | $SiO_2$ | [55] |
| GQDs (source: graphene oxide) | | Photoluminescence quenching | Aniline | [85] |
| CQDs@$NH_2$ | Fluorescence quenching | 2,4,6- trinitrotoluene | Ethylenediamine | [86] |
| CQDs/$Ag_3PO_4$ | Fluorescence quenching | 2,4- dinitrotoluene | | [87] |
| CQDs/$SiO_2$ aerogels | Fluorescence quenching | $NO_2$ | Branched polyethylenimine | [49] |

## 5.7 The future scope and challenges of CQDs for the preferential detection of environmental pollutants

CQDs have proved to be efficient and reliable NMs for the early detection of toxic metals in water supply, sewage treatment plants, and pollutants in air. The use of CQDs in various applications is so adaptable that they have be preferred in comparison to other NMs for the sensing of toxic metals and will continue to be preferred in the coming years. The efficiency of CQDs is so accepted that it has opened a new avenue for designing devices and complex environmentally benign processes.

## 5.8 Conclusion

In the current chapter we have complied the current advancements and new avenues of zero-dimensional carbon NMs including the most recent and modern applications. To date, many fabrication techniques have been designed to produce pure and doped GQDs and CQDs with distinct characteristics and morphologies, which accelerate the ongoing optimization of analytical activity for real-world applications. The validity of 0D carbon NMs in the sensing domain is significantly less developed. As table 5.2 shows, there have been numerous studies related to the specific sensing of various metal ions. Mostly, these listed NMs work based on the fluorescence intensity quenching of GQDs and CQDs. However, key questions, such as how this quenching works and how sensor selectivity is achieved, have remained topics of interest. Furthermore, CQDs prepared from the same source have also exhibited sensing applications for multiple ions. Clearly, in future studies there is much need to understand selectivity. Surface states and doping are crucial, but we appear to be a long way from understanding these complex systems and how CQDs and GQDs designed via a simple technique and from a similar source demonstrate such specific activity. This presents an excellent opportunity for scientists as the initial outcomes have been somewhat hopeful. Furthermore, such understanding is crucial for designing a commercial device because minor changes to materials and processes can modify the characteristics and output of the final sensor; otherwise a device cannot be manufactured dependably. The problem of selectivity has started to be resolved by tailoring the surface with distinct functionalization and also the expansion of biosensors based on GQDs and CQDs. To conclude, CQDs and GQDs are at a very early stage of evolution, and their potential is highly promising.

## Acknowledgments

We are very grateful to Professor Pratima Solanki for giving us the opportunity of writing the current chapter.

## References

[1] He C, Cheng J, Zhang X, Douthwaite M, Pattisson S and Hao Z 2019 Recent advances in the catalytic oxidation of volatile organic compounds: a review based on pollutant sorts and sources *Chem. Rev.* **119** 4471–568

[2] Chong M N, Jin B, Chow C W K and Saint C 2010 Recent developments in photocatalytic water treatment technology: a review *Water Res.* **44** 2997–3027

[3] Shannon M A, Bohn P W, Elimelech M, Georgiadis J G, Mar̃ias B J and Mayes A M 2008 Science and technology for water purification in the coming decades *Nature* **452** 301–10

[4] Carneiro P, Morais S and Pereira M C 2019 Nanomaterials towards biosensing of Alzheimer's disease biomarkers *Nanomaterials* **9** 1663

[5] Pirzada M and Altintas Z 2019 Nanomaterials for healthcare biosensing applications *Sensors* **19** 5311

[6] Barberis A *et al* 2015 Simultaneous amperometric detection of ascorbic acid and antioxidant capacity in orange, blueberry and kiwi juice, by a telemetric system coupled with a fullerene- or nanotubes-modified ascorbate subtractive biosensor *Biosens. Bioelectron.* **67** 214–23

[7] Yan Y *et al* 2019 Recent advances on graphene quantum dots: from chemistry and physics to applications *Adv. Mater.* **31** 1808283

[8] Robidillo C J T *et al* 2019 Ratiometric detection of nerve agents by coupling complementary properties of silicon-based quantum dots and green fluorescent protein *ACS Appl. Mater. Interfaces* **11** 33478–88

[9] Ramanathan S, Gopinath S C B, Arshad M K Md and Poopalan P 2019 Multidimensional (0D-3D) nanostructures for lung cancer biomarker analysis: comprehensive assessment on current diagnostics *Biosens. Bioelectron.* **141** 111434

[10] Xu Q, Li W, Ding L, Yang W, Xiao H and Ong W J 2019 Function-driven engineering of 1D carbon nanotubes and 0D carbon dots: mechanism, properties and applications *Nanoscale* **11** 1475–504

[11] Guo C X, Yang H B, Sheng Z M, Lu Z S, Song Q L and Li C M 2010 Layered graphene/quantum dots for photovoltaic devices *Angew Chem. Int. Ed.* **49** 3014–7

[12] Zhang X, Rajaraman B R S, Liu H and Ramakrishna S 2014 Graphene's potential in materials science and engineering *RSC Adv.* **4** 28987–9011

[13] Liang R, Wei M, Evans D G and Duan X 2014 Inorganic nanomaterials for bioimaging, targeted drug delivery and therapeutics *Chem. Commun.* **50** 14071–81

[14] Das R *et al* 2017 Recent advances in nanomaterials for water protection and monitoring *Chem. Soc. Rev.* **46** 6946–7020

[15] Chung S, Revia R A and Zhang M 2021 Graphene quantum dots and their applications in bioimaging, biosensing, and therapy *Adv. Mater.* **33** 1904362

[16] Sun H, Wu L, Wei W and Qu X 2013 Recent advances in graphene quantum dots for sensing *Mater. Today* **16** 433–42

[17] Zhu S, Song Y, Zhao X, Shao J, Zhang J and Yang B 2015 The photoluminescence mechanism in carbon dots (graphene quantum dots, carbon nanodots, and polymer dots): current state and future perspective *Nano Res.* **8** 355–81

[18] Pathan S, Jalal M, Prasad S and Bose S 2019 Aggregation-induced enhanced photoluminescence in magnetic graphene oxide quantum dots as a fluorescence probe for As (III) sensing *J. Mater. Chem. A* **7** 8510–20

[19] Qian Z S, Shan X Y, Chai L J, Chen J R and Feng H 2015 A fluorescent nanosensor based on graphene quantum dots-aptamer probe and graphene oxide platform for detection of lead (II) ion *Biosens. Bioelectron.* **68** 225–31

[20] Molaei M J 2019 A review on nanostructured carbon quantum dots and their applications in biotechnology, sensors, and chemiluminescence *Talanta* **196** 456–78

[21] Fan Q *et al* 2020 Functional carbon quantum dots for highly sensitive graphene transistors for $Cu^{2+}$ ion detection *ACS Appl. Mater. Interfaces.* **12** 4797–803

[22] Yang G, Wan X, Su Y, Zeng X and Tang J 2016 Acidophilic S-doped carbon quantum dots derived from cellulose fibers and their fluorescence sensing performance for metal ions in an extremely strong acid environment *J. Mater. Chem.* A **4** 12841–9

[23] Sun H, Zhou Y, Ren J and Qu X 2018 Carbon nanozymes: enzymatic properties, catalytic mechanism, and applications *Angew. Chem. Int. Ed.* **57** 9224–37

[24] Jaiswal S, Singh R, Singh K, Fatma S and Prasad B B 2019 Enantioselective analysis of D- and L-serine on a layer-by-layer imprinted electrochemical sensor *Biosens. Bioelectron.* **124–25** 176–83

[25] Wang H H, Li M J, Wang H J, Chai Y Q and Yuan R 2019 p–n-sensitized heterostructure $Co_3O_4$/fullerene with highly efficient photoelectrochemical performance for ultrasensitive DNA detection *ACS Appl. Mater. Interfaces* **11** 23765–72

[26] Chong Y *et al* 2014 The *in vitro* and *in vivo* toxicity of graphene quantum dots *Biomaterials* **35** 5041–8

[27] Yuan X, Liu Z, Guo Z, Ji Y, Jin M and Wang X 2014 Cellular distribution and cytotoxicity of graphene quantum dots with different functional groups *Nanoscale Res. Lett.* **9** 108

[28] Zhou Y, Sun H, Wang F, Ren J and Qu X 2017 How functional groups influence the ROS generation and cytotoxicity of graphene quantum dots *Chem. Commun.* **53** 10588–91

[29] Ruan J *et al* 2018 graphene quantum dots for radiotherapy *ACS Appl. Mater. Interfaces* **10** 14342–55

[30] Popek E P 2018 *Sampling and Analysis of Environmental Chemical Pollutants: a Complete Guide* 2nd edn (New York: Elsevier) pp 1–414

[31] Ling L L, Liu W J, Zhang S and Jiang H 2017 Magnesium oxide embedded nitrogen self-doped biochar composites: fast and high-efficiency adsorption of heavy metals in an aqueous solution *Environ. Sci. Technol.* **51** 10081–9

[32] Merkoçi A, Pumera M, Llopis X, Pérez B, Del Valle M and Alegret S 2005 New materials for electrochemical sensing VI: carbon nanotubes *TrAC—Trends Anal Chem.* **24** 826–38

[33] Joseph L *et al* 2013 Removal of bisphenol A and 17$\alpha$-ethinyl estradiol by combined coagulation and adsorption using carbon nanomaterials and powdered activated carbon *Sep. Purif. Technol.* **107** 37–47

[34] Liu Y, Liu H, Zhou Z, Wang T, Ong C N and Vecitis C D 2015 Degradation of the common aqueous antibiotic tetracycline using a carbon nanotube electrochemical filter *Environ. Sci. Technol.* **49** 7974–80

[35] Wang C, Feng C, Gao Y, Ma X, Wu Q and Wang Z 2011 Preparation of a graphene-based magnetic nanocomposite for the removal of an organic dye from aqueous solution *Chem. Eng. J.* **173** 92–7

[36] Nikolaou A, Meric S and Fatta D 2007 Occurrence patterns of pharmaceuticals in water and wastewater environments *Anal. Bioanal. Chem.* **387** 1225–34

[37] Rathee G, Singh N and Chandra R 2020 Simultaneous elimination of dyes and antibiotic with a hydrothermally generated NiAlTi layered double hydroxide adsorbent *ACS Omega* **5** 2368–77

[38] Rathee G, Kohli S, Awasthi A, Singh N and Chandra R 2020 MoS42-intercalated NiFeTi LDH as an efficient and selective adsorbent for elimination of heavy metals *RSC Adv.* **10** 19371–81

[39] Rathee G, Awasthi A, Sood D, Tomar R, Tomar V and Chandra R 2019 A new biocompatible ternary layered double hydroxide adsorbent for ultrafast removal of anionic organic dyes *Sci Rep.* **9** 16225

[40] Dordio A V, Duarte C, Barreiros M, Carvalho A J P, Pinto A P and da Costa C T 2009 Toxicity and removal efficiency of pharmaceutical metabolite clofibric acid by *Typha* spp.—potential use for phytoremediation? *Bioresour. Technol.* **100** 1156–61

[41] Feng Y, Yang L, Liu J and Logan B E 2016 Electrochemical technologies for wastewater treatment and resource reclamation *Environ. Sci.: Water Res. Technol.* **2** 800–31

[42] Encyclopedia of Membranes 2020 *Encyclopedia of membranes* (Cham: Springer) pp 1–72

[43] Cheng Q, Shou L, Chen C, Shi S and Zhou M 2017 Application of ultra-high-performance liquid chromatography coupled with LTQ-Orbitrap mass spectrometry for identification, confirmation and quantitation of illegal adulterated weight-loss drugs in plant dietary supplements *J. Chromatogr.* B **1064** 92–9

[44] Skaltsas T, Goulielmaki M, Pintzas A, Pispas S and Tagmatarchis N 2017 Carbon quantum dots/block copolymer ensembles for metal-ion sensing and bioimaging *J. Mater. Chem.* **B 5** 5397–402

[45] Li M, Chen T, Gooding J J and Liu J 2019 Review of carbon and graphene quantum dots for sensing *ACS Sens.* **4** 1732–48

[46] Liu M, Xu Y, Niu F, Gooding J J and Liu J 2016 Carbon quantum dots directly generated from electrochemical oxidation of graphite electrodes in alkaline alcohols and the applications for specific ferric ion detection and cell imaging *Analyst* **141** 2657–64
[47] Gaddam R R, Vasudevan D, Narayan R and Raju K V S N 2014 Controllable synthesis of biosourced blue-green fluorescent carbon dots from camphor for the detection of heavy metal ions in water *RSC Adv.* **4** 57137–43
[48] Pooja D, Saini S, Thakur A, Kumar B, Tyagi S and Nayak M K 2017 A 'turn-on' thiol functionalized fluorescent carbon quantum dot based chemosensory system for arsenite detection *J. Hazard. Mater.* **328** 117–26
[49] Wang R, Li G, Dong Y, Chi Y and Chen G 2013 Carbon quantum dot-functionalized aerogels for $NO_2$ gas sensing *Anal. Chem.* **85** 8065–9
[50] Liu Q, Ma C, Liu X P, Wei Y P, Mao C J and Zhu J J 2017 A novel electrochemiluminescence biosensor for the detection of microRNAs based on a DNA functionalized nitrogen doped carbon quantum dots as signal enhancers *Biosens. Bioelectron.* **92** 273–9
[51] Li H, Sun C, Vijayaraghavan R, Zhou F, Zhang X and MacFarlane D R 2016 Long lifetime photoluminescence in N, S co-doped carbon quantum dots from an ionic liquid and their applications in ultrasensitive detection of pesticides *Carbon NY* **104** 33–9
[52] Hallaj T, Amjadi M, Manzoori J L and Azizi N 2017 A novel chemiluminescence sensor for the determination of indomethacin based on sulfur and nitrogen co-doped carbon quantum dot–$KMnO_4$ reaction *Luminescence* **32** 1174–9
[53] Zhai Q, Li J and Wang E 2017 Recent advances based on nanomaterials as electrochemiluminescence probes for the fabrication of sensors *ChemElectroChem* **4** 1639–50
[54] Liu Z, Zhang F, Cui L, Wang K and Zhan H 2017 Fabrication of a highly sensitive electrochemiluminescence chlorpromazine sensor using a Ru(bpy)32+ incorporated carbon quantum dot-gelatin composite film *Anal. Methods* **9** 1011–7
[55] Xi J *et al* 2016 Pd nanoparticles decorated N-doped graphene quantum dots@N-doped carbon hollow nanospheres with high electrochemical sensing performance in cancer detection *ACS Appl. Mater. Interfaces* **8** 22563–73
[56] Zhang Z, Sun W and Wu P 2015 Highly photoluminescent carbon dots derived from egg white: facile and green synthesis, photoluminescence properties, and multiple applications *ACS Sustain. Chem. Eng.* **3** 1412–8
[57] Aslandaş A M, Balci N, Arik M, Şakiroıļu H, Onganer Y and Meral K 2015 Liquid nitrogen-assisted synthesis of fluorescent carbon dots from Blueberry and their performance in $Fe^{3+}$ detection *Appl. Surf. Sci.* **356** 747–52
[58] Zhu W, Zhang J, Jiang Z, Wang W and Liu X 2014 High-quality carbon dots: synthesis, peroxidase-like activity and their application in the detection of $H_2O_2$, $Ag^+$ and $Fe^{3+}$ *RSC Adv.* **4** 17387–92
[59] Yang Q, Wei L, Zheng X and Xiao L 2015 Single particle dynamic imaging and $Fe^{3+}$ sensing with bright carbon dots derived from bovine serum albumin proteins *Sci. Rep.* **5** 17727
[60] Ge S, He J, Ma C, Liu J, Xi F and Dong X 2019 One-step synthesis of boron-doped graphene quantum dots for fluorescent sensors and biosensor *Talanta* **199** 581–9
[61] Xu P, Wang C, Sun D, Chen Y and Zhuo K 2015 Ionic liquid as a precursor to synthesize nitrogen- and sulfur-co-doped carbon dots for detection of copper(II) ions *Chem. Res. Chinese Univ.* **31** 730–5

[62] Shao X, Gu H, Wang Z, Chai X, Tian Y and Shi G 2013 Highly selective electrochemical strategy for monitoring of cerebral $Cu^{2+}$ based on a carbon dot-TPEA hybridized surface *Anal. Chem.* **85** 418–25

[63] Wang Y *et al* 2016 Ratiometric fluorescent paper sensor utilizing hybrid carbon dots-quantum dots for the visual determination of copper ions *Nanoscale.* **8** 5977–84

[64] Zhang C, Hu Z, Song L, Cui Y and Liu X 2015 Valine-derived carbon dots with colour-tunable fluorescence for the detection of $Hg^{2+}$ with high sensitivity and selectivity *New J. Chem.* **39** 6201–6

[65] Gui R *et al* 2015 Room-temperature phosphorescence logic gates developed from nucleic acid functionalized carbon dots and graphene oxide *Nanoscale* **7** 8289–93

[66] Tan X W, Romainor A N B, Chin S F and Ng S M 2014 Carbon dots production via pyrolysis of sago waste as potential probe for metal ions sensing *J. Anal. Appl. Pyrolysis* **105** 157–65

[67] Chen J, Liu J, Li J, Xu L and Qiao Y 2017 One-pot synthesis of nitrogen and sulfur co-doped carbon dots and its application for sensor and multicolor cellular imaging *J. Colloid Interface Sci.* **485** 167–74

[68] Gogoi N, Barooah M, Majumdar G and Chowdhury D 2015 Carbon dots rooted agarose hydrogel hybrid platform for optical detection and separation of heavy metal ions *ACS Appl. Mater. Interfaces* **7** 3058–67

[69] Zhang L, Chen S, Zhao Q and Huang H 2015 Carbon dots as a fluorescent probe for label-free detection of physiological potassium level in human serum and red blood cells *Anal. Chim. Acta* **880** 130–5

[70] Gu J *et al* 2015 Carbon dot cluster as an efficient ‘off–on’ fluorescent probe to detect Au(III) and glutathione *Biosens. Bioelectron.* **68** 27–33

[71] Zhang Z *et al* 2014 Quinoline derivative-functionalized carbon dots as a fluorescent nanosensor for sensing and intracellular imaging of $Zn^{2+}$ *J. Mater. Chem.* B **2** 5020–7

[72] Shi J, Lu C, Yan D and Ma L 2013 High selectivity sensing of cobalt in HepG2 cells based on necklace model microenvironment-modulated carbon dot-improved chemiluminescence in Fenton-like system *Biosens. Bioelectron.* **45** 58–64

[73] Fan Z *et al* 2014 Surrounding media sensitive photoluminescence of boron-doped graphene quantum dots for highly fluorescent dyed crystals, chemical sensing and bioimaging *Carbon* **70** 149–56

[74] Qu F, Wang S, Liu D and You J 2015 Differentiation of multi-metal ions based on fluorescent dual-emission carbon nanodots *RSC Adv.* **5** 82570–5

[75] Chen J *et al* 2016 Cyclam-functionalized carbon dots sensor for sensitive and selective detection of copper(II) ion and sulfide anion in aqueous media and its imaging in live cells *Sensors Actuators* B **224** 298–306

[76] Fang L, Zhang L, Chen Z, Zhu C, Liu J and Zheng J 2017 Ammonium citrate derived carbon quantum dot as on–off–on fluorescent sensor for detection of chromium(VI) and sulfites *Mater. Lett.* **191** 1–4

[77] Mohd Yazid S N A, Chin S F, Pang S C and Ng S M 2013 Detection of Sn(II) ions via quenching of the fluorescence of carbon nanodots *Microchim. Acta* **180** 1–2

[78] Sharma V, Saini A K and Mobin S M 2016 Multicolour fluorescent carbon nanoparticle probes for live cell imaging and dual palladium and mercury sensors *J. Mater. Chem.* B **4** 2466–76

[79] Xiong C *et al* 2016 *In situ* electro-polymerization of nitrogen doped carbon dots and their application in an electrochemiluminescence biosensor for the detection of intracellular lead ions *Chem. Commun.* **52** 5589–92

[80] Devi P *et al* 2017 Ultrasensitive and selective sensing of selenium using nitrogen-rich ligand interfaced carbon quantum dots *ACS Appl. Mater. Interfaces* **9** 13448–56

[81] Devi P, Kaur G, Thakur A, Kaur N, Grewal A and Kumar P 2017 Waste derivitized blue luminescent carbon quantum dots for selenite sensing in water *Talanta* **170** 49–55

[82] Mohapatra S, Sahu S, Nayak S and Ghosh S K 2015 Design of $Fe_3O_4$@$SiO_2$@carbon quantum dot based nanostructure for fluorescence sensing, magnetic separation, and live cell imaging of fluoride ion *Langmuir* **31** 8111–20

[83] Li Z *et al* 2015 Highly luminescent nitrogen-doped carbon quantum dots as effective fluorescent probes for mercuric and iodide ions *J. Mater. Chem.* C **3** 1922–8

[84] Huang H, Weng Y, Zheng L, Yao B, Weng W and Lin X 2017 Nitrogen-doped carbon quantum dots as fluorescent probe for 'off–on' detection of mercury ions, l-cysteine and iodide ions *J. Colloid Interface Sci.* **506** 373–8

[85] Ghosh T, Chatterjee S and Prasad E 2015 Photoinduced electron transfer from various aniline derivatives to graphene quantum dots *J. Phys. Chem.* A **119** 11783–90

[86] Tian X, Peng H, Li Y, Yang C, Zhou Z and Wang Y 2017 Highly sensitive and selective paper sensor based on carbon quantum dots for visual detection of TNT residues in groundwater *Sensors Actuators* B **243** 1002–9

[87] Chen Q, Wang Y, Wang Y, Zhang X, Duan D and Fan C 2017 Nitrogen-doped carbon quantum dots/$Ag_3PO_4$ complex photocatalysts with enhanced visible light driven photocatalytic activity and stability *J. Colloid Interface Sci.* **491** 238–45

# Chapter 6

# The potential of zero-dimensional carbon nanomaterials as biosensors for clinical biomarker detection

**Anas Saifi, Kamal Arora, Charu Negi, Vikas Chaudhary and Saurabh Srivastava**

Cancer and cardiovascular disease are the leading causes of human deaths reported worldwide. The early and precise detection of a disease is the most reliable approach to increase the survival of patients, for which the detection of biomarkers in body fluids by employing biosensing platforms has garnered great interest recently. Thus zero-dimensional (0D) carbon nanomaterials including fullerenes, carbon quantum dots, and graphene quantum dots, due to their unique physiochemical properties, have attracted enormous attention from researchers in designing and developing ultra-sensitive biosensors. In this chapter, a summary of some recent developments in the design, properties, and applications of 0D carbon nanomaterials for clinical biomarker detection is presented.

## 6.1 Introduction

Carbon (meaning coal in Latin) is a non-metal element which is present in every living organism; thus it can be said that carbon is omnipresent in each aspect of life. As the chief component of various compounds, carbon can be hard, i.e. diamond, or soft, i.e. graphite. The applications of these unique elements are broad from drilling to beauty products. They are also employed to eliminate the presence of other poisonous substances. Among all the elements which exist as free elements, carbon has the highest melting point. If carbon is the mother tree, carbon nanomaterials are its chemical fruits. Carbon nanomaterials have drawn great attention as a consequence of their unique structure in addition to their exceptional electrical, mechanical, optical, and chemical properties. The carbon-based nanomaterials can be classified as zero-dimensional (0D), one-dimensional (1D), and two-dimensional (2D). Fullerene, quantum dots, and nanodiamonds fall into the category of 0D materials, whereas nanofibers and nanotubes are 1D, and

doi:10.1088/978-0-7503-4048-9ch6  

graphene and carbon nanowalls are 2D. Looking over the historical literature, in 1952 hollow carbon fibers with a diameter of 50 nm were synthesized [1]. In 1960 the multi-walled carbon nanotube (MWCNT) structure was reported, using an electron microscope. In addition to this, the friction properties of MWCNTs were evaluated as a consequence of the rolling property of graphene sheets [2]. In 1976 carbon fibers of nanometer-scale were grown using chemical vapor deposition (CVD) by Oberlin *et al* [3]. In 1985 the discovery of fullerenes by Kroto *et al* was a breakthrough [4], it was the first allotrope of carbon discovered in the twentieth century [5].

The carbon precursors which are utilized to make carbon nanomaterials (CNs), are crucial for large-scale production. Most traditional synthesis procedures for the manufacture of 2D graphene require the use of expensive carbon precursors such as hydrocarbons and graphite [6–9]. Nanostructures based on carbon such as carbon nanotubes, graphene oxide, graphene [10–12], carbon fibers [13], and candle soot [14] have been utilized as precursors to create 0D fullerene, quantum dots (QDs), nanodiamonds (NDs), and other carbon nanostructures. The modification in nanoparticle size through their conversion into 0D structures will impart unique properties to the carbon nanomaterials compared to their higher dimensional counterparts [15]. In contrast to higher dimension materials, 0D materials are spherically shaped nanoparticles with a diameter $< 100$ nm [16]. Due to their novel properties 0D nanomaterials are actively employed for biomedical applications [17]. The number of studies on 0D nanomaterials for potential applications based on biosensing has increased recently [18]. However, only a few provide an extensive review of this application as it is a crucial component in the development of 0D nanomaterials for biosensing applications.

### 6.1.1 Fullerenes

Kroto *et al* discovered fullerene, a chemical allotrope of carbon, in 1985. C60, the most common fullerene, possesses a truncated octahedral symmetry consisting of five to six $sp^2$ hybridized carbon rings [19]. Fullerenes have a strong electron affinity, a large surface-to-volume ratio, and a stable structure [20]. The structure of fullerene includes pentagonal and hexagonal polyhedral structures, where carbons are placed at their respective corners. Buckyball refers to a fullerene with s spherical structure that resembles a soccer ball. An architect, Richard Buckminster Fuller, noted for designing geodesic domes that look like spherical fullerenes, was the inspiration for the name. In 1985, during laser spectroscopy tests, a researcher from Rice University discovered fullerene as an unexpected surprise. Sir Harold W Kroto, Robert F Curl Jr, and Richard E Smalley received the Nobel prize in 1996 'for their discovery of fullerenes', according to the Nobel Prize archives [21].

#### *6.1.1.1 Synthesis and applications of fullerenes*

Fullerenes are employed in a wide range of applications ranging from electronics, to biology and medicine as a result of their exceptional properties. Fullerenes also have strong biocompatibility and inertia, as well as good affinity for diverse organic compounds, making them suitable for applications in biosensors [22]. Also, fullerenes have been employed as chemical and biological sensing materials for ultra-sensitive cancer detection and DNA biomarkers [23]. Initially, laser vaporization of carbon in an inert atmosphere was employed to yield a minuscule amount of fullerene. [24]. Kratschmer *et al* used an electric arc to evaporate graphite and create isolable quantities of C60 for the first time in 1990 [25]. Fullerene production in an arc discharge was examined in depth by Alekseyev and Dyuzhev [26], who unraveled this topic from calculations of an arc discharge using a molecule assembly of fullerene. In the last two decades, the chemistry of fullerene has exploded in popularity. A number of fullerene derivatives with distinct characteristics were created, and different methods for manufacturing them in larger quantities were proposed. In 1991 and 1992, the fullerenes C60 and C70 were discovered by Howard *et al* [27, 28], in benzene/oxygen flames, and they established a combustion synthesis method for fullerenes. Using microwave plasma in a low-pressure argon environment, the fullerenes C60 and C70 were produced by Xie *et al* [29] using chloroform. Various perchlorinated carbon clusters, perchlorinated intermediates of fullerenes, and fullerene are easily synthesized using the large-scale, low-cost production microwave plasma synthesis process. Taylor *et al* [30] pioneered the 1000 °C pyrolysis of naphthalene to produce C60 and C70. The benefits include a continuous procedure that does not require rod replacement and the ability to make closed fullerene cages from well-defined aromatic pieces. Ragged single walled carbon nanotubes (SWCNTs) upon treatment with an organic solvent followed by calcination, were used to synthesize fullerene through a pyrolysis approach [31]. A good yield quality of C60 was reported in 2009, by employing carbon dioxide reduction by magnesium carbonate or metallic lithium at 700 °C and a pressure of 100 MPa [32]. Although there is low yield in the above approach in the synthesis of C60, it solves the problem of contamination common in the other synthesis methods.

### 6.1.2 Quantum dots

The elements in periodic table groups II–VI form quantum dots. In 2015 Wegner and Hildebrandt found that CdTe, CdSe, and InP based quantum dots have potential uses in bioimaging, biosensing, and treatment [33]. Carbon dots are nanoparticles made of carbon that have unique qualities, such as wavelength-dependent photoluminescence, optical stability, and chemical inertness [34]. Carbon quantum dots (CQDs), are quasi-spherical nanoparticles comprising of carbon, hydrogen, oxygen, nitrogen, and other elements with a diameter $< 10$ nm. Depending on the synthesis techniques, oxidized CQDs contain 5% to 50% oxygen. The absorption spectrum of CQDs exhibits two bands attributed to the $\pi \rightarrow \pi^*$ and $n \rightarrow \pi^*$ transitions of C=C and C=O, respectively [35]. Graphene quantum dots (GQDs) are a conjugated single sheet of dimensions 2–10 nm and are a type of CQDs possessing excellent optoelectronic and magnetic properties [36, 37].

*6.1.2.1 The synthesis and applications of quantum dots*

Recently, many new synthesis methods for CQDs have been proposed, which can be modified during preparation, including the top-down and bottom-up methods. The preparation of CQDs has three drawbacks: (i) carbonization led carbonaceous aggregation; (ii) uniformity and size control; and (iii) solubility based essential surface qualities. Laser ablation, microwave irradiation, electrochemical carbonization, chemical ablation, and hydrothermal treatment are the best methods for the preparation of CQDs. A facile technique to synthesize luminous CQD in an aqueous solution was reported by Peng and Travas-Sejdic [38]. This includes dehydration of carbohydrates by concentrated $H_2SO_4$, followed by splitting of the resultant material into CQDs by reacting with $HNO_3$, and finally, passivation by amine-terminated compounds. Carbonization of low molecular weight alcohols by an electrochemical method is another approach to synthesize CQDs. On an adjustable Luggin capillary, working and counter electrodes are formed by platinum sheets whereas the reference electrode is a calomel electrode [39]. Sun and colleagues created CQDs by irradiating carbon with a laser beam at 900 °C in an argon atmosphere [40]. Pyrolysis of citric acid with different amine compounds using microwave mediation was used by Liu *et al* to prepare luminous CQDs [41]. A one-step hydrothermal method employing orange juice was used by Mohapatra *et al* to make photoluminescent CQDs [42]. Under basic circumstances, the alcohols were electrochemically carbonized to generate CQDs. Owing to their large absorption coefficient, photobleaching resistance, improved brightness, size tunable light emission, and limited emission spectrum, QDs have been employed to build different bioluminescent, chemiluminescent, and fluorescent sensors [43]. Due to their hydrophilicity and cell permeability, CQDs have surpassed metal-QDs in several potential applications, which include drug targeting, photocatalysis, and photovoltaics. Moreover, intracellular biomolecules can be detected by CQDs [44]. Selenoproteins are important in several biological functions and have been linked to a number of human disorders which include cardiovascular disease and cancer. A novel fluorescent nanoprobe consisting of CQDs was developed by Wang *et al* for *in vitro* fluorescent imaging of selenol [45]. The fabricated nanoprobe consisted of 2,4-dinitrobenzenesulphonyl chloride as a segment along with CQDs, which split upon treatment with selenocysteine, resulting in a yellow–green fluorescence. CQDs have been used to diagnose cancer successfully. An ultra-sensitive fluorescent sensor platform based on N-doped CQDs was developed by Lu and colleagues for the detection of $\beta$-glucuronidase, an invasive tumor biomarker [46].

### 6.1.3 Nanodiamonds

These are crystalline nanoparticles based on carbon with outstanding qualities such as chemical inertness, clarity, hardness, and optical properties that inherit the diamond structure at a nanoscale [47]. Truncated octahedron and spherical diamond with $sp^3$-bonded predominant carbon can be termed as an ultra-gem of all gemstones of other materials due to its outstanding qualities, such as electrical insulation, conductivity, hardness, thermal expansion, chemical corrosiveness, and biocompatibility. 'Nanodiamond' refers to a variety of materials which include diamond particles and pure-phase diamond films, along with their structural assemblages whose average size is only 5 nm.

*6.1.3.1 Synthesis and applications of nanodiamonds*

The detonation of carbon explosive materials and high-temperature–pressure transformation of graphite are two basic methods for the preparation of nanodiamond. Bundy *et al* claimed to be the first researchers who successfully synthesized diamond using a high-temperature–high-pressure technique. However, in 1980, it was suggested that diamond could be synthesized by detonating explosives in a steel container with a negative oxygen balance under vacuum. Explosive detonation is still frequently utilized; however, the detonator explosion process is incredibly rapid and complicated. Furthermore, there are several drawbacks to the detonation approach. It is difficult to remove a fraction of surface atoms from the bulk in nanodiamonds, even after the purification procedure. Low cost, low energy, controlled-sized, ease of control, few by-product, and large-scale synthesis procedures, i.e. microwave plasma chemical vapor deposition [48], hot filament chemical vapor deposition [49], pulse laser ablation [50], electron irradiation [51], and high-energy x-ray diffraction [52], have been reported for the facile synthesis of nanodiamonds.

For many years, additives of nanodiamond have been employed in electrolytic and electroless metal plating. They have been applied recently in additional applications such as mass spectrometry, diamond film chemical vapor deposition, magnetic resonance imaging, chromatography, and proteomics. Carbon 'onions', formed by graphitizing nanodiamonds, have exhibited promising applications in catalysis, composites, and energy storage, while boron-doped nanodiamonds can be employed in electrochemical double-layer capacitors, batteries, and electroanalysis, due to their conducting nature [53]. Moreover, in electrochemical environments, undoped non-conducting nanodiamond exhibits redox activity.

## 6.2 Clinical biomarkers

Our better understanding of molecular biology and genomics has resulted in an enhancement in our knowledge on the association between biomolecules, genomes, and disease, leading to an all-time high interest in biomarkers. Although numerous research studies have been conducted on the validation of diagnostic and prognostic biomarkers, this field is still relatively new in terms of clinical application. The term 'biomarker' is a portmanteau of 'biological marker', objective indicators of the health condition of any patient that can be perceived from outside, i.e. used to quantify the presence or severity of any disease and treatment effect. These are measurable and do not indicate how a person feels or functions. Biomarkers are considered as medical signs, different to medical symptoms, which are limited to those indicators of health or sickness experienced by patients. The use of biomarkers in clinical practice, particularly biomarkers measured in the laboratory, is still relatively new, and the best techniques are being refined in the developmental phase [54].

The International Programme on Chemical Safety conducted by the World Health Organization along with the United Nations and the International Labor Organization in a joint endeavor on chemical safety, described a biomarker as, 'an indicator of a particular disease state or some other physiological state of an organism which maybe chemical, physical, or biological in nature, and the measurement may be functional, physiological, biochemical, cellular, or molecular' [55]. Clinical trials might be shortened and product development and regulatory approval could be accelerated by using a biomarker-driven strategy.

An example well-known to many people is the use of blood glucose levels to determine the efficacy of diabetic treatment. Here, glucose is the biomarker for the detection of diabetes.

### 6.2.1 Types of biomarkers

A wide range of clinical biomarkers is being used these days because every biological system has its specific biomarkers. The Biomarkers, Endpoints, and other Tools (BEST) resource has divided biomarkers into several categories according to their role in research and clinical practice (figure 6.1) [56].

#### *6.2.1.1 Diagnostic biomarkers*

The precise detection of disease is a necessity of medical practice. Diagnostic biomarkers are essential for the detection and confirmation of a disease or a person's condition. For example, the variation in hemoglobin A1c and blood sugar levels are considered as biomarkers for diagnosis of Type 2 diabetes mellitus patients.

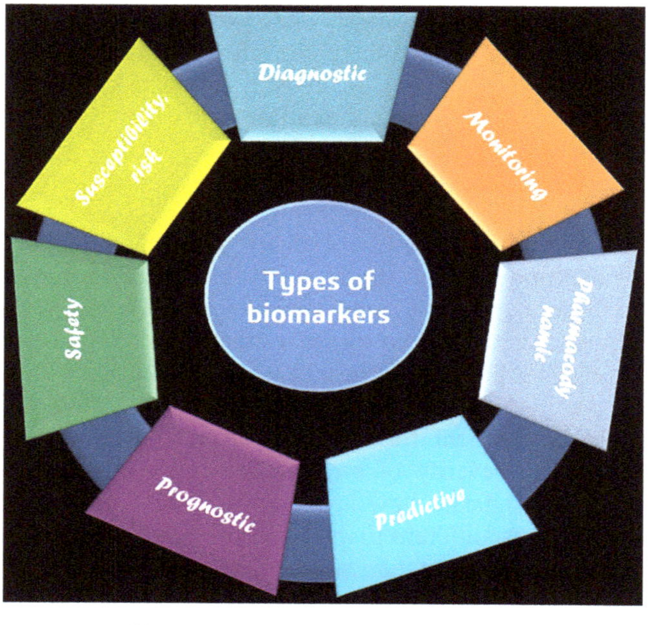


**Figure 6.1.** Different types of biomarkers.

*6.2.1.2 Monitoring biomarkers*

These are measured over time to determine the condition of disease as well as to investigate when the patient is subjected to any medication or environmental factor. For example, cancer antigen 125 (CA-125) is used as a monitoring biomarker to determine the progression of ovarian cancer and the effect of treatment—whether the response of treatment is positive or negative.

*6.2.1.3 Pharmacodynamic/response biomarkers*

These are biomarkers that indicate the occurrence of a biological response after being subjected to any medication or ecological factor. For example, chloride in sweat can be used as a responsive biomarker to measure a patient's response to cystic fibrosis transmembrane regulator (CFTR) potentiating drugs.

*6.2.1.4 Predictive biomarkers*

A predictive biomarker is a biomarker that can be used to determine whether or not a person is susceptible or immune to any medication or environmental factor, i.e. how likely an individual will respond to treatment. One of the best examples of predictive biomarkers is the mutations in the BRCA1/2 genes for sensitivity to ionizing radiation. Exposure to ionizing radiation may impair the function of the gene's protein products in the repair of double-stranded DNA breaks.

*6.2.1.5 Prognostic biomarkers*

A biomarker that demonstrates the chance of a future clinical event, recurrence of a disease, or progression in a certain population is termed as a prognostic biomarker. For example, the C-reactive protein (CRP) level may be utilized to identify patients with unstable angina or a previous history of acute myocardial infarction (AMI), who are more susceptible for recurrent coronary artery disease events.

*6.2.1.6 Safety biomarkers*

Safety biomarkers are used to analyze the condition of a patient before and after exposure to medical products or environmental agents in order to determine the susceptibility, presence, or severity of toxicity as an undesirable consequence of medical treatment. For example, patients consuming drugs that affect kidney function can be evaluated using serum creatinine as a safety biomarker.

*6.2.1.7 Susceptibility/risk biomarkers*

The susceptibility biomarkers are also known as risk biomarkers and are utilized when the person at present does not have any illness but is likely to develop a possible disease. For example, the susceptibility/risk of developing Alzheimer's can be analysed by the variation in the apolipoprotein E (APOE) gene [57].

### 6.2.2 Cardiovascular disease (CVD) and its related biomarkers

CVD, related to heart and blood vessels, is one of the major causes of death worldwide, with an approximate 17.9 million deaths annually. The CVDs include rheumatic heart disease, cerebrovascular disease, coronary heart disease, etc. One third of deaths have been reported to occur before the age of 70, of which strokes and heart attacks account for most CVD related fatalities [58].When a myocardial injury occurs, a cardiac biomarker, i.e. a protein component of cell structures, is discharged into the blood circulation. Cardiac biomarkers play a crucial role in early diagnosis, risk detection, and treatment of patients with acute exacerbations of heart failure, acute coronary syndrome (ACS), and chest discomfort [59]. The cardiac biomarkers were first documented in 1954, where an increase in blood aspartate aminotransferase concentration was observed 3–4 h after suffering from an AMI [60].

Cardiovascular biomarkers are categorized based on the pathological processes they represent. This method may lead to over-simplification because most biomarkers represent several pathophysiological processes, for example, extracardiac disease (e.g. renal failure) [61, 62]. Table 6.1 provides an overview of different cardiovascular biomarkers based on their pathological processes.

**Table 6.1.** Different cardiovascular biomarkers depending on their pathological processes.

| Biomarkers | Overview | References |
|---|---|---|
| | **Myocardial necrosis** | |
| cTn | Troponin is a striated muscle protein complex comprising troponin C, T, and I. For a given assay, above 99th percentile cardiac troponin concentration between two readings it is mandatory to diagnose AMI. | [159, 160] |
| hs-cTn | The hs-cTn assay improves the accuracy of AMI diagnosis. This can also help in assessing the risk level of the patient. | [161] |
| H-FABP | FABs are intercellular proteins with a molecular weight of 15 kDa, produced in tissues with active fatty acid metabolism. Under non-pathological conditions it is not present in circulation, while in the case of AMI it is released rapidly. | [162, 163] |
| | **Inflammation** | |
| HsCRP | In tissue injury or inflammation, a substantial amount of this pattern recognition molecule is increased. In recent research, it is very evident that CRP protein may have a direct impact on the inflammatory processes which causes myocardial damage. | [164, 165] |
| IL-6 | In response to trauma and infections, T-cells and macrophages secrete IL-6 which is encoded by the IL-6 gene, resulting in the release of CRP. | [166] |
| GDF-15 | GDF-15 is an excellent anticipator of cardiovascular events and mortality from any disease. In clinical trials, GDF-15 can be helpful in accessing the level of risk to the patient. | [167] |

| | | |
|---|---|---|
| Fibrinogen | During acute inflammation, the level of an acute-phase protein, i.e. fibrinogen, which is synthesized in the liver, exceeds 7 mg $ml^{-1}$. Several investigations state that the concentration of fibrinogen could be risky for coronary artery disease, heart stroke, and mortality. | [168] |
| | **Plaque instability** | |
| PAPP-A | PAPP-A causes insulin-derived growth factor-1 (IGF-1) to be activated, which causes lipid uptake and inflammation, both of which can lead to atherogenesis and plaque instability. | [169] |
| MPO | An enzyme that catalyzes the conversion of hydrogen and chloride into hyperchlorite. During an inflammatory response, it is secreted by active neutrophils and macrophages. | [170] |
| MMP-9 | Matrix metalloproteinase-9 (MMP-9) plays a major role in the degradation of the extracellular matrix. MMP-9 levels are higher in susceptible atherosclerotic plaques. | [171, 172] |
| | **Myocardial stress** | |
| ST2 | ST2 belongs to the IL-1 receptor family. There are two different forms in which it exists, i.e. a transmembrane receptor (ST2L) and a soluble decoy receptor (ST2). According to several studies, higher plasma ST2 concentrations have been correlated with enhanced risk of mortality and nonfatal adverse cardiac disease in patients with AMI, AHF, and CHF. | [173–175] |
| ET-1 | ET-1, secreted by the vascular endothelial cells, is a peptide hormone consisting of 21 amino acids. It is a pro-fibrotic hormone and a strong vasoconstrictor whose levels are associated with pulmonary artery pressure and shear stress. | [176] |
| Gal-3 | In 2010 the FDA approved Gal-3 as a novel biomarker for risk stratification of HF. Gal-3 is produced by activated cardiac macrophages. | [167] |
| NRG-1 | NRG-1 is produced by endothelial cells, and in turn binds to adjacent ErbB receptors on cardiac myocytes, which enhances the survival of the cell, its growth, and maintenance. | [177] |
| miRNAs | miRNAs are non-coding RNA molecules with a short length of 22 nucleotides. In the circulation, miRNAs are present, persistent, and detectable. After an MI, early detection of several cardiac miRNAs in the bloodstream can be performed, possibly reducing the time it takes to diagnose. | [177] |

### 6.2.3 Cancer

Cancer is an uncontrollable growth of abnormal cells. According to WHO reports, 9.6 million cancer deaths were reported in 2018, making it the second leading cause of fatality around the world. Men suffer mostly with lung, rectal, prostate, stomach, and liver cancer, while the most common types of cancer found in women are breast, cervical, thyroid, colorectal, and lung cancer [63]. A cancerous tumor occurs due to mutations in distinct critical genes [64]. Cancer includes about 100 different diseases with multiple risk factors and epidemiology that arise from various types of cells and organs of the human body. The pathogenesis of all tumors is thought to be the same [65].

#### *6.2.3.1 Breast cancer*

Breast cancer, in which the breast cells grow uncontrollably, is the most common cancer and the leading cause of death related to cancer in women worldwide. The statistics report around 2.26 million new breast cancer cases in women with 68 500 global deaths in 2020. By the end of 2020, 7.8 million women diagnosed with breast cancer in the previous five years were alive, thereby making it the most widespread cancer in the world [66]. The major cause of breast cancer is genetic mutations, in particular BRCA1 and BRCA2. The risks other than genetic mutations associated with the development of breast cancer in females are reproductive factors, age, environmental factors, and a personal or family history of breast disease [67, 68]. Regular screening for breast cancer is critical to diagnose cases where no symptoms have been observed [69]. Some of the conventional screening techniques for breast cancer detection are breast ultrasound, breast MRI, mammograms, etc

##### ***6.2.3.1.1 Breast cancer biomarkers***

Estrogen receptor alpha (ER$\alpha$) and epidermal growth factor 2 (ERBB2, known as HER2 or HER2/neu previously) have been identified as the two major molecular targets responsible for the development of breast cancer [70]. ER$\alpha$, a steroid hormone receptor, is expressed in nearly 70% of invasive cancers. It is a transcription factor which activates oncogenic growth pathways in breast cancer cells when it is activated by estrogen. Expression of the steroid hormone progesterone (PR), which is closely related, also acts as an ER$\alpha$ marker [71]. The tumors with the expression of either ER or PR in at least 1% of tumor cells are classified as HR+ [72].

In the epidermal growth factor family, ERBB2 is a transmembrane receptor tyrosine kinase which is overexpressed and magnified in 20% of breast cancers. In the absence of systemic therapy, it is often associated with poor diagnosis [73]. The amplification or overexpression of ERBB2 in tumors is ERBB2+ and the patients can benefit from ERBB2-targeted therapy, which includes anti-ERBB2 antibodies. Nearly 15% of all breast cancers are triple negative, which is identified as the absence of the molecular targets ER, PR, and ERBB2 [74].

So far, research efforts have been made to create convenient and non-invasive biomarkers which can diagnose the onset of breast cancer in the early stages. To improve the present approaches in the clinical diagnosis of breast cancer, non-invasive body fluid-based tests could prove to be beneficial. These non-invasive body fluids include circulating tumor cells (CTCs), carcinoma antigens, micro-RNAs, sweat, urine, nipple aspirate fluids (NAF), tears, circulating extracellular vesicles

(EVs) in the peripheral blood, circulating cell-free tumor nucleic acids (DNA or RNA), and volatile organic compounds in exhaled breath [75].

### *6.2.3.2 Lung cancer*

Lung cancer occurs in the lungs with a possibility of spreading into other body parts such as the brain or lymph nodes, etc. Cancer which progresses to the bodily organs also has the potential of spreading to the lungs as well. Lung cancers can be small cell (SCLC) or non-small cell (NSCLC). The growth of lung cancer can vary and thus the treatment methods are different as well [76, 77]. Smoking is one of the major cause of lung cancer and in the year 2020, 2.21 million cases of lung cancer were reportedly diagnosed and global deaths of 1.80 million were reported [63, 78]. Some of the conventional screening techniques for lung cancer detection are chest computed tomography, chest radiography, magnetic resonance imaging, PET/CT scans, mediastinoscopy, fine-needle aspiration, bronchoscopy, etc [78, 79].

#### ***6.2.3.2.1 Lung cancer biomarkers***

Studies report many serum tumor markers for the clinical diagnosis of lung cancer, which include carcinoembryonic antigen (CEA), CYFRA 21–1, CA-125, neuron-specific enolase (NSE), chromogranin A, squamous cell carcinoma antigen, $\alpha$1-antitrypsin, and retinol-binding protein (RBP). However, no single blood tests have been reported for lung cancer detection. CEA is an extensively studied tumor marker and has been reported to be amplified in 0%–38% of SCLC patients with limited disease and in 40%–65% of those with extensive disease. In patients with NSCLC, the CEA is amplified in 30%–65% of patients. Muley *et al* reported that among 153 NSCLC patients, the ones with higher levels of CEA or CFRA 21–1 had lower chances of survival compared to normal patients. Pre-operative CEA levels greater than 5 ng $mL^{-1}$ in patients with stage IA disease is associated with lower chances of disease-free survival (22.2 versus 75%). However, relatively few patients fall into this category, although this may help in identifying a subgroup of early-stage patients requiring aggressive treatment. Early-stage cancer patients compared to the ones at the metastatic stage are reported to have lower CEA and CA-125 levels. In another study, which included 37 patients with advanced NSCLC, the ones with a detailed radiological response had a decrease in the levels of these markers.

Lung cancer has been classified into the following two categories:

(i) Gene changes (mutations) or rearrangements that promote the growth of cancer cells.

*Gene mutation biomarkers*: Epidermal growth factor (EGFR), anaplastic lymphoma kinase (ALK), ROS1, RET, HER2, proto-oncogene, KRAS, discoidin domain receptor tyrosine kinase 2 (DDR2), NTRK1, BRAF, MET proto-oncogene PIK3CA, and fibroblast growth factor receptor (FGFR).

(ii) Immune system biomarkers that predict how well your cancer will respond to immunotherapy.

*Immune system biomarkers*: CTLA-4, PD-L1, and PD-1 [80–83].

*6.2.3.3 Colorectal cancer*

Colorectal cancer (CRC), the second most deadly cancer worldwide, is the uncontrollable growth of cells in the colon or the rectum of the large intestine and is also known as colon cancer, rectal cancer, or bowel cancer. A total of 1.93 million new CRC cases were reported in the year 2020 with around 935 000 worldwide deaths [63] and the estimated increase in cases will rise to nearly 2.5 million in 2035 [84]. Several methods of screening CRC have been reported, which include both visual examinations as well as high-sensitivity stool-based tests. The visual examinations are carried out at a healthcare facility and they include flexible sigmoidoscopy, computed tomographic colonography (CTC), and colonoscopy. The high-sensitivity stool-based tests include multitargeted stool DNA tests, the high-sensitivity-guaiac-based fecal occult blood test (gFOBT), and the fecal immunochemical test (FIT)[85].

***6.2.3.3.1 Colon cancer biomarkers***

CRC is regarded as a heterogeneous disease developed due to the growth of epigenetic and genetic mutations [86]. The most commonly mutated genes in CRC patients are APC (about 80%–82% of cases), TP53 (48%–59%), KRAS (40%–50%), and PIK3CA (14%–18%) [87]. Lately, in order to aid in disease diagnosis, examination, and treatment, new guidelines have been established to test the molecular biomarkers in CRC tumor tissues. Some biomarkers, including NRAS, KRAS, and BRAF, are based on the genetic mutations important in the development of CRC or are related to the defects in the DNA mismatch repair system (MMR). The microsatellite instability (MSI) status is determined by these defects [88]. For the diagnosis, prognostication, and therapy response in CRC patients, these biomarkers have been further divided for their possible use as genetic and epigenetic biomarkers:

(i) *Tissue-based biomarkers*: MSI, KRAS gene mutations, and BRAF gene mutations.
(ii) *Blood-based biomarkers*: CEA and carbohydrate antigen 19–9 (CA19–9) [89].
(iii) *Epigenetic biomarker*: miRNA, adenomatous polyposis coli (APC) mutations, and CpG island methylator phenotype.
(iv) *Other promising biomarkers*: Chromosome 18q loss of heterozygosity, loss of PTEN, TP53 expression, loss of NDST4, IGFR-IR, and phosphatidylinositide-3-kinases (PI3K) mutation.

*6.2.3.4 Prostate cancer*

The prostate gland is exclusively present in men. It produces the fluid which is found in semen and help in its nourishment and transport. When the growth of prostate gland cells is uncontrollable, prostate cancer starts to develop. The majority of prostate cancers develop slowly. Almost all prostate cancers are adenocarcinomas [90]. Risk factors include age, environmental factors, family history, etc [91]. In 2020, 1.41 million prostate cases were reported worldwide [63]. Some of the conventional screening techniques for prostate cancer detection include digital rectal examination

(DRE) which is followed by transrectal ultrasound (TRUS) guided biopsy, prostate-specific antigen (PSA) blood test, CT of the pelvis, and MRI [91–93].

#### *6.2.3.4.1 Prostate cancer biomarkers*

PSA is the most prevailing serum biomarker used for the diagnosis of prostate cancer. Since PSA has been introduced as a biomarker, it has led to an enhancement of early-stage and localized prostate cancer detection [94]. However, due to its potential to provide false positive and negative results along with the inconsistent clinical trial results, the use of PSA has become more controversial [95].

Existing biomarkers

(a) *Serum/urine biomarkers*: ProPSA and PHI (Prostate Health Index), the 4Kscore test, TMPRSS2-ERG, PCa antigen 3, SelectMDx, and SChLAP1 [96].

(b) *Tissue biomarker*: Oncotype DX PCa assay, ConfirmMDx, Decipher, Prolaris, and ELAVL1 [96, 97].

Next-generation biomarkers

(a) *Serum/urine biomarkers*: PCA3, SCHLAP1, TMPRSS2–ETS fusion, PTEN, and AR-V7 [98].

(b) *Tissue biomarker*: PCA3, SCHLAP1, TMPRSS2–ETS fusion, and PTEN [98].

## 6.3 Biosensors for the detection of clinical biomarkers

### 6.3.1 Introduction to biosensors

A biosensor is a self-supporting intrinsic device which is able to provide distinct quantitative or semi-quantitative analytical results using a biological element. In a biosensor, physiochemical variation due to the peculiar interactions between an analyte and a bio-recognition element is quantified using a suitable transducer. The bio-recognition component may be an enzyme, antibody, nucleic acid, receptor protein, or cell/tissue. Recently, biosensors have received increasing attention in the environmental monitoring and healthcare sector owing to their high sensitivity, specificity, portability, speed, and low cost. The history of biosensors began in 1962 when the very first enzyme-based glucose biosensor was developed by Clark and Lyons [99, 100]. The device comprised a thin layer of glucose oxidase (GOx) that was captured on an oxygen electrode using a semipermeable dialysis membrane. The experiment was based on monitoring the oxygen utilized by the following enzyme-catalyzed reaction:

$$\text{glucose} + O_2 \xrightarrow{\text{glucose oxidase}} \text{gluconic acid} + H_2O_2.$$

Although the recent types of biosensors are much more advanced, the instability of the desired biomolecules in a variety of conditions is still challenging.

Recently, carbon nanomaterials have attracted much research interest in biosensors due to their large surface area, biocompatibility, distinctive electro-optical properties, and high mechanical strength. Additionally, owing to their large

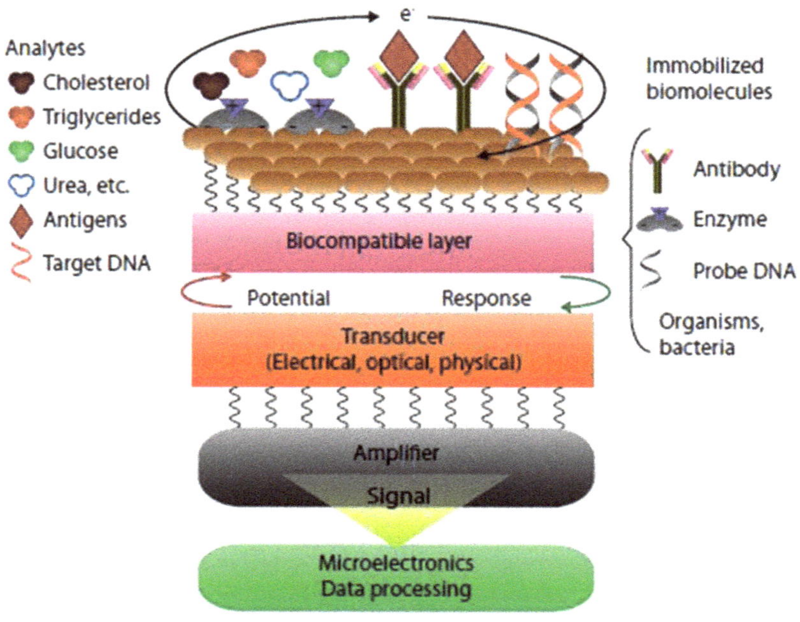


**Figure 6.2.** Schematic of a biosensor and its components. (Reproduced with permission from [154]. Copyright 2011 Springer Nature.)

electrochemical current, greater electrocatalytic activity, and ability to reveal direct electron transfer for a few proteins, carbon nanomaterial based platforms have indicated their potential in highly sensitive biosensors.

A biosensor has three main components: (i) a biological element or bio-recognition element including an enzyme, antibody, DNA, cells, etc, which recognizes an analyte, also known as a bioreceptor; (ii) a matrix for immobilizing biomolecules, including nanomaterials, conducting polymers, self-assembled monolayers, and sol–gel films; and (iii) a transducer which converts a biochemical reaction product into a measurable signal (figure 6.2). Biosensors are divided into two categories, the first is based on the biological recognition element and the second category is based on the transducers implemented. Usually, the biological recognition elements are enzymes, DNAs, antibodies, cell receptors, micro-organisms, etc. The next classification of biosensors depends on the method of transduction used. They are again separated into three types: (i) optical biosensors; (ii) electrochemical biosensors; and (iii) mass-based biosensors.

### 6.3.2 The advantages of biosensors over conventional techniques in the detection of clinical biomarkers

Biosensors are beneficial because of their ease-of-use, high sensitivity, rapid detection, and inexpensive method of detecting clinical biomarkers compared to the conventional techniques, which helps in predicting an effective treatment. Other advantages are their small sample volume requirement, shorter assay time than conventional techniques, i.e. ELISA, low energy consumption, ease of transport, and multiplexing ability [101]. Biosensors provide as rapid a return of information to the clinician as conventional techniques, although conventional techniques take more time for detection of the clinical biomarker. Biosensors are a very innovative tool for bedside care for patients, in particular in an emergency situation. Many methods are available for clinical biomarker detection but the major challenge is the identification of a single biomarker in a sample and it may not be adequate for clinicians to reach some conclusion [102]. Moreover, the biomarker level does not reach the limit of detection (LOD) of conventional techniques. A biosensor provides a multiplexed detection platform, which is capable of rapid and multiple biomarker detection. Also, biosensors offer a precise diagnosis which is very important before any therapeutic intervention [103].

Biosensors have an ultra-sensitive configuration for quantitative detection of analytes closer to the patients [104]. The most commonly used conventional techniques are ELISA or lateral flow-based, which are costly and provide only qualitative information. In contrast, a biosensor helps to reduce the test costs and to generate digital signals that can be combined with existing devices such as android mobiles. Micro- and nanotechnology-based biosensors are capable of complicated molecular assays which are required for the diagnosis of many infectious diseases [105]. In parallel, biosensors have significance in understanding the pathogen genomics and proteomics and their interplay with the host. Biosensors have high sensitivity as well as selectivity for biomarkers.

### 6.3.3 Optical biosensors

Optical biosensors, the most studied class of biosensors, identifies any change in the property of light, for instance, fluorescence, absorption, refractive index, or scattering of light arising due to the interaction between an analyte and a receptor. This leads to the generation of a signal corresponding to the concentration of a substance which is measured using biological materials, including antigens, enzymes, receptors, cells, nucleic acids, and complete tissues [121]. They have proven to be to be superior to the conventional analytical techniques by offering greater adaptability and sensitivity, enabling the real-time, faster, label-free, and efficient detection of various chemical and biological materials. Numerous multidisciplinary approaches, such as micromechanical systems, microelectronics, biotechnology, chemistry, and molecular biology, along with advanced concepts, have been utilized for the fabrication of new optical biosensors. The research and methodological development of optical biosensors have accomplished significant growth over the past decade [121–123].

Optical biosensing can be label-free or label-based. The detected signal in the label-free mode is directly generated by the interaction of the analyte with the transducer, whereas in case of a label-based sensor, a label is used, which in turn generates an optical signal by a fluorescent, luminescent, or colorimetric method. An optical biosensor (figure 6.3) is a compact analytical device that consists of a bio-recognition sensing element combined with an optical transducer system.

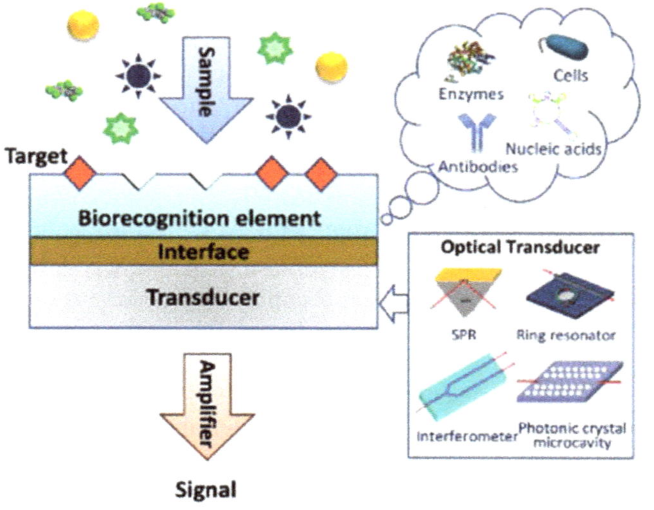


**Figure 6.3.** Biological detection based on optical biosensors. (Reproduced with permission from [155]. Copyright 2019 Elsevier.)

The optical biosensor is used to generate a signal corresponding to the analyte's concentration. Optical waveguide interferometry, evanescent wave fluorescence, surface plasmon resonance (SPR), and localized surface plasmon resonance (LSPR) employ the field adjacent to the surface of the biosensor to identify the interaction of a biological element with an analyte [122, 123]. Optical biosensors thus possess a great advantage over conventional tools in detecting biological systems, thereby promoting significant progress in clinical diagnosis and drug discovery [124].

### 6.3.4 Electrochemical biosensors

Specific binding is the basic principle for the detection of various biomarkers, which permits capturing a specific analyte from the sample. Electrochemistry based biomarker detection techniques are one of the most widely used technologies in the area of food safety, environment protection, human deficiency detection, and diagnosis. Based on their mechanism, electrochemical biosensors are divided into two categories, i.e. label-based and label-free. In label-based electrochemical biosensors, a secondary antibody is used along with the electroactive labels, i.e. biocompatible nanoparticles. The secondary antibody is bound with the target analyte with an intermediate primary antibody, which is immobilized onto the electrode surface through specific binding, and a change in current according to the analyte concentration is observed at the electrode [106]. In label-free electrochemical biosensors, the signal is generated due to the oxidation/reduction reactions of bioconjugation on the electrode surface, which offers high sensitivity and a rapid method of diagnosis [107].

The properties of electrochemical sensors have been recently improved via two strategies. First, enhancing the capture efficiency on the surface of the devices

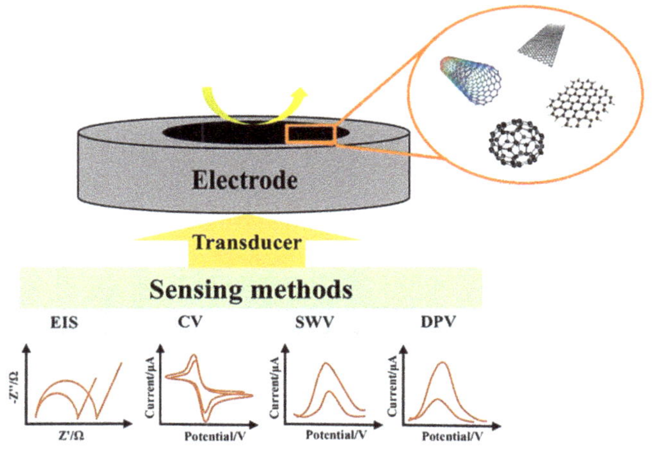


**Figure 6.4.** The working principle of an electrochemical biosensor with different components. (Reproduced with permission from [156]. Copyright 2021 Elsevier.)

through reducing the effects of inevitable factors, such as impurities in the buffer solution, changes in the environmental temperature and humidity, etc. An ideal biosensor captures specific biomarkers and does not capture any other impurities. However, the impacts of impurities and environmental effects can be reduced by building a barrier, for example, deposition of a bovine serum albumin (BSA) layer [108]. Second, the output signal can be increased by using some specific labels at the same concentration of analyte. These specific labels amplify the signals; for example fluorescent labels, which are detected through charge-coupled devices, generate larger electric signals [109]. The proportion of label on the second antibody's surface is accountable for the generation of a high output signal even at low analyte concentration, which improves the resolution as well as the detection limit.

An electrochemical biosensor is a combination of three components (figure 6.4): (a) a recognition element which interacts with the analyte; (b) a transducer which generates a measurable signal from the biological changes; and (c) an electronic readout circuit for the management of data. An electrochemical biosensor can be classified on the basis of the type of transducer into voltametric, amperometric, potentiometric, conductometric, impedimetric, and field effect transistor (FET) [110].

#### *6.3.4.1 Voltametric techniques*

Voltametric techniques are categorized into two- and three-electrode systems. In a three-electrode voltametric system, a voltage is applied between a working electrode and a reference electrode (i.e. Ag/AgCl). The current is measured as an output signal between the working electrode and the counter electrode. In a two-electrode system, the reference electrode is shorted with a counter electrode. The input voltage can be applied in many modes, such as square wave, linear sweep, stripping, cyclic, and differential pulse. Among the different applied voltage modes, the differential pulse and square-wave modes are mostly employed due to their high sensitivity. In both two- and three-electrode systems, the substrate on the surface of the biosensor should be designed with care for specific recognition of the analyte. The limit of detection differs, which is dependent on the substrate, which can be Au nanoparticles, CNTs, magnetic particles, and quantum dots.

#### *6.3.4.2 Amperometric techniques*

Generally, amperometry is the most regularly used technique, in which a change in current as the output signal is measured with respect to time at a constant applied potential. The change in current at a fixed potential is observed due to the Redox activity of an electroactive species in a reaction. In amperometric biosensors, an antibody is labeled with an electrochemically active material that easily binds with the analyte and generates a signal as a change in current. The limit of detection of amperometric biosensors is also dependent on the electrode properties and the resolution of the electronic configuration. In chronoamperometry, a potential is applied in the square-wave mode to the working electrode and the output current is measured as a function of time.

#### *6.3.4.3 Impedimetric techniques*

Impedimetric techniques, also known as electrochemical impedance spectroscopy (EIS), are used for the electrochemical characterization of a surface. EIS is used to observe the electrode kinetics and substrate–analyte binding characteristics. More importantly, this technique is used for real-time detection of biomarkers, which has increased its significance in the area of biosensors. In electrochemical impedance spectroscopy, the alternating current is measured at different frequencies to overserve an impedance spectrum. The analyte–surface interactions (i.e. antigens–antibodies) on the electrode surfaces generates a capacitance and interfacial resistances, which can be used as an output response for biomarker detection.

EIS is the most frequently used impedance-based technique because it requires just 5 mV or 10 mV excitation voltage, which is much small than for voltametric methods (−200 mV–600 mV excitation voltage). It operates at a low excitation voltage, thus generating low heat, which makes it a long-lasting method. Additionally, EIS offers multiple parameters for sensing, such as double-layer capacitance ($C_{PE}$), charge transfer resistance ($R_{ct}$), Warburg impedance ($Z_w$), and electrolyte resistance ($R_S$).

#### *6.3.4.4 Potentiometric techniques*

In potentiometric techniques, an electric potential of the working electrode is measured when the applied current is zero. In this approach a potential is generated at a working electrode due to the electrochemical reactions, which is related to the concentration of analyte. The correlation between the concentration and the potential is given by the Nernst equation. It offers a low limit of detection which is significant for biomarker detection at the lower concentration in the sample. Another potentiometric technique is known as chronopotentiometry and measures the potential as a function of time in response to a constant or square-wave current.

#### *6.3.4.5 Conductometric techniques*

Conductometric biosensors can be beneficial for the design and development of an advanced technique for biomarker detection. Their unique advantages, such as no requirement for a reference electrode, operating at a low driving voltage, and large-scale production, make it more suitable.

#### *6.3.4.6 Field effect transistor (FET)*

The FET is comprised of three terminal devices which control the flow of electrons between the source and drain through an electric field at gate in a material that is semiconducting. At the gate, the electric field is controlled by altering the potential relative to the source and drain electrode. Many types of FET are available, for example, enzyme FET (EnFET) and ion-selective field effect transistor (ISFET).

## 6.4 A biosensing platform based on zero-dimensional carbon nanomaterials for clinical biomarkers

Carbon-based nanomaterials have gained extensive popularity in the field of nanotechnology due to their bulk production at a low cost, low toxicity, and ease of surface functionalization [111]. They have potential in biology-based applications, for example, drug delivery, sensing, and imaging. In this context, zero-dimensional carbon nanomaterials have become an ideal candidate for the production of nano-sensors, possessing greater performance and lower consumption of power due to their excellent optoelectronic properties [112].

GQDs, CQDs, and fullerenes are zero-dimensional carbon nanomaterials which possess exceptional properties, such as a variable energy bandgap, non-toxicity, large surface area, chemical resistance, and zero-blinking.

### 6.4.1 Graphene quantum dot based biosensors for clinical biomarkers

#### *6.4.1.1 Optical detection of clinical biomarkers using graphene quantum dots*

GQDs are zero-dimensional nano-sized derivatives of graphene exhibiting excellent optoelectronic properties, which are necessary for biosensors for diagnosis of clinical diseases (figure 6.5) [113]. The excellent and tunable photoluminescent properties of GQDs make them highly applicable in bioimaging and biosensing. The edges of GQDs are occupied by oxygen-rich functional groups which provides them good water solubility and biocompatibility.

Bhatnagar *et al* designed a sensitive amine-functionalized GQD based fluorescent sensor (figure 6.6) conjugated with the antibody anti-cTnI to detect traces of cTnI in blood for early diagnosis of heart attacks. The detection was based on fluorescence resonance energy transfer (FRET), which utilizes the synergy between the conjugate and graphene. The amine-functionalized GQDs and anti-cTnI nanoprobe emitted blue light when illuminated with ultraviolet light. The deposited graphene on the nanoprobe acted as a quencher and the sensor no longer fluoresced. Upon addition of the cTnI antigen to the fabricated senor, the fluorescence was restored as a consequence of the complex formation between the antigen and the cTnI antibody. The sensor was highly sensitive when tested with clinical samples as well. The time of

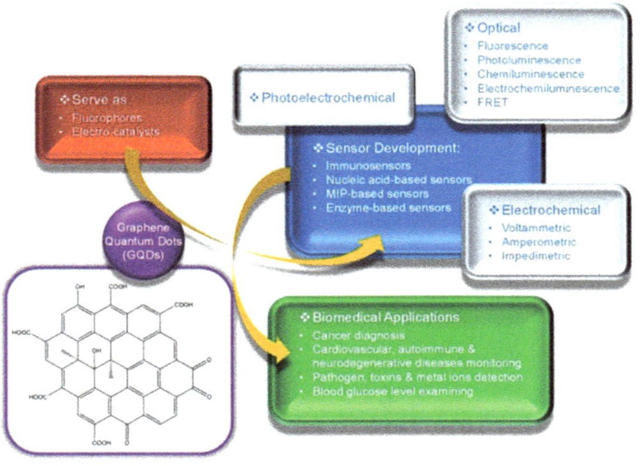


**Figure 6.5.** Graphene quantum dots as optical and electrochemical biosensors. (Reproduced from [157]. Copyright 2020 the Author(s).)

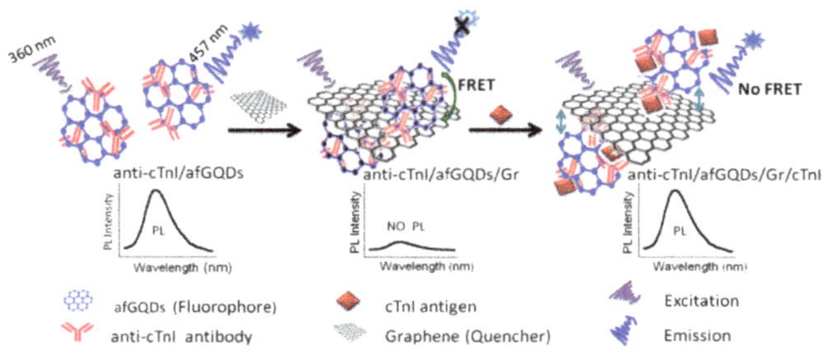


**Figure 6.6.** Illustration of a fluorescent amine-functionalized GQDs based biosensor for the detection of cTnI. (Reproduced with permission from [114]. Copyright 2016 Elsevier.)

detection was 10 min with a limit of detection of 0.192 pg $mL^{-1}$. This could, therefore, lead to its promising application in the real-time analysis of cTnI in biological samples [114].

MicroRNA (miRNA), an endogenous non-coding RNA, performs an important role in regulating the functions and aging of the heart. For example, miRNA-22 stimulates aging and activates the cardiac fibroblasts in the aging heart [115, 116]. In a study by Boon *et al*, it was shown that the expression of miRNA-34a was affected by a senescing heart. Silencing or deleting this miRna-34a could decrease the death of cardiac muscle cells, also known as cardiomyocytes, thereby helping in recovering myocardial function [117]. Recognition of the aging-related biomarker miRNA-34a could thus provide newer perspectives on the investigation and early detection of cardiovascular diseases. Based on this fact, Sun *et al* developed a unique heterodimer probe with a comet-like structure, based on Au nanoflowers and GQDs which could detect miRNA-34a. The designed probe worked on the principle of FRET with a lower detection limit of 0.1 femtomolar. This was nearly two times lower than the existing sensing techniques, which could thus provide early detection of cardiovascular diseases and propose the required therapeutic protocols [116]. Biosensors based on GQDs exhibit unique electrical and chemical properties, and their ease of interaction with biomolecules via electrostatic bonding or $\pi$–$\pi$ interaction renders them an ideal tool for cardiovascular biosensing applications [118].

The diagnosis of cancer in its early stages is significant in increasing the chances of treatment success and reducing the number of cancer-related deaths. The recent advancement of nanobiotechnology has revolutionized cancer detection by allowing small-scale biosensors for prompt and efficient cancer detection [119, 120]. Nanobiosensors exhibit great potential over previous methods such as low limits of detection, real-time analysis, a small amount of sample to be analyzed, high-throughput screening, and label-free detection [121]. Carbon-based nanomaterials, which are characterized by the presence of graphene structures, present significant advantages, such as ultrasensitivity and high selectivity, for biosensor devices as a result of their excellent physiochemical and optoelectronic properties. For cancer detection, the biosensors are fabricated such as to recognize specific cancer analytes and convert them into signals. These analytes include antibodies, carcinoma antigens, nucleic acids, proteins, and cancer-related by-products, for example, proteins exposed on the exterior of cancer cells, $H_2O_2$, and changes in the pH values [122].

In addition to the detection of small molecules, GQD based biosensors have been reported as tools to diagnose cancer [123]. The production of lactic acid by tumors leads to the hydrolysis of adenosine triphosphate (ATP) in anaerobic conditions. Thus, the value of pH becomes lower than that of healthy tissues, which makes this characteristic an efficient tool for the diagnosis of cancer. Fan *et al* fabricated a pH-responsive fluorescent GQD probe to differentiate tumors from normal tissues. The probe had minimal toxicity and exhibited an intense fluorescence transition from green at pH below 6.8 to blue above pH 6.8, which is similar to the acidic biological microenvironment in solid tumors. This unique and reversible fluorescent property helped in distinguishing the tumors from healthy tissues. The fluorescence intensity was reported to be unchanged even after 24 h of illumination, which demonstrated the excellent stability of the fabricated fluorescent GQD based sensor, which in turn can have great potential applications as a universal fluorescent probe for the detection of tumors [124]. In another study, Cui *et al* developed a GQD based fluorescent magnetic biosensor consisting of magnetite ($Fe_3O_4$) and molybdenum disulfide ($MoS_2$) nanosheets (figure 6.7). The photoluminescent GQDs were synthesized using an electrochemical approach and were functionalized with epithelial cell adhesion molecule (EpCAM) receptors (an aptamer). The fabricated biosensor was employed for the rapid and efficient detection and sensitive separation of CTC. Apart from this, the biosensor exhibited low cell toxicity and an average capture efficiency of 90% which was reported to be higher than that of the other circular tumor cell separation methods reported so far. The fluorescent magnetic biosensor was rapid (15 min) and was reported to be extremely specific and sensitive in

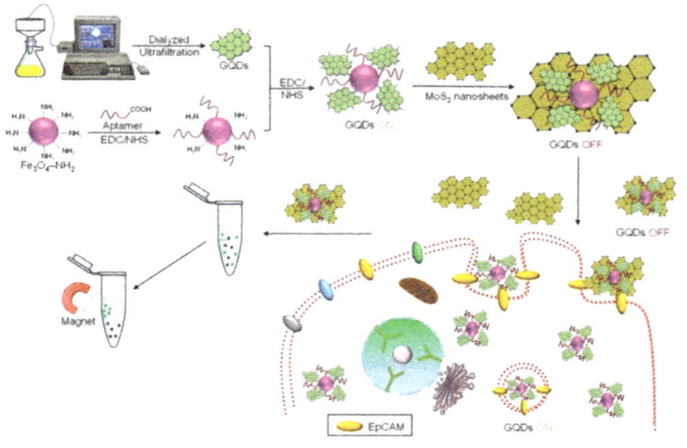


**Figure 6.7.** Fluorescent magnetic GQD, magnetite ($Fe_3O_4$), and molybdenum disulfide ($MoS_2$) nanosheet based biosensor for clinical detection of circulating tumor cells. (Reproduced with permission from [125]. Copyright 2019 Springer.)

detection due to the presence of aptamers which aided the capture of both low and high expressing circular tumor cells, highlighting its potential for early detection of cancer [125].

Among the various nano-sized substrates that have been investigated so far for the clinical detection of cancer biomarkers, the zero-dimensional GQDs present great potential due to their exceptional electro-optical properties, the presence of various active sites allowing greater chemical functionalization, and their large surface area. Their bonding with biomolecules and the ability to recognize cancer biomarkers or diagnose the tumor cells makes them a versatile tool for the fast and accurate sensing of cancer [125].

Biosensors based on GQDs can be used to detect other biomarkers, such as ascorbic acid, amino acids, and dopamine as well [112]. Recently, an exceptionally sensitive GQD based fluorescence turn-on nanosensor was fabricated. The nanosensor exhibited orange emission which could efficiently detect ascorbic acid. The interaction of ascorbic acid with hydroxide radicals led to the fluorescence recovery of the GQDs, which was otherwise quenched by o-benzoquinone. This fluorescence switching provided high selectivity and a broad range of detection to the sensor. Apart from this, the biosensor was biocompatible since it did not involve the use of heavy metal in the system. The limit of detection of the fabricated biosensor was 0.32 micromoles for ascorbic acid, which was quite a lot lower than the other fluorescent biosensors reported so far [126]. Monitoring glutathione (GSH) has drawn appreciable attention for its crucial role in human diseases. A GQD-$MnO_2$ nanosheet based biosensor was fabricated by Yan *et al*, which could ultra-sensitively detect the presence of GSH in living cells. The FRET between GQDs and $MnO_2$ resulted in quenching of the fluorescence intensity of the GQDs. This fluorescence 'turn–off–on' biosensor upon entering the cell promoted the reduction of $MnO_2$ nanosheets to $Mn^{2+}$ by GSH, thereby releasing the GQDs which developed a recovered fluorescence signal. An ultra-sensitive response to GSH along with an extremely low detection limit of 150 nanomolars was achieved using the fabricated biosensor [127].

#### *6.4.1.2 Electrochemical detection of clinical biomarkers using graphene quantum dots*

GQDs show good electron mobility and electron-accepting properties similar to peroxidase. They also exhibit electrochemical catalytic properties toward hydrogen peroxide which allow its detection in living cells. Moreover, GQDs can be considered as an alternative to graphene, GO, or rGO in biosensors because they have better electronic conductivity than graphene oxide and more functional groups can be immobilized for biomolecule attachment than on pure graphene. Pure graphene can be functionalized with carboxylic and/or amine groups but it affects their electronic properties. Moreover, the presence of oxygenated and peripheral carboxyl groups with the aromatic structure of graphene also affects their electrochemical properties.

An electrochemical GQD based biosensor was reported for the clinical detection of cTnI, where 1-ethyl-3-(3-dimethylaminopropyl) carbodiimide/N-hydrogysuccinimide was used as a coupling agent for the GQDs and an amino-functionalized gold electrode. The 1-ethyl-3-(3-dimethylaminopropyl) carbodiimide was again used for coupling 4-aminothiophenol and poly-amidoamine dendrimer on the GQDs [128]. This sensor was used to detect troponin-I with a lower detection limit of 20 fg mL $^{-1}$ and 25 fg mL$^{-1}$ for the differential pulse voltammetry (DPV) and cyclic voltammetry (CV) techniques, respectively. These zero-dimensional carbon materials can be used in screen-printed technology to enhance sensitivity. An enzyme-free electrochemical based biosensor for cTnI detection was developed using modified screen-printed gold electrodes (Au-SPEs), GQDs, and gold nanoparticles (AuNPs). AuNPs and GQDs are powerful nanomaterials for biosensor applications because of their good biocompatibility and high electrocatalytic activity, and as they enhance the surface area of the electrode and provide active sites for chemical modification [129]. In this sensor, the GQDs and AuNPs were implanted onto the Au-SPEs and then anti-cTnI was immobilized for cTn-I detection. Finally, the concentration of cTn-I in the sample could be determined utilizing amperometry, EIS, CV and square-wave voltammetry (SWV).

In another study, a label-free impedimetric immune biosensor was fabricated for diagnosis of cMyoglobin, a cardiac biomarker [130]. In this impedimetric biosensor, GQDs were employed as a template on SPEs for a sensing platform which was conjugated with anti-myoglobin antibodies. The values of $R_{ct}$ were observed as a function of antigen concentration. The detection of cMyo was studied in the presence of other proteins with a 0.02 ng mL$^{-1}$ limit of detection, comparable to standard ELISA techniques.

Lakshaman Kumar *et al* designed and developed GQDs functionalized with acetic acid which served as an interface over a gold electrode for the electrochemical detection of cTnI, a highly sensitive and specific biomarker for AMI detection [131]. The interaction of the biomarker with the functionalized GQDs was investigated using amperometry and cyclic voltammetry. The study reported 0.17 to 3 ng $mL^{-1}$ as the range of detection with a 0.02 ng $mL^{-1}$ limit of detection, which rendered the biosensor as highly sensitive and highly stable. The study proposed an antibody-free tool for the electrochemical detection of cTnI providing an efficient, low cost, and stable method of detection [131].

Recently, Mansuriya *et al* fabricated an electrochemical nano-immunosensor comprising a screen-printed Au electrode modified with GQDs and AuNPs for efficient detection of cTnI [132]. This involved the incorporation of anti-cTnI antibody onto the surface of the gold NP–GQD nanocomposites. The biomarker was then determined using CV, EIS, amperometry, and SWV. Interestingly, the nano-immunosensor showed excellent sensitivity towards the biomarker with a 0.5 pg $mL^{-1}$ lower detection limit. This study first reported the successful detection and investigation of an analyte using four different electrochemical techniques. Also, the immunosensor presented a highly sensitive, specific, and rapid detection of the biomarker, thereby demonstrating its potential use in real-time analysis [133].

### 6.4.2 Carbon quantum dot based biosensors for clinical biomarkers

#### *6.4.2.1 Optical detection of clinical biomarkers using carbon quantum dots*

CQDs, a new type of 0D carbon nanomaterial with a particle size below 10 nm, have provoked enormous interest in the area of biosensors due to their unique photo-luminescent and photo-physical properties [134]. In addition, CQDs can be produced easily and also have some advantages, such as easy functionalization with amine and carboxyl groups similar to GQDs. Carboxyl functionalized CQDs show excellent water solubility and good compatibility with organic and inorganic species. CQDs are promising alternatives for GQDs because they are cost-effective, easy to prepare, and highly biocompatible due to the absence of toxic metal impurities. The CQDs can be prepared by various methods such as microwave, acidic oxidation, electrochemical oxidation, ultrasonication, arc discharge, and hydrothermal methods.

CQDs have poorer crystallinity compared to GQDs as a result of greater surface defects and a lower content of crystalline $sp^2$ carbon [135]. CQDs have exceptional optical properties in chemiluminescence, electrochemiluminescence, and fluorescence [136, 137]. The synthesis and functionalization of CQDs is quick and facile, similar to that of GQDs. The sensitivity of CQDs as biosensors can be improved by functionalization of their surfaces or by doping to enhance their optoelectronic properties, surface activity, and biocompatibility [138].

Chen *et al* designed a sensitive CQD and deoxyriobonuclease-I based biosensor for the detection of myoglobin. Myoglobin is an essential cardiac biomarker and plays an important role in the detection of AMI. Their study showed the specific recognition of myoglobin which could lead to the recovery of the fluorescence which was otherwise quenched by the $\pi$–$\pi$ interaction of the myoglobin aptamer and CQDs in the absence of myoglobin. The sensitive and rapid sensing of myoglobin was achieved with a 20 pg $mL^{-1}$ limit of detection and in a range between 50 and 100 ng $mL^{-1}$. Also, the sensitive and specific detection in real-time biological samples makes the sensor useful for detecting AMI [139]. Tan *et al* reported s ratiometric fluorescent immunoassay of cTnI using ELISA. The study used palladium–iridium nanocubes as a label for the secondary antibody with CQDs as the reference fluorophore for the detection of the biomarker. o-phenylenediamine was used as a non-fluorescent substrate which was catalyzed by the nanocubes to a yellow fluorescent product in the presence of $H_2O_2$ with an emission peak at 570 nm upon 400 nm excitation. The fluorescent product quenched the fluorescence of the CQDs by being adsorbed on their surface. Consequently, upon the increase of cTnI, the fluorescence peak for the yellow fluorescent product at 579 nm increased with a decrease for the 450 nm of CQDs. This resulted in ratiometric fluorescent signals which could be used for a quantitative assay of cTnI. The changes in two fluorescence emission peaks made the detection results more reliable and sensitive, making ELISA significantly promising in clinical diagnosis [140].

The highly efficient charge transfer quenching of fluorescence emissions has rendered CQDs excellent and efficient sensors [141]. Miao *et al* presented an environmentally friendly strategy of fabricating CQDs from tomato juice as a label-free sensitive assay for the detection of CEA. The CQDs showed excellent photostability, high fluorescent quantum yield, and non-toxicity. The study reported the adsorption and desorption of aptamers by the surface of carboxyl group rich CQDs via a competitive mechanism. The CEA aptamer was attached to the CQDs surface by $\pi$–$\pi$ stacking leading to quenching of fluorescence. Due to the presence of CEA, the CEA aptamer detached from the surface of the CQDs, which was a result of their strong affinity, thereby restoring the fluorescence. The detection range was reported from 1 to 0.5 ng $mL^{-1}$ with a lower limit of detection of 0.3 ng $mL^{-1}$. The strategy used in this study could be applied in a recyclable manner, broadening the scope of CQDs in sensing-based applications, along with their biocompatibility [142]. The same concept of detection was used by Motaghi *et al*, who devised AS1411, a nucleolin aptamer and CQD based sensor to detect cancer cells (figure 6.8). Mouse breast 4T1, human breast MCF7, and human cervical HeLa cancer cells were selected as the target cells, whereas the human foreskin fibroblast cells HFF-P16 were selected as the control cells. The target cancer cells were spectrofluorometrically detected in the presence of control cells after incubating them in carbon dot–aptamer solutions, followed by subsequent centrifugation and measurement of their fluorescent intensities. The AS1411 aptamers specifically targeted the cancer cells with more nucleolin on their surface than MCF7 or 4T1 cells, leading to the release of carbon dots and an enhancement in the fluorescent intensity. This study presented a facile and cost-effective method of cancer cell detection with sufficient sensitivity and a nearly 100 cells $mL^{-1}$ limit of detection [143].

An ultra-sensitive N-doped CQD based fluorescent sensor was developed by Lu *et al* for efficient detection of $\beta$-glucuronidase. The doped CQDs exhibited green photoluminescence where 4-nitrophenyl-$\beta$-D-glucuronide was used as a substrate in which p-nitrophenol served as a strong absorber to quench the fluorescence of the N-CQDs. This biosensor proved to be highly sensitive to $\beta$-glucuronidase with a detection limit of 0.3 $\mu$U $L^{-1}$, thereby leading to a newer technique for the design and development of selective and sensitive tumor biomarkers by fluorescent QDs [46].

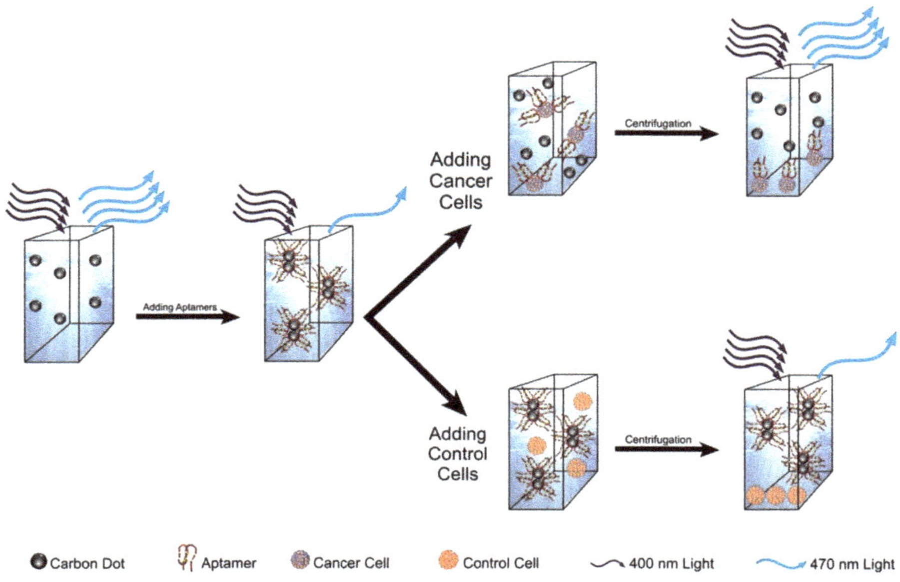


**Figure 6.8.** Schematic illustration of cancer cell detection utilizing the CQDs-AS1411 aptamer nanoconjugates. (Reproduced from [157]. Copyright the author(s).)

A bidimensional optical sensor designed and developed by Liu *et al* combined the benefits of fluorescence and colorimetry for the detection of arginine (Arg). For this, the CQDs were synthesized in ethylene glycol as the stabilizing and reducing agent. The influence of arginine on the growth of AuNP–CQD composites was monitored by colorimetric detection with a lower detection limit of 37 nM and a detection limit of 450 nM obtained spectrofluorometrically. The fabricated dual-signal biosensor exhibited excellent selectivity towards arginine over other amino acids [144].

#### *6.4.2.2 Electrochemical detection of clinical biomarkers using carbon quantum dots*

Unlike other carbon allotropes, CQDs have not been explored much for electrochemical biosensor based applications. The few applications of CQDs in electrochemical biosensors were mainly focused on its electrocatalytic activity towards oxygen and hydrogen peroxide reduction and the sensing of dopamine [145], 2,4,6-trinitrotoluene [146], patulin [147], and glucose [148].

The electrochemical signals of a biosensor can be enhanced by modifying a glassy carbon electrode (GCE) with CQDs, for highly sensitive detection of etoposide (ETO) up to 5 nM ETO. Modification of the surface is an effective approach to changing the morphological properties of materials according to specific applications. The surface functionalization of CQDs can be made through various approaches, i.e. changing the surface chemistry or interactions. It is a well-known fact that the surface of CQDs is rich with oxygen-containing groups, which allow the high possibility of modifying the surface of CQDs [149].

### 6.4.3 A fullerene based biosensor for clinical biomarkers

*6.4.3.1 Optical detection of clinical biomarkers using fullerenes*

Fullerenes, due to their exceptional physiochemical properties, can be used to construct highly sensitive biosensors [22]. Surface functionalization of fullerenes can be performed in order to improve their luminescent properties. Peng *et al* presented a simple one-pot strategy to fabricate core–shell manganese–fullerene nanocomposites for GSH-activable magnetic resonance/fluorescence bimodal imaging functions (figure 6.9). The nanocomposites exhibited a quenched fluorescence induced by FRET, which upon exposure to GSH resulted in multicolor fluorescence recovery due to reconstruction of the nanoprobes. This nanoprobe based on manganese–fullerene can potentially be applicable for the detection and regulation of the biological processes of GSH-associated diseases [150].

The design and fabrication of fullerenes and their derivatives play a vital role in developing their applications, particularly in sensing, imaging, and marking in biomedical fields. Based on this, Ma *et al* fabricated fluorescent C60-based nanoparticles which utilized thiol-ene chemistry and folic acid modification. The fluorescent C60 nanoparticles were reported to have a relatively high quantum yield of 26% and showed intense blue luminescence. Also, they were found to be highly photostable and resistant to photobleaching under constant UV illumination for five hours. The as-synthesized fluorescent C60 nanoparticles were used as probes for folic acid detection, targeting, and cancer cell imaging. They exhibited high specificity and sensitivity to folic acid with a 0–80 micromolar detection range and a 0.24 micromolar limit of detection. The folic acid modification in the nanoparticles facilitated their penetration into the cancer cells with higher expression of folate receptors, which therefore was applicable in targeted cellular imaging of cancer cells [151].

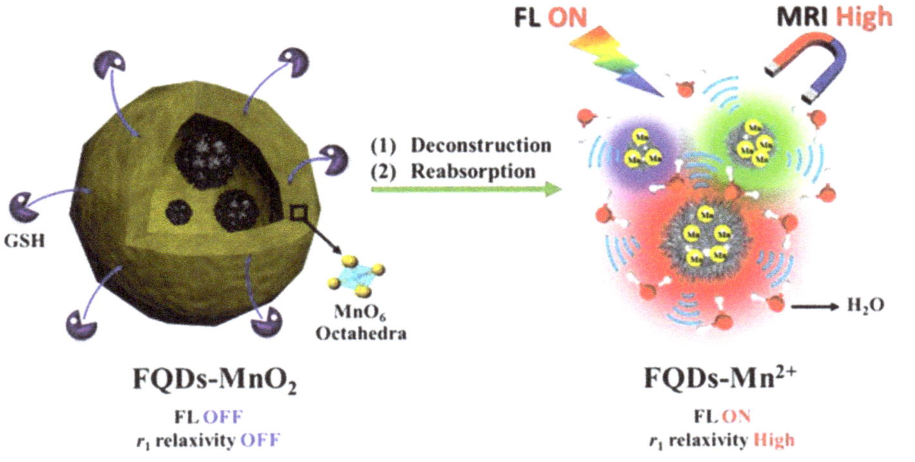


**Figure 6.9.** Schematic illustration of manganese–fullerene based core–shell nanoprobes. (Reproduced with permission from [149]. Copyright 2019 American Chemical Society.)

*6.4.3.2 Electrochemical detection of clinical biomarkers using fullerenes*

Fullerenes also find potential applications in cancer diagnosis, like GQDs and CQDs. Supramolecular chemistry has been widely utilized to synthesize functionalized C60, which would preserve its original structure as well as its integral electrochemical properties [152]. For example, Yuan *et al* designed and developed an electrochemical biosensor based on amino group-functionalized fullerene (n-C60) coupled Pd–Pt bimetallic nanocrystal, and 4-mercaptophenylboronic acid (4-MPBA) to detect $\alpha$2,3-sialylated glycans ($\alpha$2,3-sial-Gs), which are important tumor biomarkers for cancer diagnosis in the early stages. The load capacity, sensitivity, and conductivity of the biosensor was improved by the incorporation of n-C60 on the surface of the PdPt bimetallic nanocrystal. The detection was observed to be within a wide range between 10 fg $mL^{-1}$ to 100 ng $mL^{-1}$ with a lower detection limit of 3 fg $mL^{-1}$. The sensor exhibited great stability and a good recovery rate, thereby indicating its potential in scientific research [153].

Prussian blue carried Au nanoparticles reportedly amplify the electrochemical signals and they possess excellent sensing activity, redox biocompatibility, and electrochemical activity. In a study by Han *et al*, these nanoparticles were decorated onto the surface of n-C60 and thiol group-functionalized fullerenes. The fabricated nanoparticles possessed good water solubility along with large surface-active sites and they were labeled by detection aptamers and alkaline phosphatase as a tracer. A sandwich-type electrochemical biosensor was thus designed, wherein the Au nanoparticles along with onion-like mesoporous graphene sheets were employed as the biosensor. This utilized the multi-labeled functionalized nanohybrid of C60 as a tracing tag. Platelet-derived growth factor B-chain, a cancer-related protein, which is responsible for the growth, transformation, and progression of tumors, was used as the model target to be detected by the fabricated sensor. The linear detection range was reported to be in the range of 2–40 $\mu$M with a 0.6 pM limit of detection, indicating the high sensitivity of the aptasensor with a potential application in electrochemical bioanalysis [152].

## 6.5 Conclusion and future outlook

To conclude, we have summarized the potential applications of zero-dimensional carbon nanomaterial based biosensors for the clinical detection of cardiovascular and cancer biomarkers. Studies show that zero-dimensional carbon nanomaterials can be exploited commercially in different areas and compete with traditional semiconductor zero-dimensional materials. Due to their unique qualities, zero-dimensional carbon nanomaterials have demonstrated vast applicability in clinical biomarker detection via a biosensing protocol. They are likely to become more widely used in the coming years. Although they have been found to be very interesting, the synthesis techniques of some of these materials are costly along with their low yield. Therefore, there is scope to develop new cost-effective strategies for high yields. For the high efficacy of biosensor functionalization and the bio-functionalization of zero-dimensional carbon nanomaterials, proper immobilization of the desired biomolecule is very important. Therefore, there is a need to develop new strategies for the structural and chemical tailoring of these materials in order to enhance their surface functionalization and bio-functionalization, leading to high sensitivity, stability, and specificity of detection of a particular biomarker. The low toxicity and cost-effectiveness of these materials can also lead to the substitution of conventional materials for biomedical applications in the near future.

## Acknowledgements

SS acknowledges financial assistance from the DST, New Delhi, India (Grant No. DST/INSPIRE/04/2014/002540), and Dr A P J Aktu, Uttar Pradesh, India (Grant under Visvesvaraya Research Promotion Scheme).

## References

[1] Monthioux M and Kuznetsov V L 2006 Who should be given the credit for the discovery of carbon nanotubes? *Carbon* **44** 1621–3

[2] Bollmann W and Spreadborough J 1960 Action of graphite as a lubricant *Nature* **186** 29–30

[3] Oberlin A, Endo M and Koyama T 1976 Filamentous growth of carbon through benzene decomposition *J. Cryst. Growth* **32** 335–49

[4] Kroto H, Health J R, O'Brien S C, Curl R F and Smalley R E 1985 C60: Buckminsterfullerene *Nature* **318** 1985

[5] Subramoney S 1998 Novel nanocarbons—structure, properties, and potential applications *Adv. Mater.* **10** 1157–71

[6] Oza G *et al* 2016 Camphor-mediated synthesis of carbon nanoparticles, graphitic shell encapsulated carbon nanocubes and carbon dots for bioimaging *Sci. Rep.* **6** 1–9

[7] Shehab M, Ebrahim S and Soliman M 2017 Graphene quantum dots prepared from glucose as optical sensor for glucose *J. Lumin.* **184** 110–6

[8] Hu Y, Yang J, Tian J, Jia L and Yu J-S 2014 Waste frying oil as a precursor for one-step synthesis of sulfur-doped carbon dots with pH-sensitive photoluminescence *Carbon* **77** 775–82

[9] Zhao H *et al* 2016 Egg yolk-derived phosphorus and nitrogen dual doped nano carbon capsules for high-performance lithium ion batteries *Mater. Lett.* **167** 93–7

[10] Wei S, Zhang R, Liu Y, Ding H and Zhang Y-L 2016 Graphene quantum dots prepared from chemical exfoliation of multiwall carbon nanotubes: an efficient photocatalyst promoter *Catal. Commun.* **74** 104–9

[11] Minati L, Torrengo S, Maniglio D, Migliaresi C and Speranza G 2012 Luminescent graphene quantum dots from oxidized multi-walled carbon nanotubes *Mater. Chem. Phys.* **137** 12–6

[12] Shinde D B and Pillai V K 2012 Electrochemical preparation of luminescent graphene quantum dots from multiwalled carbon nanotubes *Chem.–A Eur. J.* **18** 12522–8

[13] Peng J *et al* 2012 Graphene quantum dots derived from carbon fibers *Nano Lett.* **12** 844–9

[14] Liu H, Ye T and Mao C 2007 Fluorescent carbon nanoparticles derived from candle soot *Angew. Chem.* **119** 6593–5

[15] Liang R, Wei M, Evans D G and Duan X 2014 Inorganic nanomaterials for bioimaging, targeted drug delivery and therapeutics *Chem. Commun.* **50** 14071–81

[16] Sondhi P, Maruf M H U and Stine K J 2020 Nanomaterials for biosensing lipopolysaccharide *Biosensors* **10** 2

[17] Yao J, Li P, Li L and Yang M 2018 Biochemistry and biomedicine of quantum dots: from biodetection to bioimaging, drug discovery, diagnostics, and therapy *Acta Biomater.* **74** 36–55

[18] Chen J B *et al* 2020 Nanostructured architectures for biomolecular detection inside and outside the cell *Adv. Funct. Mater.* **30** 1907701

[19] Carneiro P, Morais S and Pereira M C 2019 Nanomaterials towards biosensing of Alzheimer's disease biomarkers *Nanomaterials.* **9** 1663

[20] Sun H, Zhou Y, Ren J and Qu X 2018 Carbon nanozymes: enzymatic properties, catalytic mechanism, and applications *Angew. Chem. Int. Ed.* **57** 9224–37

[21] Mostofizadeh A, Li Y, Song B and Huang Y 2011 Synthesis, properties, and applications of low-dimensional carbon-related nanomaterials *J. Nanomater.* **2011** 685081

[22] Afreen S, Muthoosamy K, Manickam S and Hashim U 2015 Functionalized fullerene (C60) as a potential nanomediator in the fabrication of highly sensitive biosensors *Biosens. Bioelectron.* **63** 354–64

[23] Miyazawa K 2015 Synthesis of fullerene nanowhiskers using the liquid–liquid interfacial precipitation method and their mechanical, electrical and superconducting properties *Sci. Technol. Adv. Mater.* **16** 013502

[24] Kroto H W, Heath J R, O'Brien S C, Curl R F and Smalley R E 1985 C60: buckminsterfullerene *Nature* **318** 162–3

[25] Krätschmer W, Fostiropoulos K and Huffman D 1990 The infrared and ultraviolet absorption spectra of laboratory-produced carbon dust: evidence for the presence of the C60 molecule *Chem. Phys. Lett.* **170** 167–70

[26] Alekseyev N and Dyuzhev G 2003 Fullerene formation in an arc discharge *Carbon* **41** 1343–8

[27] Howard J B, McKinnon J T, Makarovsky Y, Lafleur A L and Johnson M E 1991 Fullerenes C60 and C70 in flames *Nature* **352** 139–41

[28] Howard J B, Lafleur A L, Makarovsky Y, Mitra S, Pope C J and Yadav T K 1992 Fullerenes synthesis in combustion *Carbon* **30** 1183–201

[29] Xie S-y, Huang R-b, Yu L-j, Ding J and Zheng L-s 1999 Microwave synthesis of fullerenes from chloroform *Appl. Phys. Lett.* **75** 2764–6

[30] Taylor R, Langley G J, Kroto H W and Walton D 1993 Formation of C 60 by pyrolysis of naphthalene *Nature* **366** 728–31

[31] Koshio A, Yudasaka M, Ozawa M and Iijima S 2002 Fullerene formation via pyrolysis of ragged single-wall carbon nanotubes *Nano Lett.* **2** 995–7

[32] Chen C and Lou Z 2009 Formation of C60 by reduction of $CO_2$ *J. Supercrit. Fluid.* **50** 42–5

[33] Xiao M *et al* 2019 Rationally engineered nucleic acid architectures for biosensing applications *Chem. Rev.* **119** 11631–717

[34] Hong W *et al* 2018 *In vivo* toxicology of carbon dots by 1H NMR-based metabolomics *Toxicol. Res.* **7** 834–47

[35] Yue X *et al* 2014 Green synthesis of fluorescent carbon quantum dots for detection of $Hg^{2+}$ *Chin. J. Anal. Chem.* **42** 1252–8

[36] Cayuela A, Soriano M, Carrillo-Carrión C and Valcárcel M 2016 Semiconductor and carbon-based fluorescent nanodots: the need for consistency *Chem. Commun.* **52** 1311–26

[37] Wang S, Cole I S and Li Q 2016 The toxicity of graphene quantum dots *RSC Adv.* **6** 89867–78

[38] Peng H and Travas-Sejdic J 2009 Simple aqueous solution route to luminescent carbogenic dots from carbohydrates *Chem. Mater.* **21** 5563–5

[39] Deng J *et al* 2014 Electrochemical synthesis of carbon nanodots directly from alcohols *Chem.–A Eur. J.* **20** 4993–9

[40] Sun Y-P *et al* 2006 Quantum-sized carbon dots for bright and colorful photoluminescence *JACS* **128** 7756–7

[41] Zhai X *et al* 2012 Highly luminescent carbon nanodots by microwave-assisted pyrolysis *Chem. Commun.* **48** 7955–7

[42] Sahu S, Behera B, Maiti T K and Mohapatra S 2012 Simple one-step synthesis of highly luminescent carbon dots from orange juice: application as excellent bio-imaging agents *Chem. Commun.* **48** 8835–7

[43] Ma Y, Zhang Y and William W Y 2019 Near infrared emitting quantum dots: synthesis, luminescence properties and applications *J. Mater. Chem.* C **7** 13662–79

[44] Zhang Q Q, Chen B B, Zou H Y, Li Y F and Huang C Z 2018 Inner filter with carbon quantum dots: a selective sensing platform for detection of hematin in human red cells *Biosens. Bioelectron.* **100** 148–54

[45] Wang Q, Zhang S, Zhong Y, Yang X-F, Li Z and Li H 2017 Preparation of yellow–green-emissive carbon dots and their application in constructing a fluorescent turn-on nanoprobe for imaging of selenol in living cells *Anal. Chem.* **89** 1734–41

[46] Lu S *et al* 2016 Facile and ultrasensitive fluorescence sensor platform for tumor invasive biomaker $\beta$-glucuronidase detection and inhibitor evaluation with carbon quantum dots based on inner-filter effect *Biosens. Bioelectron.* **85** 358–62

[47] Zhang X *et al* 2010 Biodistribution and toxicity of nanodiamonds in mice after intra-tracheal instillation *Toxicol. Lett.* **198** 237–43

[48] Purohit V, Jain D, Sathe V, Ganesan V and Bhoraskar S 2007 Synthesis of nanocrystalline diamonds by microwave plasma *J. Phys. D: Appl. Phys.* **40** 1794

[49] Hao T, Zhang H, Shi C and Han G 2006 Nano-crystalline diamond films synthesized at low temperature and low pressure by hot filament chemical vapor deposition *Surf. Coat. Technol.* **201** 801–6

[50] Yang L, May P, Yin L, Smith J and Rosser K 2007 Growth of diamond nanocrystals by pulsed laser ablation of graphite in liquid *Diam. Relat. Mater.* **16** 725–9

[51] Hawelek L, Brodka A, Dore J C, Honkimaki V, Tomita S and Burian A 2008 Structural studies of nanodiamond by high-energy x-ray diffraction *Diam. Relat. Mater.* **17** 1186–93

[52] Amans D *et al* 2009 Nanodiamond synthesis by pulsed laser ablation in liquids *Diam. Relat. Mater.* **18** 177–80

[53] Mochalin V N, Shenderova O, Ho D and Gogotsi Y 2012 The properties and applications of nanodiamonds *Nat. Nanotechnol.* **7** 11–23

[54] Strimbu K and Tavel J A 2010 What are biomarkers? *Curr. Opin. HIV AIDS* **5** 463–6

[55] International Programme on Chemical Safety 1993 Biomarkers and risk assessment: concepts and principlesJoint Sponsorship of the United Nations Environment Programme, the International Labour Organisation, and the World Health Organization http://inchem.org/documents/ehc/ehc/ehc155.htm

[56] FDA-NIH Biomarker Working Group 2016 *BEST (Biomarkers, EndpointS, and other Tools) Resource* (Silver Spring, MD: US Food and Drug Administration) http://inchem.org/documents/ehc/ehc/ehc155.htm

[57] Weaver O and Leung J W T 2017 Biomarkers and imaging of breast cancer *Am. J. Roentgenol.* **210** 271–8

[58] World Health Organization 2021 Cardiovascular diseases (CVDs) https://who.int/news-room/fact-sheets/detail/cardiovascular-diseases-(cvds)

[59] Singh V, Martinezclark P, Pascual M, Shaw E S and O'Neill W W 2010 Cardiac biomarkers—the old and the new: a review *Coronary Artery Dis.* **21** 244–56

[60] Danese E and Montagnana M 2016 An historical approach to the diagnostic biomarkers of acute coronary syndrome *Ann. Transl. Med.* **4** 194

[61] Lyngbakken M N, Myhre P L, Røsjø H and Omland T 2018 Novel biomarkers of cardiovascular disease: applications in clinical practice *Crit. Rev. Clin. Lab. Sci.* **56** 33–60

[62] Wang J, Tan G-J, Han L-N, Bai Y-Y, He M and Liu H-B 2017 Novel biomarkers for cardiovascular risk prediction *J. Geriatr. Cardiol.* **14** 135–50

[63] World Health Organization 2022 Cancer https://who.int/health-topics/cancer#tab=tab_1

[64] Ames B N, Gold L S and Willett W C 1995 The causes and prevention of cancer *Proc. Natl Acad. Sci. USA* **92** 5258–65

[65] Stratton M R, Campbell P J and Futreal P A 2009 The cancer genome *Nature* **458** 719–24

[66] World Health Organization Breast cancer https://who.int/news-room/fact-sheets/detail/breast-cancer

[67] Shah R, Rosso K and Nathanson S D 2014 Pathogenesis, prevention, diagnosis and treatment of breast cancer *World J. Clin. Oncol.* **5** 283–98

[68] Sun Y-S *et al* 2017 Risk factors and preventions of breast cancer *Int. J. Biol. Sci.* **13** 1387–97

[69] American Cancer Society What is breast cancer? https://cancer.org/cancer/breast-cancer/about/what-is-breast-cancer.html

[70] Waks A G and Winer E P 2019 Breast cancer treatment: a review *JAMA* **321** 288–300

[71] Joshi H and Press M F 2018 Molecular oncology of breast cancer *The Breast* ed K I Bland, E M Copeland, V S Klimberg and W J Gradishar 5th edn (Amsterdam: Elsevier) ch 22 pp 282–307

[72] Hammond M E H *et al* 2010 American Society of Clinical Oncology/College of American Pathologists Guideline recommendations for immunohistochemical testing of estrogen and progesterone receptors in breast cancer *J. Clin. Oncol.* **28** 2784–95

[73] Piccart-Gebhart M J *et al* 2005 Trastuzumab after adjuvant chemotherapy in HER2-positive breast cancer *New Engl. J. Med.* **353** 1659–72

[74] Harbeck N *et al* 2019 Breast cancer *Nat. Rev. Dis. Primers.* **5** 66

[75] Li J *et al* 2020 Non-invasive biomarkers for early detection of breast cancer *Cancers* **12** 2767

[76] NIH-National Cancer Institute Lung cancer https://cancer.gov/types/lung

[77] Centers for Disease Control and Prevention What is lung cancer? https://cdc.gov/cancer/lung/basic_info/what-is-lung-cancer.htm

[78] Cersosimo R J 2002 Lung cancer: a review *Am. J. Health-Syst. Pharm.* **59** 611–42

[79] American Cancer Society Tests for lung cancer https://cancer.org/cancer/lung-cancer/detection-diagnosis-staging/how-diagnosed.html

[80] Healthline Lung cancer biomarkers: what to know https://healthline.com/health/lung-cancer/lung-cancer-biomarkers

[81] Villalobos P and Wistuba I I 2017 Lung cancer biomarkers *Hematol. Oncol. Clin. North Am.* **31** 13–29

[82] Pennell N A, Arcila M E, Gandara D R and West H 2019 Biomarker testing for patients with advanced non–small cell lung cancer: real-world issues and tough choices *Am. Soc. Clin. Oncol. Educ. Book.* **39** 531–42
[83] Lindeman N I *et al* 2018 Updated molecular testing guideline for the selection of lung cancer patients for treatment with targeted tyrosine kinase inhibitors: guideline from the College of American Pathologists, the Int. Association for the Study of Lung Cancer, and the Association for Molecular Pathology *J. Thorac. Oncol.* **142** 321–46
[84] Dekker E, Tanis P J, Vleugels J L A, Kasi P M and Wallace M B 2019 Colorectal cancer *Lancet* **394** 1467–80
[85] American Cancer Society Colorectal cancer facts and figures 2020–22 https://cancer.org/research/cancer-facts-statistics/colorectal-cancer-facts-figures.html
[86] Sepulveda A R *et al* 2017 Molecular biomarkers for the evaluation of colorectal cancer: guideline from the American Society for Clinical Pathology, College of American Pathologists, Association for Molecular Pathology, and the American Society of Clinical Oncology *J. Clin. Oncol.* **35** 1453–86
[87] Carethers J M and Jung B H 2015 Genetics and genetic biomarkers in sporadic colorectal cancer *Gastroenterology* **149** 1177–90
[88] Koncina E, Haan S, Rauh S and Letellier E 2020 Prognostic and predictive molecular biomarkers for colorectal cancer: updates and challenges *Cancers* **12** 319
[89] Gonzalez-Pons M and Cruz-Correa M 2015 Colorectal cancer biomarkers: where are we now? *BioMed Res. Int.* **2015** 149014
[90] American Cancer Society What is prostate cancer? https://cancer.org/cancer/prostate-cancer/about/what-is-prostate-cancer.html
[91] Mazhar D and Waxman J 2002 Prostate cancer *Postgrad. Med. J.* **78** 590–5
[92] Schiebler M L *et al* 1993 Current role of MR imaging in the staging of adenocarcinoma of the prostate *Radiology* **189** 339–52
[93] Descotes J-L 2019 Diagnosis of prostate cancer *Asian J. Urol.* **6** 129–36
[94] Rosser C J and Parker A 2005 RE: the prostate specific antigen era in the united states is over for prostate cancer: what happened in the last 20 years? *J. Urol.* **174** 1154–5
[95] Pinsky P F, Prorok P C and Kramer B S 2017 Prostate cancer screening—a perspective on the current state of the evidence *N. Engl. J. Med.* **376** 1285–9
[96] McGrath S *et al* 2016 Prostate cancer biomarkers: are we hitting the mark? *Prostate Int.* **4** 130–5
[97] Litwin M S and Tan H J 2017 The diagnosis and treatment of prostate cancer: a review *JAMA* **317** 2532–42
[98] Koo K M, Mainwaring P N, Tomlins S A and Trau M 2019 Merging new-age biomarkers and nanodiagnostics for precision prostate cancer management *Nat. Rev. Urol.* **16** 302–17
[99] Clark L C Jr and Lyons C 1962 Electrode systems for continuous monitoring in cardiovascular surgery *Ann. NY Acad. Sci.* **102** 29–45
[100] Malhotra B D and Chaubey A 2003 Biosensors for clinical diagnostics industry *Sensors Actuators* B **91** 117–27
[101] Parkash O and Shueb R H 2015 Diagnosis of dengue infection using conventional and biosensor based techniques *Viruses* **7** 5410–27
[102] Diamandis E P 2012 The failure of protein cancer biomarkers to reach the clinic: why, and what can be done to address the problem? *BMC Med.* **10** 87

[103] Wang S *et al* 2020 Bioinspired poly(vinyl alcohol) film actuator powered by water evaporation under ambient conditions *Macromol. Mater. Eng.* **305** 2000145

[104] Patel S, Nanda R, Sahoo S and Mohapatra E 2016 Biosensors in health care: the milestones achieved in their development towards lab-on-chip-analysis *Biochem. Res. Int.* **2016** 3130469

[105] Sin M L Y, Mach K E, Wong P K and Liao J C 2014 Advances and challenges in biosensor-based diagnosis of infectious diseases *Expert Rev. Mol. Diagn.* **14** 225–44

[106] Kerman K, Saito M, Tamiya E, Yamamura S and Takamura Y 2008 Nanomaterial-based electrochemical biosensors for medical applications *TrAC, Trends Anal. Chem.* **27** 585–92

[107] Hai X, Li Y, Zhu C, Song W, Cao J and Bi S 2020 DNA-based label-free electrochemical biosensors: from principles to applications *TrAC, Trends Anal. Chem.* **133** 116098

[108] Rahdar S, Rahdar A, Ahmadi S and Trant J F 2019 Adsorption of bovine serum albumin (BSA) by bare magnetite nanoparticles with surface oxidative impurities that prevent aggregation *Can. J. Chem.* **97** 577–83

[109] Sharma A *et al* 2018 Designed strategies for fluorescence-based biosensors for the detection of mycotoxins *Toxins* **10** 197

[110] Thévenot D R, Toth K, Durst R A and Wilson G S 2001 Electrochemical biosensors: recommended definitions and classification *Biosens. Bioelectron.* **16** 121–31

[111] Cha C, Shin S R, Annabi N, Dokmeci M R and Khademhosseini A 2013 Carbon-based nanomaterials: multifunctional materials for biomedical engineering *ACS Nano* **7** 2891–7

[112] Wang Z, Hu T, Liang R and Wei M 2020 Application of zero-dimensional nanomaterials in biosensing *Front. Chem.* **8** 320

[113] Alagarsamy K N *et al* 2021 Carbon nanomaterials for cardiovascular theranostics: promises and challenges *Bioact. Mater.* **6** 2261–80

[114] Bhatnagar D, Kumar V, Kumar A and Kaur I 2016 Graphene quantum dots FRET based sensor for early detection of heart attack in human *Biosens. Bioelectron.* **79** 495–9

[115] Jazbutyte V *et al* 2013 MicroRNA-22 increases senescence and activates cardiac fibroblasts in the aging heart *Age* **35** 747–62

[116] Sun J *et al* 2018 Comet-like heterodimers 'gold nanoflower @graphene quantum dots' probe with FRET 'off' to DNA circuit signal 'on' for sensing and imaging microRNA in vitro and *in vivo Anal. Chem.* **90** 11538–47

[117] Boon R A *et al* 2013 MicroRNA-34a regulates cardiac ageing and function *Nature* **495** 107–10

[118] Foo M E and Gopinath S C B 2017 Feasibility of graphene in biomedical applications *Biomed. Pharmacother.* **94** 354–61

[119] Mahato K, Kumar A, Maurya P K and Chandra P 2018 Shifting paradigm of cancer diagnoses in clinically relevant samples based on miniaturized electrochemical nano-biosensors and microfluidic devices *Biosens. Bioelectron.* **100** 411–28

[120] Afreen S, He Z, Xiao Y and Zhu J-J 2020 Nanoscale metal–organic frameworks in detecting cancer biomarkers *J. Mater. Chem.* B **8** 1338–49

[121] Chauhan N, Maekawa T and Kumar D N S 2017 Graphene based biosensors—accelerating medical diagnostics to new-dimensions *J. Mater. Res.* **32** 2860–82

[122] Iannazzo D, Espro C, Celesti C, Ferlazzo A and Neri G 2021 Smart biosensors for cancer diagnosis based on graphene quantum dots *Cancers* **13** 3194

[123] Xi J *et al* 2016 Pd nanoparticles decorated N-doped graphene quantum dots@N-doped carbon hollow nanospheres with high electrochemical sensing performance in cancer detection *ACS Appl. Mater. Interfaces* **8** 22563–73
[124] Fan Z, Zhou S, Garcia C, Fan L and Zhou J 2017 pH-responsive fluorescent graphene quantum dots for fluorescence-guided cancer surgery and diagnosis *Nanoscale* **9** 4928–33
[125] Cui F *et al* 2019 A novel magnetic fluorescent biosensor based on graphene quantum dots for rapid, efficient, and sensitive separation and detection of circulating tumor cells *Anal. Bioanal. Chem.* **411** 985–95
[126] Liu H, Na W, Liu Z, Chen X and Su X 2017 A novel turn-on fluorescent strategy for sensing ascorbic acid using graphene quantum dots as fluorescent probe *Biosens. Bioelectron.* **92** 229–33
[127] Yan X *et al* 2016 Graphene quantum dot–$MnO_2$ nanosheet based optical sensing platform: a sensitive fluorescence 'turn off–on' nanosensor for glutathione detection and intracellular imaging *ACS Appl. Mater. Interfaces* **8** 21990–6
[128] Campuzano S, Yáñez-Sedeño P and Pingarrón J M 2019 Carbon dots and graphene quantum dots in electrochemical biosensing *Nanomaterials* **9** 634
[129] Mansuriya B D and Altintas Z 2020 Applications of graphene quantum dots in biomedical sensors *Sensors* **20** 1072
[130] Tuteja S K *et al* 2016 A label-free electrochemical immunosensor for the detection of cardiac marker using graphene quantum dots (GQDs) *Biosens. Bioelectron.* **86** 548–56
[131] Lakshmanakumar M, Nesakumar N, Sethuraman S, Rajan K S, Krishnan U M and Rayappan J B B 2019 Functionalized graphene quantum dot interfaced electrochemical detection of cardiac troponin I: an antibody free approach *Sci. Rep.* **9** 17348
[132] Mansuriya B D and Altintas Z 2021 Enzyme-free electrochemical nano-immunosensor based on graphene quantum dots and gold nanoparticles for cardiac biomarker determination *Nanomaterials* **11** 578
[133] Wang W, Cheng L and Liu W 2014 Biological applications of carbon dots *Sci. China Chem.* **57** 522–39
[134] Pirzada M and Altintas Z 2019 Nanomaterials for healthcare biosensing applications *Sensors* **19** 5311
[135] Atabaev T S 2018 Doped carbon dots for sensing and bioimaging applications: a minireview *Nanomaterials* **8** 342
[136] Molaei M J 2019 Carbon quantum dots and their biomedical and therapeutic applications: a review *RSC Adv.* **9** 6460–81
[137] Molaei M J 2019 A review on nanostructured carbon quantum dots and their applications in biotechnology, sensors, and chemiluminescence *Talanta* **196** 456–78
[138] Chen J *et al* 2019 A fluorescent biosensor for cardiac biomarker myoglobin detection based on carbon dots and deoxyribonuclease I-aided target recycling signal amplification *RSC Adv.* **9** 4463–8
[139] Tan X, Zhang L, Tang Q, Zheng G and Li H 2019 Ratiometric fluorescent immunoassay for the cardiac troponin-I using carbon dots and palladium-iridium nanocubes with peroxidase-mimicking activity *Microchim. Acta* **186** 280
[140] Liang W, Bunker C E and Sun Y-P 2020 Carbon dots: zero-dimensional carbon allotrope with unique photoinduced redox characteristics *ACS Omega* **5** 965–71
[141] Miao H, Wang L, Zhuo Y, Zhou Z and Yang X 2016 Label-free fluorimetric detection of CEA using carbon dots derived from tomato juice *Biosens. Bioelectron.* **86** 83–9

[142] Motaghi H, Mehrgardi M A and Bouvet P 2017 Carbon dots-AS1411 aptamer nano-conjugate for ultrasensitive spectrofluorometric detection of cancer cells *Sci. Rep.* **7** 10513

[143] Liu T *et al* 2017 A colorimetric and fluorometric dual-signal sensor for arginine detection by inhibiting the growth of gold nanoparticles/carbon quantum dots composite *Biosens. Bioelectron.* **87** 772–8

[144] Zheng W *et al* 2018 A molecularly-imprinted-electrochemical-sensor modified with nano-carbon-dots with high sensitivity and selectivity for rapid determination of glucose *Anal. Biochem.* **555** 42–9

[145] Guo W, Pi F, Zhang H, Sun J, Zhang Y and Sun X 2017 A novel molecularly imprinted electrochemical sensor modified with carbon dots, chitosan, gold nanoparticles for the determination of patulin *Biosens. Bioelectron.* **98** 299–304

[146] Zhang L, Han Y, Zhu J, Zhai Y and Dong S 2015 Simple and sensitive fluorescent and electrochemical trinitrotoluene sensors based on aqueous carbon dots *Anal. Chem.* **87** 2033–6

[147] Huang Q *et al* 2013 Carbon dots and chitosan composite film based biosensor for the sensitive and selective determination of dopamine *Analyst* **138** 5417–23

[148] Nguyen H V *et al* 2016 Electrochemical sensing of etoposide using carbon quantum dot modified glassy carbon electrode *Analyst* **141** 2665–75

[149] Peng Y *et al* 2019 Activatable core–shell metallofullerene: an efficient nanoplatform for bimodal sensing of glutathione *ACS Appl. Mater. Interfaces* **11** 46637–44

[150] Ma Y, Fu S, Tan Y and Zhang A 2021 Design and synthesis of highly fluorescent and stable fullerene nanoparticles as probes for folic acid detection and targeted cancer cell imaging *Nanotechnology* **32** 195501

[151] Han J *et al* 2013 Multi-labeled functionalized $C_{60}$ nanohybrid as tracing tag for ultra-sensitive electrochemical aptasensing *Biosens. Bioelectron.* **46** 74–9

[152] Yuan Q *et al* 2018 Sandwich-type biosensor for the detection of $\alpha$2,3-sialylated glycans based on fullerene-palladium-platinum alloy and 4-mercaptophenylboronic acid nano-particle hybrids coupled with Au-methylene blue-MAL signal amplification *Biosens. Bioelectron.* **102** 321–7

[153] Solanki P R, Kaushik A, Agrawal V V and Malhotra B D 2011 Nanostructured metal oxide-based biosensors *NPG Asia Mater.* **3** 17–24

[154] Chen Y, Liu J, Yang Z, Wilkinson J S and Zhou X 2019 Optical biosensors based on refractometric sensing schemes: a review *Biosens. Bioelectron.* **144** 111693

[155] Du H, Xie Y and Wang J 2021 Nanomaterial-sensors for herbicides detection using electrochemical techniques and prospect applications *TrAC, Trends Anal. Chem.* **135** 116178

[156] Mansuriya B D and Altintas Z 2020 Applications of graphene quantum dots in biomedical sensors *Sensors* **20** 1072

[157] Motaghi H, Mehrgardi M A-O and Bouvet P 2017 Carbon dots-AS1411 aptamer nano-conjugate for ultrasensitive spectrofluorometric detection of cancer cells *Sci. Rep.* **7** 10513

[158] Keller T *et al* 2009 Sensitive troponin I assay in early diagnosis of acute myocardial infarction *New Engl. J. Med.* **361** 868–77

[159] Apple F S and Collinson P O 2012 Analytical characteristics of high-sensitivity cardiac troponin assays *Clin. Chem.* **58** 54–61

[160] Korley Frederick K and Jaffe Allan S 2013 Preparing the United States for high-sensitivity cardiac troponin assays *J. Am. Coll. Cardiol.* **61** 1753–8

[161] Glatz J F and van der Vusse G J 1996 Cellular fatty acid-binding proteins: their function and physiological significance *Progr. Lipid Res.* **35** 243–82

[162] Tambara K, Fujita M, Miyamoto S, Doi K, Nishimura K and Komeda M 2005 Pericardial fluid level of heart-type cytoplasmic fatty acid-binding protein (H-FABP) is an indicator of severe myocardial ischemia *Cardiology* **93** 281–4

[163] Landry A, Docherty P, Ouellette S and Cartier L J 2017 Causes and outcomes of markedly elevated C-reactive protein levels *Can. Fam. Physician* **63** e316–23

[164] Ridker P M, Cushman M, Stampfer M J, Tracy R P and Hennekens C H 1997 Inflammation, aspirin, and the risk of cardiovascular disease in apparently healthy men *New Engl. J. Med.* **336** 973–9

[165] Ridker P M, Rifai N,, Stampfer M J and Hennekens C H 2000 Plasma concentration of interleukin-6 and the risk of future myocardial infarction among apparently healthy men *Circulation* **101** 1767–72

[166] Wang J, Tan G J, Han L N, Bai Y Y, He M and Liu H B 2017 Novel biomarkers for cardiovascular risk prediction *J. Geriatr. Cardiol.* **14** 135–50

[167] Danesh J *et al* 2005 Plasma fibrinogen level and the risk of major cardiovascular diseases and nonvascular mortality: an individual participant meta-analysis *JAMA* **294** 1799–809

[168] Sangiorgi G *et al* 2006 Pregnancy-associated plasma protein-A is markedly expressed by monocyte-macrophage cells in vulnerable and ruptured carotid atherosclerotic plaques: a link between inflammation and cerebrovascular events *J. Am. Coll. Cardiol.* **47** 2201–11

[169] Hochholzer W, Morrow D and Giugliano R P 2010 Novel biomarkers in cardiovascular disease: update 2010 *Am. Heart J.* **160** 583–94

[170] Sluijter J P *et al* 2006 Matrix metalloproteinase 2 is associated with stable and matrix metalloproteinases 8 and 9 with vulnerable carotid atherosclerotic lesions *Stroke* **37** 235–9

[171] Newby A C 2015 Metalloproteinases promote plaque rupture and myocardial infarction: a persuasive concept waiting for clinical translation *Matrix Biol.* **44–46** 157–66

[172] Januzzi J L Jr 2013 ST2 as a cardiovascular risk biomarker: from the bench to the bedside *J. Cardiovasc. Transl. Res.* **6** 493–500

[173] Dhillon O S *et al* 2013 Pre-discharge risk stratification in unselected STEMI: is there a role for ST2 or its natural ligand IL-33 when compared with contemporary risk markers? *Int. J. Cardiol.* **167** 2182–8

[174] Demyanets S *et al* 2014 Soluble ST2 and interleukin-33 levels in coronary artery disease: relation to disease activity and adverse outcome *PLOS One* **9** e95055

[175] Davenport A P *et al* 2016 Endothelin *Pharmacol. Rev.* **68** 357–418

[176] Lee K F *et al* 1995 Requirement for neuregulin receptor erbB2 in neural and cardiac development *Nature* **378** 394–8

[177] Wang G K *et al* 2010 Circulating microRNA: a novel potential biomarker for early diagnosis of acute myocardial infarction in humans *Eur. Heart J.* **31** 659–66

# Chapter 7

# Zero-dimensional carbon nanomaterials for cancer diagnosis

**Stephen Rathinaraj Benjamin, Eli José Miranda Ribeiro Júnior, Geanne Matos de Andrade and Reinaldo Barreto Oriá**

In recent decades, the development of zero-dimensional (0D) nanomaterials, which include graphene quantum dots (GQDs) and carbon quantum dots (CQDs), has led to their extensive application in cancer detection. The development of 0D nanomaterials with specific charge, particle size, defined shape, ligand binding, etc, has improved their biocompatibility and the active targeting of cancer cells. In addition, a novel approach for cancer diagnosis and the real-time monitoring of biological processes is required. In this chapter, we provide a brief description of current research and present future prospects for cancer diagnosis using 0D nanomaterials and their composites.

## 7.1 Introduction

During the last few years, cancer has become one of the most common causes of death for humans. According to the World Health Organization, roughly 18 million new cases of cancer and approximately 9.6 million cancer-related fatalities were registered in 2018, and the yearly occurrence of cancer cases is predicted to increase to 22 million in the next two decades. These ominous predictions have renewed researchers' focus on developing and implementing novel methods for cancer treatment [1, 2]. Nanotechnology is a fast developing field that is vital in the current development of functional biomaterials. The development of revolutionary nanomaterials with novel and unique features is essential in order to maintain the current rate of technological innovation. Carbon nanostructures (CNSs) and their derivatives have attracted considerable attention owing to their exceptional features and prospective uses in a wide range of existing and new areas [3]. Carbon nanostructures include a diverse range of carbon allotropes, as well as a diverse range of

doi:10.1088/978-0-7503-4048-9ch7  

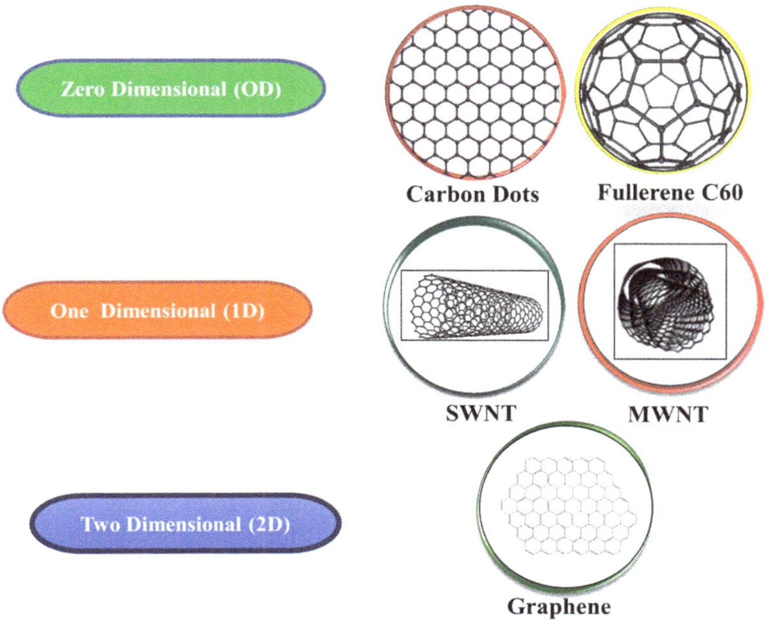


**Figure 7.1.** The most common types of carbon nanostructures.

forms and sizes. Carbon can form incredibly low-dimensional structures by bonding to itself in a unique design, including 0D (fullerenes), 1D (nanotubes), 2D (graphene sheets), and 3D (diamond-like carbon) (figure 7.1). In recent decades, nanotechnology has been a rapidly developing area. Several nanomaterials with variations in their mechanical and physico-chemical characteristics can be produced via chemical modification, atomic organization, and nanomaterial dimensions. Due to the obvious intrinsic structural characteristics, including their small dimensions and high surface-to-volume ratios, 0D nanomaterials include more active surface areas according to unit weight. 0D nanomaterials have enhanced or new characteristics due to edge and quantum confinement phenomena. Furthermore, in zero-dimensional (0D) nanomaterials, all dimensions are at the nanoscale (no dimension is greater than 100 nm). 0D nanomaterials are nanoparticles. Various zero-dimensional nanomaterials have been studied widely up to this point. For example, carbon quantum dots (CQDs), graphene quantum dots (GQDs), fullerenes, magnetic nanoparticles (MNPs), inorganic quantum dots, noble metal nanoparticles, up-conversion nanoparticles (UCNPs), and polymer dots (Pdots) are all typical 0D nanomaterials with practical benefits in the science of materials and in photovoltaic devices. A newly produced structure may provide unique features by varying its size and decreasing it to zero-dimensions. Most 0D nanoparticles are spherical or quasi-spherical, featuring a diameter smaller than 100 nm. 0D nanomaterials have unique characteristics that allow them to be used in biomedical applications such as biosensors and biochips. They are chemically inert, have cellular permeability, and are biocompatible. Biosensors can detect antigens, biomolecules, antibodies, DNA, viruses, and bacteria. It is difficult to improve biosensors' effective surface area, electrochemical interaction, conduction, or optical performance. 0D nanomaterials are utilized as strong sensing materials to improve biosensor sensitivity and analytical efficiency. Many studies on 0D nanomaterials have been published recently, but few describe this use in detail, which is essential for 0D nanomaterial biosensor development. In addition, the most often employed 0D nanomaterials in biosensing for cancer detection will be discussed in this chapter.

## 7.2 0D nanomaterials in cancer diagnosis employing electrochemical biosensing

### 7.2.1 Carbon quantum dots

Carbon quantum dots have a potential future in biological applications [4] due to their minimal toxicity, photothermal stability, enormous production capability, and low cost [5]. Their ultra-small size (less than 10 nm) enhances the blood circulation time. Furthermore, heteroatom doping (metal and nonmetal doping) can provide quantum dots with new functionalities and increase their optical characteristics [6].

### 7.2.2 Graphene quantum dots

Graphene quantum dots (GQD), zero-dimensional graphene compounds, have enormous potential in imaging, sensing, and biomedical therapy owing to their interesting physico-chemical and optical features [7]. They are generally biocompatible and are eliminated slowly by the bodies of mice, resulting in no appreciable toxicity even when administered in a multidosing regimen [8]. GQDs allow a steady, powerful, and adjustable imaging technique for cancer surgery and diagnosis because of the substantial fluorescence activity caused by quantum confinement with edge effects [9].

### 7.2.3 Prostate specific antigens

Prostate cancer is the greatest risk factor for mortality in men over the age of 50, along with lung cancer. It is critical to conduct clinical assessments of cancer proteins or biomarkers in order to detect cancer early and track disease development [10]. Biomarkers have gained in clinical importance as a result of their broad variety of uses in cancer detection, prevention, early diagnosis, and therapy. Biomarkers are critical in the detection and treatment of cancers in their early phases [11]. Prostate specific antigen (PSA) is a semen mutual protein that may be transported to the bloodstream. PSA is a significant prostate cancer biomarker that may be utilized for a variety of purposes, including detection, therapy monitoring, and recurrence monitoring after treatment [12]. Men with prostate cancer have an increase in PSA values as a result of prostatic dysfunction. In human blood, the normal PSA range is 4.0 to 10 ng $mL^{-1}$. However, the detection of PSA is required for prostate cancer diagnosis [13]. Many types of metal sulfide QDs, such as PbS, ZnS, and CdS, have been used as labels with very sensitive electrochemical biosensors. Yang *et al* [14] utilized QDs–graphene sheets (GSs) as labels to detect PSA. The 1-pyrenebutanoic acid, succinimidyl ester (PBSE) was initially conjugated to the GS by stacking ($\pi$–$\pi$). After that, via amidation between both the PBSE succinimidyl ester and anti-PSA Ab1 amine groups, the GS-PBSE solution attaches to the glassy carbon electrode (GCE) surface. The immunosensor has an excellent linear range concentration between 0.005–10 ng $mL^{-1}$ and demonstrated a limit of detection (LOD) of 3 pg $mL^{-1}$ in human serum samples. In order to detect Ab2 and modify the electrode surface, researchers used electrochemically active QDs and magnetized metal–organic frameworks, and the sandwich test was completed with CdTe:Ni QD-labeled Ab2 solutions. Their electrochemical signaling is related to Cd or Ni ions, in addition to being correlated with cadmium ions because the redox peaks of nickel are positive (in particular at pH 7.0). In addition, the constructed immunosensor has a linearity of 1–100 ng $mL^{-1}$ and an LOD of 0.45 pg $mL^{-1}$ [15].

Farschi and colleagues published a study describing an immunosensor for PSA detection that uses a citrate-AgNP coated GQD nano-ink enhanced flexible paper sheet [16]. The cysteamine functionalized gold nanoparticles altered the texture of the cellulose paper (ivory sheets/photographic paper). It has a considerable amount of biotinylated anti-PSA Ab1 surface area. The immunosensor was treated with PSA antigen for 30 min at 4 °C. Consequently, the immunosensor achieved a quantification limit of 0.07 $\mu$g $L^{-1}$, and a linearity range between 0.07 and 60 $\mu$g $L^{-1}$. Yang *et al* [14] used QD-functionalized graphene sheets (FGSs) to detect PSA antibodies in human blood samples. The electrochemical performance of GS is enhanced due to the high affinity of PSA antibodies for the QDs. In their study, they found that the lower detection value was 3 pg $mL^{-1}$ and that the concentration range value was between 0.005–10 ng $mL^{-1}$. Label-free, sensitive PSA testing remains a significant problem in the detection of male prostate cancer. Using label-free PSA aptasensors and gold nanorod (AuNR)-modified screen-printed electrodes, the authors compared the two methods. The composite GQD–AuNRs show a fast transit of electron and catalytic capabilities. Anti-PSA aptamer immobilized electrodes were printed on a customized screen. In ideal circumstances, both sensors provide an LOD of 0.14 ng $mL^{-1}$. The PSA detection and anti-PSA results of the aptasensor were tested in real samples and found to be effective [17]. In addition, the potential to functionalize and significantly alter graphene-based quantum dots has enabled application-specific tailoring. The aptasensor based on N-GQDs had the best effectiveness with an LOD of 1.54 pM in the technique of electrical impedance spectroscopy and minimal interference in mixed samples. The performance of this electrode was verified with great recovery, selectivity, and sensitivity in spiked human blood samples, making it a dependable and efficient aptasensor [18].

Mehrnaz *et al* [19] developed an impedimetric immunosensor using gold nanoparticles and multiwalled carbon nanotubes–graphene quantum dots for prostate cancer biomarker detection in human blood samples. The label-free immunosensor was made by coating a GCE with nanocomposite as a signal amplifier and afterwards immobilizing it as an antibody against PSA on the modified electrode. The label-free immunosensor significantly enhanced sensing capability, with a range of 1–10 000 pg $mL^{-1}$ and an LOD value of 0.48 pg $mL^{-1}$. Graphene oxide (GO)–quantum dots@silver core–shell nanocrystals were utilized to create an ultrasensitive

PSA-sensor-sandwiching nanoprobe. By integrating large numbers of GQDs into one probe, the sample-to-target ratio was improved to enable more fluorescent signals. The incorporated GQDs were released easily after oxidative etching with hydrogen peroxide damaged the silver shell. The proposed immunosensor had a linear relationship between fluorescence intensity and PSA concentration, exhibiting a lower detection value of 0.3 pg $mL^{-1}$ [20].

Early cancer diagnosis requires very sensitive tumor marker detection. In a new electrochemical sandwich-type immunosensor for the measurement of PSA, gold nanoparticles are synthesized using an easy green hydrothermal method utilizing nitrogen-doped quantum graphene dots (Au@N-GQDs). Secondary antibodies ($Ab_2$) were identified with a good biological compatibility and a specific surface area. The composition of Au, Ag, and $Cu_2O$ has high electrocatalytic activity to reduce $H_2O_2$ for better PSA detection. The immunosensor showed a linear concentration range of 0.01 pg $mL^{-1}$ to 100 ng $mL^{-1}$ and an LOD value of 0.003 pg $mL^{-1}$ [21]. An electrochemical immunosensor for detecting PSA was developed by Malekzad *et al* [22]. Using the green and biocompatible nanocomposite with a conductive matrix of poly L-cysteine (P-Cys) and dual signal amplification components of GQD/gold nanoparticles (GNPs), anti-PSA antibodies were immobilized. This led to the development of a very sensitive amperometric immunosensor based on P-Cys, GNPs, and GQDs. In this study, a new gold-surfaced transducer with a large surface area for immobilizing anti-PSA was developed entirely electrochemically. The immunosensor detected PSA at physiological pH, the PSA concentration calibration was in the range 2 to 9 pg $mL^{-1}$, and the LOD was 1.8 pg $mL^{-1}$. Wu *et al* [23] developed an ultrasensitive electrochemiluminescent (ECL) biosensor employing GQDs (LOD = 0.29 pg $mL^{-1}$). Au/Ag-reduced graphene oxide was employed to develop two GQDs (aminated and carboxyl) to enhance the electrochemiluminescence and increase the electrode surface area. Gan and colleagues [24] developed the magnetic nanoparticles CdTe-QDs–graphene (CdTe-G) and dextran–$Fe_3O_4$ for the CA 19–9 ECL immunosensor. CA 19–9 has been proven to differentiate between benign and malignant pancreatic diseases. Captured samples for the enrichment of CA 19–9 were produced on dextran—$Fe_3O_4$ by immobilizing Ab1. Signal samples (CA 19–9 Ab2/CdTe-G) may generate an ECL signal when $Ab_2$ is bound to the surface of the CdTe-G. The ECL signal was observed to increased linearly from 0.005 to 100 pg $mL^{-1}$ at CA 19–9 concentration, with an LOD value of 0.002 pg $mL^{-1}$. The development of ECL biosensor QDs may also lead to their use as signal labels for highly fluorescent and stable semi-conductors. ECL is a very valuable method to identify biomolecules, particularly tumor biomarkers, linked to disease by employing ECL-active QDs as signaling labels.

### 7.2.4 Carcinoembryonic antigen (CEA)

CEA is expressed abundantly in gastric, colorectal, ovarian, breast, and lung cancers [25]. The normal cell levels of CEA are 3–5 ng $mL^{-1}$, and malignancies exhibit up to 10 ng $mL^{-1}$ [26]. In another study Hasanzadeh *et al* [27] utilized a gold nanosphere (AuNS) electrochemical reaction integrated onto functionalized

GQDs to detect CA15–3 in MCF-7 cell lysates and human plasma. The thiol groups were added to the GCE/GQDs by amide attaching GQD carboxyl to CysA amine groups. Subsequently, GCE/GQDs/CysA was used for AuNS electrochemical processes. The GQDs/CysA/AuNSs hybrid interface efficiently immobilized anti-CA15–3 Abs, determining their bioactivity and stability. The CA15–3 tumor marker was identified at 0.11 U $mL^{-1}$. The immunosensor's high sensitivity is due to the enormous loading of CA15–3 Abs on the GQDs/CysA/AuNSs hybrid interface. CEA was identified with an N, S-graphene QDs@au-polyaniline composite impedimetric immunosensor. In malignancies such as colon cancer, CEA serum levels increase, making it a valuable tumor marker in clinical trials. As a bi-functional sensor, N,S-GQDs were coated onto Au-polyaniline (PANI) nanowires with Au-thiol linkage to enhance electrochemical properties and attach anti-CEA. The high stability of the CEA platform was developed to identify impedimetric immunoassays in human blood samples that used N,S-GQD@Au-PANI nanowires and CEA antibody–antigen bioconjugates, which were generated via the incorporation of CEA tumor markers. In addition, the label-free immunosensor has a linearity range of concentration 0.5–1000 ng $mL^{-1}$ and an LOD value of 0.01 ng $mL^{-1}$ [28]. Yang *et al* [29] utilized GQDs doped with nitrogen conjugated with Pt–Pd bimetal nanomaterials to detect CEA in the range 5–50 ng $mL^{-1}$ with a lower detection value of 2 fg $mL^{-1}$. Ji *et al* [30] designed an ultrasensitive immunosensor based on assessed CdSeTe@ZnS–$SiO_2$ QDs for detection of CEA, building the graded-gap CdSeTe@ZnS–$SiO_2$ QD bilayers on a glutaraldehyde-activated electrode. As a consequence, bilayers had four times the ECL enhancement of the reference samples. The linear detection range was from 1 pg $mL^{-1}$ to 200 ng $mL^{-1}$ with a lower detection value of 0.4 pg $mL^{-1}$. Nie *et al* [31] used P5FIn/erGO composites and AuNPs coated GQDs (GQDs@AuNPs) to develop an ECL immunosensor for specific and sensitive CEA detection. The P5FIn/erGo composites improved the transport of ions through electrochemical reactions and immobilized Ab1 effectively. This enhanced the electron transfer capacity with a steady ECL intensity using GQDs@AuNPs as labels. The ECL immunosensor has a linearity range of 0.1 pg $mL^{-1}$–10 ng $mL^{-1}$ and an LOD value of 3.78 fg $mL^{-1}$. An enzyme-free electrochemical immunosensor for CEA was created using peroxidase-like GQDs. In this study, the electrodes were modified using SWCNTs@GQDs and reduced graphene oxide (rGO)-Au nanoparticles, that have a high specific surface area and strong conductivity, resulting in dual amplification. Electrochemical immunosensors are capable of detecting CEA concentrations ranging between 50 pg $mL^{-1}$ to 650 pg $mL^{-1}$ with a lower limit of detection of 5.3 pg $mL^{-1}$ [32]. Gold nanoparticles (AuNPs) were redox-hybridized with graphene quantum dots utilizing hydrazine-modified (HM) graphene quantum dots. AuNPs are useful for biosensors because of their propensity to self-assemble with particular biomolecules. Detection using an ECL immunosensor with carcinoembryonic antigens was accomplished by utilizing HM-GQD-AuNPs on a GCE. The proposed immunosensor has an excellent linear response of 0.02 to 80 ng $mL^{-1}$, with a low detection limit of 0.01 ng $mL^{-1}$ [33].

### 7.2.5 TP53/tumor protein (p53)

p53 is the major biomarker in the tumor suppressor process, and it regulates cell proliferation, DNA repair, genetic integrity, and apoptosis. In reaction to DNA damage, p53 regulates cell growth and death [34]. As a consequence, highly precise detection and identification of p53 at very low concentrations is highly valuable, as it may aid in early-stage cancer screening and therapy effectiveness [35, 36]. The amount of p53 in human blood (normal mean value 150 ng $L^{-1}$) may accurately represent the tissue p53 gene and/or protein changes, allowing for a noninvasive tumor screening method. Hasanzadeh and co-workers [37] constructed a very sensitive voltammetric immunosensor for measuring p53, employing ternary signal amplification. As a dual amplifier, GQDs/AuNPs were utilized in this work to enhance the gold electrode. The surface of the Au/P-Cys/GQDs/AuNPs was electrostatically changed employing streptavidin-HRP and biotinylated anti-p53 Ab was applied to the modified electrode. For p53 sensing, 10 min of incubation with various p53 concentrations was used. This resulted in a linear range of 0.0244–0.369 pM when applying the square wave voltammetry (SWV) method, and 0.19–50 pM when applying the differential pulse voltammetry (DPV) method with a quantification value of 12.1 fM. The ECL immunosensor enabled ultrasensitive p53 biomarker detection using AuNPs to enhance the CdS-QD ECL emission in hydrogen peroxide. Both acetate and hydrogen peroxide are key ECL luminophores/coreactants; they extinguish and enhance quantum dots (QDs), respectively. In the presence of $H_2O_2$, AuNPs either extinguish or increase the ECL of CdS-QDs. The sandwich immunocomplex of anti-p53 $Ab_1$/p53 protein/anti-p53 $Ab_2$ was used to add AuNPs to CdS-QDs immobilized on GCE. The incorporation of GOs on the electrode surface in the production of immunosensors enhances ECL intensity. To distinguish CdS-QDs (ECL donor) from AuNPs, the sandwich method was employed (as ECL acceptors). This immunosensor has a linear range between 10 fg $mL^{-1}$ and 20 ng $mL^{-1}$, and an LOD value of 4 fg $mL^{-1}$ [38].

## 7.3 OD carbon based nanosystems for cancer therapy

A wide range of cancer-related drug delivery and imaging applications have been developed using graphene quantum dots, namely nanoscale diagnosis testing of enzymes highly expressed in cancer cells [39], fluorescence imaging [40], and chemo-photothermal treatment, that combines both therapeutic techniques in one approach [41]. GQDs have been employed for photodynamic treatment [42] and bacteria eradication [43]. GQDs containing COOH exhibit excellent photodynamic properties when stimulated by two NIR photons [44], suggesting that they may be useful in deep tissue cancer treatment [45]. These properties allowed for the loading and use of chemotherapeutics as well as the loading and use of radioisotopes in cancer therapy [46].

Recently, Yao *et al* [47] developed GQD-capped $Fe_3O_4/SiO_2$ magnetic mesoporous nanoparticles as a multifunctional platform for regulated drug administration, magnetic hyperthermia (MHT), and photothermal therapy (PTT). Due to the pH-dependent interaction between mesoporous $SiO_2$/GQD and pharmaceuticals, regulated drug release might be initiated in an acidic condition. PTT also made use of GQDs in the system with excellent photothermal conversion efficiency and robust NIR absorbance. Hyperthermia treatment uses $Fe_3O_4$ nanoparticles to produce heat below an alternating magnetic field. The release of the medication could be accelerated by reversing the effects of hyperthermia. *In vitro* studies indicated that the combination of chemotherapy, MHT, and PTT was more effective at killing tumor cells than any single treatment.

CD surface and/or edge modification may be used to provide targeted capabilities in photodynamic treatment (PDT), PTT, and chemotherapy. PTT is a light-triggered method that uses photothermal transduction agents (PTAs) to transform light into heat for tumor hyperthermia ablation under NIR light radiation. PTT has prompted widespread attention in the field of cancer treatment because of its improved precision and reduced adverse effects. The PTT and tumor imaging using GQD-FA performed by Li *et al* [48] included carboxyl groups around the exterior of the $sp^2$-domain. Due to the strong $\pi$–$\pi$ stacking interactions for PTT, the generated GQDs have an extraordinarily high loading capacity for IR780 (33.19%). For the first time, *in vivo* NIR fluorescence imaging and biodistribution showed that IR780/GQDs-FA could accumulate in tumors, resulting in sufficient hyperthermia to induce cancer cell necrosis and tumor apoptosis upon 808 nm laser treatment, resulting in total tumor ablation without relapse.

### 7.3.1 Drug delivery

Graphene quantum dots (GQDs) have been used extensively in bioimaging and drug delivery because of its outstanding physico-chemical properties, chemical stability, and biocompatibility. Due to their enhanced dosage tolerance and therapeutic efficiency, the compact delivery of drugs using nanostructures with multifunctional properties is currently receiving a lot of attention in cancer therapy. GQDs have various advantages, including $sp^2$ carbons, a high electron density, and a broad spectrum of functional compounds, such as hydroxy, epoxy, and carbonyl [49]. GQDs allow $\pi$–$\pi$ stacking between drug compounds and have significant potential for drug delivery via functionalizing oxygen groups. Recent studies have shown that GQDs may significantly enhance the nuclear accumulation of medications such as doxorubicin (DOX) or cisplatin, as well as significantly increase the cytotoxicity and DNA cleavage activity of these agents. GQDs are smaller and have a different chemical reactivity than other graphene nanomaterials such as GO and CNTs [50]. GQDs would be an attractive choice for a drug delivery platform because of their interactions and ease of functionalization through oxygen functional groups [49]. In recent years, resulting in increased drug performance, nanoparticle-based drug delivery applications (DDSs) have demonstrated promising effects in cancer treatment, particularly GQD–chitosan hybrid xerogels [51] and chitosan-coated GQDs [52] for sensitive drug delivery system.

As an example, Iannazzo *et al* [46] proposed a biocompatible drug delivery system and cell-traceable approach based on GQDs for the targeted delivery of doxorubicin to tumor cells. The GQDs were chemically coupled to biotin from the cancer recognition module (vitamin K/B7), which used a proactive polymeric linkage that could be initiated directly inside cells to identify biotin receptors abundantly expressing on tumor cells. DOX, the medicinal drug utilized in this investigation, was deposited onto the GQD surface, using the remarkable absorption capabilities of carbon nanoparticles. This work developed a cell-traceable drug carrier and novel biocompatible method suitable for selectively delivering the medicament agent to cancer cells while minimizing the anticancer medication's systemic toxicity and adverse reactions commonly inherent in traditional chemotherapy. Recently, Namazi *et al* [53] prepared new magnetic particles accompanied by GQD modified triazine dendrimers as a prospective carrier for a pH-dependent DOX drug delivery system. In this work, the authors developed microspheres that showed a high drug loading capacity of 63.09% (w/w%) and no sign of cytotoxicity (cell viability >80%) in the A549 cells after 72 h.

### 7.3.2 Photodynamic therapy

Photodynamic therapy (PDT) is a type of phototherapy that employs light and a chemical agent in tandem to induce cell damage through oxygen molecules (phototoxicity). It was observed that the capacity of GQDs to cause reactive oxygen species (ROS) in tumor cells was a remarkable phenomenon that needed more investigation [54]. Additionally, the up-conversion feature of GQDs enables the conversion of low-energy of NIR light into ultraviolet light, a phenomenon that has been investigated intensively for biological systems. As a result, GQDs are promising candidates for photodynamic treatment. Ahirwar *et al* [55] employed CQDs and graphene oxide quantum dots (GOQDs) for photosensitizers with a microwatt UV tube as the radiation source for a photodynamic treatment technique for skin cancer. GQDs and GOQDs had very minimal cytotoxicity against B16F10 and MCF-7 cells, and are highly absorbed and broadly distributed through cells.

Recently, GOQDs have proven to have great benefits, such as low toxicity, cheap cost, high biocompatibility, durability to photobleaching, and simplicity of production. In addition, Choi *et al* [56] synthesized up-conversion nanoparticles (UCNPs) coupled with GOQDs and studied their application in cancer cell monitoring and photodynamic treatment. The distinct features of UCNPs and GOQDs were combined to generate a unique nanostructure with desirable functionality for drug delivery and cell imaging. The anti-tumor therapeutic effects of this multifunctional nanocomposite were enhanced. Yin *et al* [57] investigated the effects of GQD cytotoxicity and oxidative stress. According to their study, GQDs have a tendency to advance into effective antioxidants capable of preventing ROS-induced cellular damage.

### 7.3.3 Photothermal therapy

Photothermal therapy (PTT) uses high-temperature radiation to kill cancer cells [58]. The near-infrared absorption adjacent to tumor cells and strong photothermal impact are critical for the effectiveness of this treatment. Phototherapy provides various benefits over traditional treatments, notably precision, adaptability, efficacy, and minimal adverse reactions on normal tissue. Carbon, semiconductor, precious metal, and nanocomposite particles are among the most often utilized photothermal components. Despite the great prospects for such nanomaterials, further research is needed to improve their photothermal effects and optical integrity. GQDs have recently attracted attention and are being investigated as a potentially effective photothermal material [59, 60]. This new photothermal compound is inexpensive, biocompatible, and has a strong photothermal performance. In *in vitro* biological investigations, GQDs have shown lower cytotoxicity, which may be further reduced with additional graphene size reduction. Additionally, GQDs have further application potential in terms of safety. According to a current study [41], GQDs may collect more near-infrared energy and convert it to photothermal energy to further efficiently destroy localized aggressive cancer cells due to their light absorption features. GQDs have the potential to be a novel photothermal treatment material.

## 7.4 Future directions

Despite significant improvements, current electrochemical biosensors remain basic and simple. Conventional research biosensors are large and non-compact, limiting their use in biomedical diagnostics, particularly in resource-constrained environments. These problems contribute to health problems in areas with low wages and resource constraints. The demand for portable, sensitive, and easy-to-use biosensors on a real-time basis has grown in recent decades. Moreover, point-of-care testing enables doctors and medical personnel to obtain lab-quality findings in minutes rather than hours. Using portable sensing equipment, point-of-care testing streamlines the diagnosis and provides patients with the most efficient and effective therapy possible using portable sensing devices.

Obviously, biosensors are constantly being improved. It is feasible to miniaturize the biosensing system using microfluidic lab-on-a-chip technology. As a detector or instrumentation interaction, smartphone-based biosensors have been studied. Healthcare, food security, environmental sensing, and biosecurity are all applications for these biosensors, which may be particularly beneficial in remote and rural areas. Electrochemistry combined with smartphone-based analysis has led to some very compact and easy-to-use solutions. The disadvantages of these systems include the necessity to connect the smartphone to electrodes, potentiostats, and circuits. This adds expense and complexity to the system. These gadgets are often linked to smartphones, increasing energy usage. The major sensing issue remains the manufacture of multiplex biosensors for the simultaneous determination of multiple molecules. Developing biosensors that can detect several disease indicators simultaneously may improve detection sensitivity while minimizing false positives. Multiple bioreceptors on a single platform are suggested for this purpose. Multifunctional nanomaterials, particularly multifunctional QDs, may be used to create multiplexed biosensors. More importantly, the establishment of long-term biological receptors will enable new analytes to be identified. Receptor modeling and engineering are important in order to optimize the components of biological recognition for stability, immunogenesis, and other biophysical characteristics. In addition, a bioreceptor must possess biophysical and biochemical characteristics which make it solid and resistant to destruction or alteration during processing and storage. In general, an ideal biosensor provides reliable and continuous evaluation in real-time of all biological molecule modifications over the course of a day. This comprises physiological biosensor rejection, downsizing, prolonged receptor and transducer stability, continuous calibration, fast stabilization, baseline variation, and reliability. The biosensor must be small enough for easy implantation and minimal discomfort.

## 7.5 Conclusion

This chapter addressed current advances in the assessment of disease-related biomolecules, in particular cancer biomarkers, with a focus on quantum dots. This research focused on the unique features and figures of merit of various biosensors but did not address other elements of electrochemical sensing systems. Despite significant progress, designing and using nanomaterial-based electrochemical biosensors remains challenging. These issues exacerbate healthcare issues in low-income and resource-constrained regions. Point-of-care diagnostics need portable and inexpensive biological sensing equipment. The findings will be used to develop equipment for point-of-care diagnostics, home clinics, and telemedicine. However, understanding the fundamental biochemistry, physiology, electrochemistry, surface science, and materials science is required to succeed in this field. It is possible that economic and legal considerations may override scientific ones in the end.

## References

[1] Sung H *et al* 2021 Global cancer statistics 2020: GLOBOCAN estimates of incidence and mortality worldwide for 36 cancers in 185 countries *CA Cancer J. Clin.* **71** 209–49

[2] Ferlay J *et al* 2021 Cancer statistics for the year 2020: an overview *Int. J. Cancer.* **149** 778–89

[3] Peng J *et al* 2012 Graphene quantum dots derived from carbon fibers *Nano Lett.* **12** 844–9

[4] Xu Q, Li W, Ding L, Yang W, Xiao H and Ong W-J 2019 Function-driven engineering of 1D carbon nanotubes and 0D carbon dots: mechanism, properties and applications *Nanoscale* **11** 1475–504

[5] Wang J, Zhu Y and Wang L 2019 Synthesis and applications of red-emissive carbon dots *Chem. Rec.* **19** 2083–94

[6] Xu Q *et al* 2016 Heteroatom-doped carbon dots: synthesis, characterization, properties, photoluminescence mechanism and biological applications *J. Mater. Chem.* B **4** 7204–19

[7] Schroeder K L, Goreham R V and Nann T 2016 Graphene quantum dots for theranostics and bioimaging *Pharm. Res.* **33** 2337–57

[8] Chong Y *et al* 2014 The *in vitro* and *in vivo* toxicity of graphene quantum dots *Biomaterials* **35** 5041–8

[9] Deng L, Liu L, Zhu C, Li D and Dong S 2013 Hybrid gold nanocube@silica@graphene-quantum-dot superstructures: synthesis and specific cell surface protein imaging applications *Chem. Commun.* **49** 2503–5

[10] Farshchi F, Hasanzadeh M and Mokhtarzadeh A 2020 A novel electroconductive interface based on $Fe_3O_4$ magnetic nanoparticle and cysteamine functionalized AuNPs: preparation and application as signal amplification element to minoring of antigen–antibody immunocomplex and biosensing of prostate cancer *J. Mol. Recognit.* **33** e2825

[11] Chikkaveeraiah B V, Bhirde A A, Morgan N Y, Eden H S and Chen X 2012 Electrochemical immunosensors for detection of cancer protein biomarkers *ACS Nano* **6** 6546–61

[12] Lilja H, Ulmert D and Vickers A J 2008 Prostate-specific antigen and prostate cancer: prediction, detection and monitoring *Nat. Rev. Cancer* **8** 268–78

[13] Etzioni R *et al* 2008 Quantifying the role of PSA screening in the US prostate cancer mortality decline *Cancer Causes Control.* **19** 175–81

[14] Yang M, Javadi A and Gong S 2011 Sensitive electrochemical immunosensor for the detection of cancer biomarker using quantum dot functionalized graphene sheets as labels *Sens. Actuators* B **155** 357–60

[15] Ehzari H, Amiri M and Safari M 2020 Enzyme-free sandwich-type electrochemical immunosensor for highly sensitive prostate specific antigen based on conjugation of quantum dots and antibody on surface of modified glassy carbon electrode with core–shell magnetic metal–organic frameworks *Talanta* **210** 120641

[16] Farschi F, Saadati A and Hasanzadeh M 2020 A novel immunosensor for the monitoring of PSA using binding of biotinylated antibody to the prostate specific antigen based on nano-ink modified flexible paper substrate: efficient method for diagnosis of cancer using biosensing technology *Heliyon* **6** e04327

[17] Srivastava M, Nirala N R, Srivastava S K and Prakash R 2018 A comparative study of aptasensor vs immunosensor for label-free PSA cancer detection on GQDs-AuNRs modified screen-printed electrodes *Sci. Rep.* **8** 1923

[18] Nxele S R and Nyokong T 2021 The effects of the composition and structure of quantum dots combined with cobalt phthalocyanine and an aptamer on the electrochemical detection of prostate specific antigen *Dye Pigment.* **192** 109407

[19] Ghanavati M, Tadayon F and Bagheri H 2020 A novel label-free impedimetric immunosensor for sensitive detection of prostate specific antigen using Au nanoparticles/MWCNTs-graphene quantum dots nanocomposite *Microchem. J.* **159** 105301

[20] Pei H, Zhu S, Yang M, Kong R, Zheng Y and Qu F 2015 Graphene oxide quantum dots@silver core–shell nanocrystals as turn-on fluorescent nanoprobe for ultrasensitive detection of prostate specific antigen *Biosens. Bioelectron.* **74** 909–14

[21] Yang Y *et al* 2018 An ultrasensitive sandwich-type electrochemical immunosensor based on the signal amplification strategy of echinoidea-shaped Au@Ag-$Cu_2O$ nanoparticles for prostate specific antigen detection *Biosens. Bioelectron.* **99** 450–7

[22] Malekzad H, Hasanzadeh M, Shadjou N and Jouyban A 2017 Highly sensitive immunosensing of prostate specific antigen using poly cysteine caped by graphene quantum dots and gold nanoparticle: a novel signal amplification strategy *Int. J. Biol. Macromol.* **105** 522–32

[23] Wu D *et al* 2016 Label-free electrochemiluminescent immunosensor for detection of prostate specific antigen based on aminated graphene quantum dots and carboxyl graphene quantum dots *Sci. Rep.* **6** 20511

[24] Gan N *et al* 2013 An ultrasensitive electrochemiluminescence immunoassay for carbohydrate antigen 19-9 in serum based on antibody labeled $Fe_3O_4$ nanoparticles as capture probes and graphene/CdTe quantum dot bionanoconjugates as signal amplifiers *Int. J. Mol. Sci.* **14** 10397–411

[25] Silsirivanit A 2019 Glycosylation markers in cancer *Adv. Clin. Chem* **89** 189–213

[26] Tang D and Ren J 2008 *In situ* amplified electrochemical immunoassay for carcinoembryonic antigen using horseradish peroxidase-encapsulated nanogold hollow microspheres as labels *Anal. Chem.* **80** 21

[27] Hasanzadeh M *et al* 2018 An innovative immunosensor for ultrasensitive detection of breast cancer specific carbohydrate (CA 15-3) in unprocessed human plasma and MCF-7 breast cancer cell lysates using gold nanospear electrochemically assembled onto thiolated graphene quantum dots *Int. J. Biol. Macromol.* **114** 1008–17

[28] Ganganboina A B and Doong R-A 2019 Graphene quantum dots decorated gold-polyaniline nanowire for impedimetric detection of carcinoembryonic antigen *Sci. Rep.* **9** 7214

[29] Yang Y *et al* 2017 A novel label-free electrochemical immunosensor based on functionalized nitrogen-doped graphene quantum dots for carcinoembryonic antigen detection *Biosens. Bioelectron.* **90** 31–8

[30] Ji J *et al* 2014 High-efficient energy funneling based on electrochemiluminescence resonance energy transfer in graded-gap quantum dots bilayers for immunoassay *Anal. Chem.* **86** 3284–90

[31] Nie G, Wang Y, Tang Y, Zhao D and Guo Q 2018 A graphene quantum dots based electrochemiluminescence immunosensor for carcinoembryonic antigen detection using poly (5-formylindole)/reduced graphene oxide nanocomposite *Biosens. Bioelectron.* **101** 123–8

[32] Luo Y *et al* 2018 SWCNTs@GQDs composites as nanocarriers for enzyme-free dual-signal amplification electrochemical immunoassay of cancer biomarker *Anal. Chim. Acta* **1042** 44–51

[33] Dong Y, Wu H, Shang P, Zeng X and Chi Y 2015 Immobilizing water-soluble graphene quantum dots with gold nanoparticles for a low potential electrochemiluminescence immunosensor *Nanoscale* **7** 16366–71

[34] Jafari H *et al* 2013 Genotyping of human papillomavirus and TP53 mutations at exons 5 to 7 in lung cancer patients from Iran *BioImpacts* **3** 135–40

[35] Hasanzadeh M, Shadjou N and de la Guardia M 2017 Current advancement in immunosensing of p53 tumor suppressor protein based on nanomaterials: analytical approach *TrAC Trends Anal. Chem.* **89** 13–20

[36] Hasanzadeh M and Shadjou N 2017 (Nano)-materials and methods of signal enhancement for genosensing of p53 tumor suppressor protein: novel research overview *Mater. Sci. Eng.* C **76** 1424–39

[37] Hasanzadeh M, Baghban H N, Shadjou N and Mokhtarzadeh A 2021 Ultrasensitive electrochemical immunosensing of tumor suppressor protein p53 in unprocessed human plasma and cell lysates using a novel nanocomposite based on poly-cysteine/graphene quantum dots/gold nanoparticle *Int. J. Biol. Macromol.* **107** 1348–63

[38] Maduraiveeran G, Sasidharan M and Ganesan V 2018 Electrochemical sensor and biosensor platforms based on advanced nanomaterials for biological and biomedical applications *Biosens. Bioelectron.* **103** 113–29

[39] Zhang H *et al* 2020 Graphene quantum dot-based nanocomposites for diagnosing cancer biomarker APE1 in living cells *ACS Appl. Mater. Interfaces* **12** 13634–43

[40] Campbell E, Hasan M T, Gonzalez Rodriguez R, Akkaraju G R and Naumov A V 2019 Doped graphene quantum dots for intracellular multicolor imaging and cancer detection *ACS Biomater. Sci. Eng.* **5** 4671–82

[41] Wang C *et al* 2020 Fabrication and characterization of novel cRGD modified graphene quantum dots for chemo-photothermal combination therapy *Sens. Actuators* B **309** 127732

[42] Du D, Wang K, Wen Y, Li Y and Li Y Y 2016 Photodynamic graphene quantum dot: reduction condition regulated photoactivity and size dependent efficacy *ACS Appl. Mater. Interfaces* **8** 3287–94

[43] Ristic B Z *et al* 2014 Photodynamic antibacterial effect of graphene quantum dots *Biomaterials* **35** 4428–35

[44] Tabish T A *et al* 2018 Biocompatibility and toxicity of graphene quantum dots for potential application in photodynamic therapy *Nanomedicine* **13** 1923–37

[45] Kuo W-S *et al* 2016 Two-photon photoexcited photodynamic therapy and contrast agent with antimicrobial graphene quantum dots *ACS Appl. Mater. Interfaces* **8** 30467–74

[46] Iannazzo D *et al* 2017 Graphene quantum dots for cancer targeted drug delivery *Int. J. Pharm.* **518** 185–92

[47] Yao X *et al* 2017 Graphene quantum dots-capped magnetic mesoporous silica nanoparticles as a multifunctional platform for controlled drug delivery, magnetic hyperthermia, and photothermal therapy *Small* **13** 1602225

[48] Li S *et al* 2017 Exceptionally high payload of the IR780 iodide on folic acid-functionalized graphene quantum dots for targeted photothermal therapy *ACS Appl. Mater. Interfaces* **9** 22332–41

[49] Su Y L *et al* 2017 Hierarchically targeted and penetrated delivery of drugs to tumors by size-changeable graphene quantum dot nanoaircrafts for photolytic therapy *Adv. Funct. Mater.* **27** 1700056

[50] Iannazzo D, Ziccarelli I and Pistone A 2017 Graphene quantum dots: multifunctional nanoplatforms for anticancer therapy *J. Mater. Chem.* B **5** 6471–89

[51] Lv O *et al* 2016 Highly fluorescent and morphology-controllable graphene quantum dots-chitosan hybrid xerogels for *in vivo* imaging and pH-sensitive drug carrier *Mater. Sci. Eng.* C **67** 478–85

[52] Sheng Y *et al* 2020 pH-sensitive drug delivery based on chitosan wrapped graphene quantum dots with enhanced fluorescent stability. *Mater. Sci. Eng.* C **112** 110888

[53] Karimi S and Namazi H 2020 Simple preparation of maltose-functionalized dendrimer/graphene quantum dots as a pH-sensitive biocompatible carrier for targeted delivery of doxorubicin *Int. J. Biol. Macromol.* **156** 648–59

[54] Hola K, Zhang Y, Wang Y, Giannelis E P, Zboril R and Rogach A L 2014 Carbon dots—emerging light emitters for bioimaging, cancer therapy and optoelectronics *Nano Today* **9** 590–603

[55] Ahirwar S, Mallick S and Bahadur D 2020 Photodynamic therapy using graphene quantum dot derivatives *J. Solid State Chem.* **282** 121107

[56] Choi S Y *et al* 2017 Synthesis of upconversion nanoparticles conjugated with graphene oxide quantum dots and their use against cancer cell imaging and photodynamic therapy *Biosens. Bioelectron.* **93** 267–73

[57] Chong Y *et al* 2016 Crossover between anti- and pro-oxidant activities of graphene quantum dots in the absence or presence of light *ACS Nano* **10** 8690–9

[58] Kumawat M K *et al* 2019 Preparation of graphene oxide-graphene quantum dots hybrid and its application in cancer theranostics *Mater. Sci. Eng.* C **103** 109774

[59] Justin R *et al* 2016 Photoluminescent and superparamagnetic reduced graphene oxide-iron oxide quantum dots for dual-modality imaging, drug delivery and photothermal therapy *Carbon* **97** 54–70

[60] Wang H *et al* 2019 Nitrogen and boron dual-doped graphene quantum dots for near-infrared second window imaging and photothermal therapy *Appl. Mater. Today.* **14** 108–17

**IOP** Publishing

Zero-dimensional Carbon Nanomaterials
Fundamentals and applications
**Pratima R Solanki and Ravindra Pratap Singh**

# Chapter 8

# Bio-imaging applications of zero-dimensional carbon nanomaterials

**Najla Bentrad**

Carbon is the most common and valuable element found in nature. Carbon nanoparticles, which have a small size and unique optical properties, have received significant attention as a result of their potential biomedical applications. Fluorescent carbon nanomaterials, as well as a new class of carbon-based nanostructures, graphene, carbon quantum dots, fullerenes, and carbon-based nanotubes have emerged. The unique properties of zero-dimensional carbon quantum dots, as well as their production, material and optical characterization, the genesis of fluorescence, as well as bio-imaging applications utilizing carbon quantum dots, are discussed in this chapter.

## 8.1 Introduction

Given the optical characteristics and critical significance of fluorescent nanoparticles, they are becoming the focus of many scientific investigations [1, 2]. Fullerenes and carbon quantum dots, for example, have been used in live cell imaging as fluorescent contrast agents since their discovery [3–5]. Carbon quantum dots (CQDs, C-dots, or CDs) are carbon nanoparticles that have a diameter of less than 10 nm. Carbon dots (CDs) were accidentally isolated using preparative electrophoresis [6]. CDs have also emerged as innovative nanoparticles that are helpful for a variety of applications because of their ability to express extensive and outstanding characteristics with biological materials. The CD nanostructures can be synthesized, which makes it simpler to achieve multiple criteria [6].

The CQDs are graphitic carbon-based semiconductor nanostructures. Because of their excellent stability and low toxicity, they are well known for their usage as tracers in imaging [7–9]. Each form of CQDs has its own physicochemical characteristics, and the origins of CQD fluorescence are a point of contention with many processes having been proposed [10, 11].

doi:10.1088/978-0-7503-4048-9ch8

A 'green chemistry' approach emphasizes the use of minimal raw materials and energy, the use of less hazardous reagents, and the production of toxicity-free functional products. The production of these 0D carbon nanostructures gets special consideration for its influence on boosting their performance in diagnostics [12]. Studying zero-dimensional carbon nanomaterials is even more intriguing since they may be made from bio-based materials using a variety of precursors, including pure chemical molecules, plant, microbial stains, or animal biomass to improve biodegradability [13–16].

It is also crucial to examine the hydrothermal pathways and procedures used to make these materials, as well as the organic solvents used in extraction approaches [17]. Advanced characterization techniques, including CQD characterization, have been achieved using microscopy and spectroscopy analyses [17–21].

Developing 0D CQDs that overcome the above unknown properties will be critical in the future. Their biocompatibility, low toxicity, and biodegradability make them a promising tool in the healthcare industry [22–25] and so can be viewed as viable alternatives in the domains of biology, the environment, and food [6]. These carbon nanoparticles have received a lot of attention since they offer remarkable characteristics in terms of their chemical, mechanical, thermal, and antibacterial activities, as well as high reinforcing capacity and thermal stability [23–25]. This chapter summarizes recent developments in carbon-polymer 0D materials and we present current research problems and the possibilities for the development of 0D carbon nanomaterials, as well as their significant imaging applications, which are more ecologically safe than current techniques.

## 8.2 The structure of zero-dimensional carbon nanomaterials

In its solid state, carbon exists in several allotropic forms, including amorphous carbon and three crystalline 0D forms (figure 8.1). Several additional synthetic allotropic forms, such as graphene sheets or nanotubes, have been discovered. For example, graphite is a two-dimensional hexagonal crystal with $sp^2$ atomic orbital hybridization, while diamond is a three-dimensional crystalline structure with an $sp^3$ hybridization. A graphene sheet is coiled in on itself to form a carbon nanotube [12].

The properties of zero-dimensional (0D) carbon nanostructures (fullerenes and carbon dots) have attracted interest in their employment in wastewater treatment via membrane technology because they are hydrophilic, which is the optimum criterion for supplying safe water [12]. An efficient technique for water treatment was reported using manufactured nanocarbons integrated into a polymer as a composite membrane.

### 8.2.1 Fullerenes

Fullerenes are simple zero-dimensional structures with a single elemental component 'carbon' and no charged or interacting functional groups; these molecules are good units for studying the basic science of self-assembly (figures 8.1 and 8.2) [5–26]. Fullerene molecules are composed of 60 carbon atoms arranged at the vertices of a

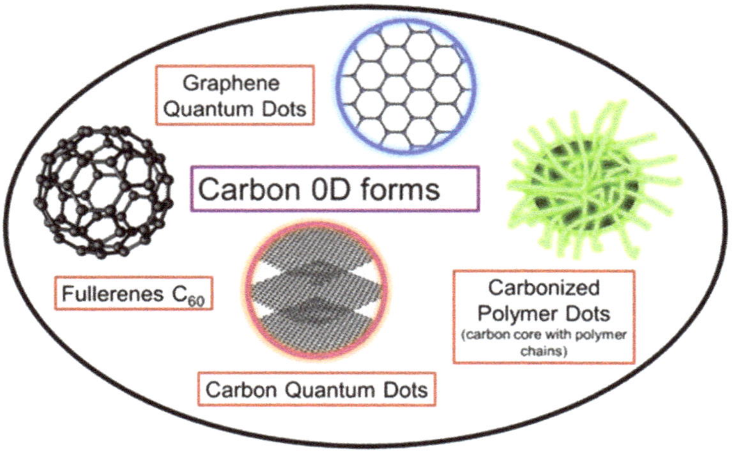


**Figure 8.1.** Schematic representation of carbon types in 0D form. (Adapted from [12]. By the authors (2020), CC BY.)

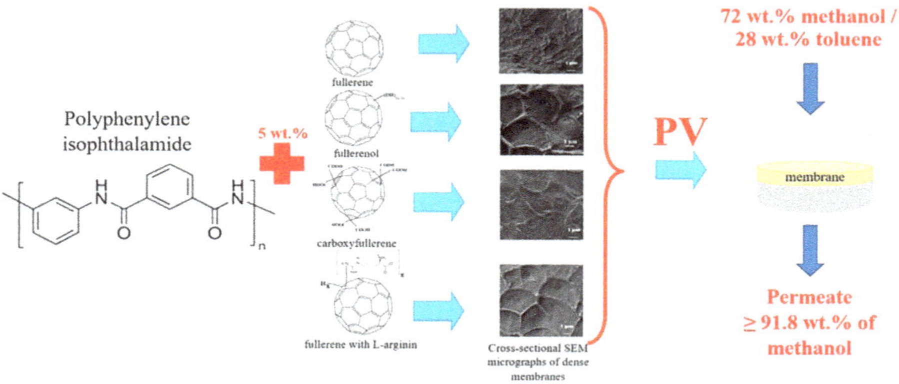


**Figure 8.2.** Aromatic polyamide membranes coated with fullerene give the best pervaporation (PV) capabilities for azeotropic methanol-toluene (72/28 wt percent) mixtures. (Adapted from [12]. By the authors (2020) CC-BY.)

regular polyhedron, a sheet of carbon atoms linked as hexagonal, pentagonal, or heptagonal rings; they then take the form of spheres, ellipsoids, and tubes [26], as well as the new concept of supramolecular differentiation [5–26].

## 8.3 Applications in bio-imaging and molecular traceability

Fullerene (C60) nanoparticles protect the liver by minimizing oxidative stress in mitochondria, for example, by scavenging reactive oxygen species (ROS), which are significant mediators in liver diseases [26]. The C60 can therefore be used as a photosensitizer, antioxidant, bio-imaging agent, and gene/drug carrier [27] due to the C–O–Si bond where the nanoparticle's photoluminescence occurs.

Bio-imaging might benefit from photoluminescent fullerene nanoparticles and nanofibers. For example, an electrospinning technique creates and characterize a novel nanofibrous material containing fullerene and poly(L-lactide) nanoparticles, and composite nanofibers [28].

Fluorescence activation and efficient singlet oxygen generation were enabled by successfully synthesized NPs by introducing a fullerene unit on pH-activated rhodamine B hydrazide for cancer imaging and therapy [29, 30]. The pH-sensitive medicine administration and bio-imaging system was made simple due to the biocompatible and water-soluble fluorescent fullerene-coated mesoporous silica nanoparticle [29]. The C60-based nanoparticles might be a promising material for cancer prevention and disorders associated with oxidative stress and additional biological applications [6, 25–31].

## 8.4 Sources and synthesis of carbon nanomaterial

Natural materials from renewable natural resources consist of mainly organic matter with significant phosphorus and nitrogen content. Examples are herbs, roots, leaves, flowers, and fruits which are used to make carbon dots with low-cost [17] that are employed in sensing, bio-imaging, drug delivery, and catalysis reactions [6, 14, 17–33].

Nitrogen-doped carbon quantum dots (N-CDs), as well as a series of N-CDs with various nitrogen concentrations, have the potential to increase the fluorescence performance and expand their detection applications [35, 36]. This is a desirable quality, as these elements have been shown to act as dopants to improve carbon dot fluorescence [35, 36].

The N-CDs exhibit a strong blue fluorescence at 401 nm when excited at 325 nm. Their quantum yield (QY) is around 14% higher than the reference material quinine sulfate. These fluorescent probes were less cytotoxic and functioned in cell bio-imaging [36]. A microwave approach may be used to produce N-CDs that brighten rapidly and easily [36].

Doping carbon dots with nitrogen (N-CDs) using a hydrothermal procedure with citric acid and polyethyleneimine has been shown to increase their quantum yield [6, 20–34] and these can be used as probes in cellular imaging due to their great photostability and minimal toxicity [17, 35], and demonstrated high water solubility [17].

Phosphorus, as well as other heteroatoms (atoms in an organic molecule that have at least one electron doublet, but are neither carbon nor hydrogen and are nonmetallic), such as boron or sulfur, can modulate and adjust the optical characteristics of carbon dots [2]. Plant-based synthesis of nitrogen-doped fluorescent carbon nanoparticles has been developed through glycine and biocompatible polyvinylpyrrolidone undergoing hydrothermal treatment.

## 8.5 Optimization of the synthesis reaction

The suggested synthesis is efficient and supports green chemistry principles even though no organic solvents are employed and no harmful by-products are produced, in this case residual resources are utilized to promote large-scale production of dots [13, 14].

Every year, the citrus processing sector generates a large quantity of by-products, a major source of citrus pectin. A fluorescent CD may be made using a sodium hydroxide-mediated citrus pectin hydrogel. Several studies have shown that microwaves can be used for CQD production, and a few of them highlight the use of this method with industrial waste as a precursor, with the focus on that issue.

The hydrogel decreases at very high temperature (hydrothermal reaction at 100 °C) that prevents the formation of visibly burnt precipitates during the synthesis process, even at temperatures as high as 180 °C. The synthesized CDs (with 27 nm width) are efficiently distributed in water and exhibit fluorescence with good photostability and biocompatibility [13, 14]. *Citrus* peels can be utilized for the creation of different CDs using a simple and low-cost carbonization method, providing an alternative strategy for future lemon germplasm utilization as a non-cytotoxic carbon precursor [13, 14].

## 8.6 Characterization techniques

Specifically, fluorescence spectrometry allows us to determine the emission ranges of the produced nanoparticles based on variation of wavelength excitation. Various reaction parameters for the intensity and type of photoluminescence are observed.

Carbon dots were observed by attenuated total reflectance (ATR) spectroscopy combined with Fourier transform infrared spectroscopy (ATR-FTIR) and x-ray diffraction (XRD) photoelectron spectroscopy. UV–visible (UV–vis) and fluorescence spectroscopy were used to investigate the optical characteristics of the produced carbon dots [17–21].

For example, UV–visible spectroscopy (UV–vis) indicates the absorbance range of the nanoparticles produced when they are subjected to incident UV rays of varying wavelengths. Indeed, ATR-FTIR spectroscopy permits the identification of many functional groups present on the nanoparticle surface and determining their size and distribution [6, 18–36].

## 8.7 Microscopes

Transmission electron microscopy (TEM) research indicates that carbon quantum dots are monodisperse and quasi-spherical, with an average size of 2.6 nm. They generate a bright blue fluorescence when they are exposed to UV light [7, 8]. A drop of 1 mg $mL^{-1}$ concentrated CQD suspension was added on a carbon grid device to prepare the samples for TEM observation. Monodisperse spherical particles with a uniform distribution, crystalline carbon, and a consistent size of less than 10 nm was observed using high-resolution transmission electron microscopy (HRTEM).

## 8.8 Technical differences

The UV–vis absorption was measured for samples synthesized using different processes to vary the pH, reaction time, and temperature. The results are documented in figure 8.3. Unlike the fluorescence response, different reaction parameters do not significantly affect the absorption spectrum of the carbon dots. However, small peaks can be observed and may be affected by the C=O bond [37].

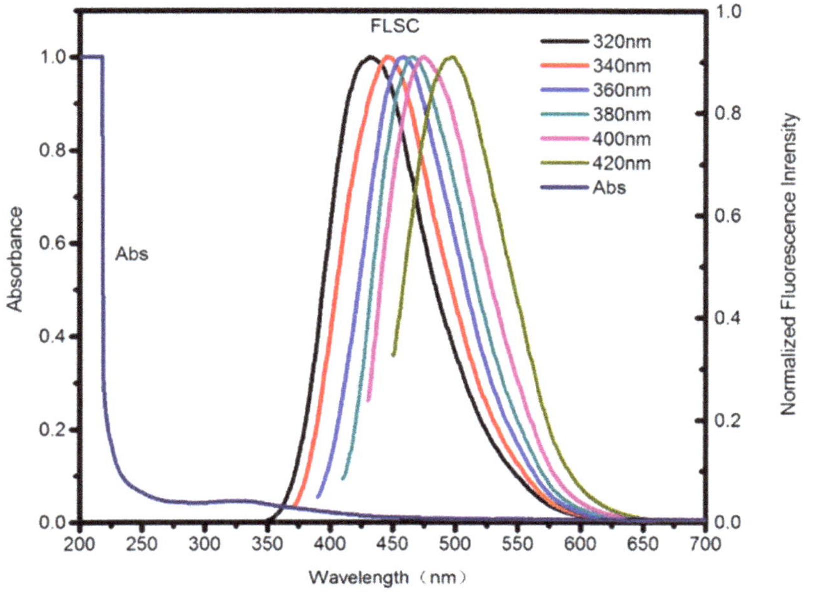


**Figure 8.3.** The functional groups present in the CDs in different samples may be seen in the FTIR spectra. (Adapted from [39]. By the authors (2020) CC-BY.)

# 8.9 Spectroscopic techniques: Fourier transform infrared spectroscopy (FTIR/ATR-FTIR)

FTIR/ATR-FTIR measurements are used to analyze the functional groups of carbon quantum dots. This analysis, therefore, validates covalent –NH, –OH, –SH, $–NH_3$, –C–O, –C=C–, and C–H bonds, and some groups such as C–N and C–S remain in the sample structure before and after synthesis of carbon quantum dots, which is very promising, as nitrogen and phosphorus can act as dopants to functionalize the CQDs [38]. As is observed in figure 8.4, the peak at 3500 $cm^{-1}$ confirms the –N–H group, while the –OH group is characterized by peaks at 3000 $cm^{-1}$.

In addition, the –C=C– bonds causes a vibration (signal) characterized at 1588 $cm^{-1}$. At higher wave numbers, a signal reflecting the –C–H bond is present at 1346 $cm^{-1}$. The $NH_3$ groups of produced carbon quantum dots are observed at 2072 $cm^{-1}$, respectively [39]. The intense band centered at 1760–1690 $cm^{-1}$ shows the presence of the carbonyl group (C=O) of a carboxylic acid chain group [2]. Similarly, a peak at 1040–909 $cm^{-1}$ can be described as a P–O bond [34]. It is shown that functional groups remain present after heat treatment, however, there is a slight reduction in the relative intensity of the peak after carbonization.

# 8.10 X-ray photoelectron spectroscopy

X-ray photoelectron spectroscopy (XPS) was used to investigate the chemical composition of CDs. $sp^2$-hybridized olefinic carbons (C=C), $sp^3$-hybridized carbons

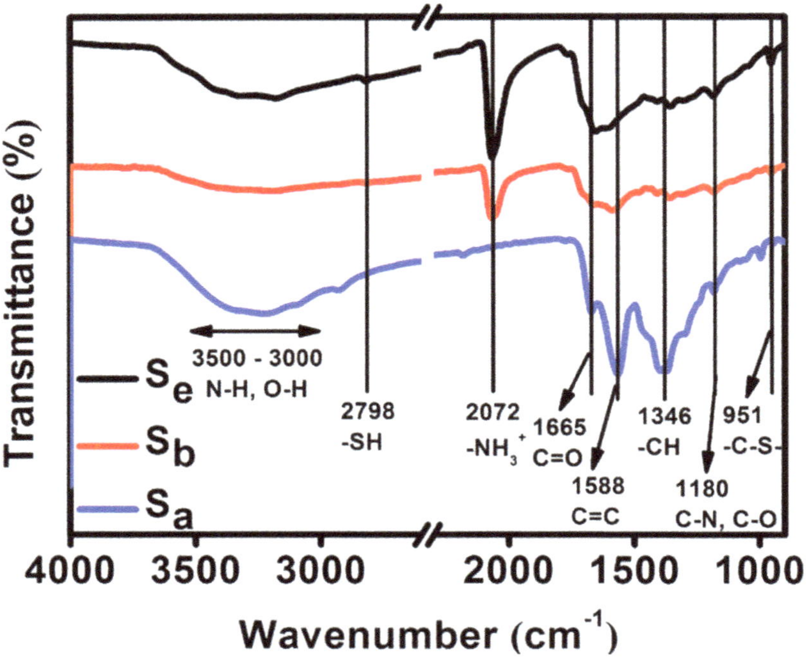


**Figure 8.4.** X-ray photoelectron spectroscopy was used to further examine the chemical composition of the C-dots. (Adapted from [39]. By the authors (2018) CC-BY.)

(C–O, C–S, and C–N), and oxidized carbon in the carboxyl group may all be fitted to the C1s spectra.

The chemical composition of the CQDs was determined using x-ray photoelectron spectroscopy. Although the P peak could not be identified due to the low amount (below the detection limit) of phosphorus, the analysis verified N, O, and C [2, 39]. According to some studies, there are three important peaks, which are attributable to C1s, N1, and O1. According to another study, some peaks can be decomposed into four peaks that can be attributed to C–C/C=C at 284.5 eV, C–O/C–N at 285.6 eV, C=O at 288.6 eV, and COOH at 290.5 eV. The N1s spectrum shows two main peaks centered at 400.5 eV and 402.5 eV, attributed to pyrrolic and graphitic nitrogen, respectively. The O1 spectra can be decomposed into two peaks at 532.0 and 534.1 eV, which are attributed to the C=O and C–O groups, respectively [2, 39].

Other research has demonstrated that employing water molecules to form hydrogen bonding with CDs and cyanuric acid successfully improves phosphorescence and optimizes ion detection based on the visual phosphorescence of C=O bonds [34].

## 8.11 Yield calculation

CDs can emit various colors of light, such as blue, green, and red photoluminescence, that have relative quantum yields (QYs) of 52.6%, 35.1%, and 12.9%, respectively. The generated N-CDs were also compared to carbon quantum dots (CQDs); the QY of the N-CDs was considerably improved by nitrogen atoms, with the highest emission efficiency of 36.3%. Various polychromatic nanostructures may be created by the protonation of nitrogen atoms into carbon dots, which is responsible for fluorescence amplification [41].

In another study, the fluorescent signal of the N-CDs was found in the lungs 12 and 24 h after injection using *in vivo* imaging devices. The potential uses of N-CDs in *in vivo* fluorescence imaging has been demonstrated [17]. The quantum yield (QY) fluorescence of N-CD glycine nitrogen doping was significantly increased from 21.43% to 47.59% compared to N-CD without doping (14.52%) [17].

The absolute QY of nitrogen-doped carbon quantum dots was 92.1% with intense blue fluorescence [25]. The solution environment interacts with the electronic structure of these CDs, altering their fluorescence response [37]. Also, shallot carbon dots (SL-CD) have a QY of 18.6%, indicating that they may be employed to detect $Cd^{2+}$ ions with high sensitivity at 15.0 nM [9].

Another ecologically friendly approach was used for synthesizing CDs with a QY over 19.8% [7]. They were employed in a $Fe^{3+}$ assay with a detection limit of $1.7 \times 10^{-9}$ mol $L^{-1}$ and for the –COOH, –OH, and $-NH_2$ groups [7].

Fluorescence intensity was observed to decline in strongly acid and basic environments; the explanation for this phenomenon is the passivation of the carbon quantum dot surface by an excess of hydroxide groups [28]. The molecular surface reduces the effective diameter of the carbon quantum dots, and it affects the fluorescence intensity [2, 6–42]. This is a positive attribute for carbon quantum dots for biomedical applications, where neutral pH compatibility is important [10]. The related folate receptor-targeted photoluminescent carbon dots have photoluminescence efficiencies of 15.4%, 12.6%, and 16.2%, respectively. Green, blue, and yellow are the converted photoluminescent characteristics in the near-infrared spectral band at 800 nm [34].

## 8.12 Fluorescence properties of carbon quantum dots

This section describes the fluorescent character of CQDs, which can emit colors with wavelengths that can approach those of the infrared. The maximum emission intensity is influenced strongly by the adopted synthesis methods [2, 6, 10, 23–43]. However, a progressive redshift can be noted when the CQD is excited with an increasingly longer wavelength. This could be explained by the uneven distribution and non-homogeneity of the formed particles [2, 6, 10, 25–43].

For this purpose, CQD passivation was performed successively by polyethylene glycol (PEG) and poly(propionylethyleneimine-co-ethyleneimine) (PPEI-EI) and subsequently the fluorescence was measured at 400–694 nm. The higher the excitation wavelength, the longer the wavelength of the perceived color, which ranges from blue to red, thus it would be a major challenge to standardize [43].

### 8.12.1 Enhanced fluorescence properties

The origin of the luminescent properties of CQDs is the subject of several theories [10]. This implies the possibility of aliphatic and aromatic group production [43]. As result, polycyclic aromatic compounds (PAH) are incorporated during the synthesis of CQD by varying their concentration each time [44]. As PAHs do not have the same energy levels, the excitation of PAHs generates a variable energy emission.

Most carbon dots' maximal emission is now confined to the blue wavelengths [41]. Citric acid and urea, for example, are pyrolyzed to generate CDs that progressively transition from blue to red light, encompassing the full light spectrum [41]. Another study looked at how the optical characteristics of CDs manufactured from common precursors such as citric acid and ethylenediamine were affected by solvent media [11]. The pH and polarity of the solvent may be changed to adjust the site of the optical transitions in CDs. The interaction with chlorinated solvents is the most important feature, with modest-resolution peaks in the absorption and PL bands in series, comparable to those found in polycyclic aromatic hydrocarbons or organic dyes [2, 6, 11–41]. The origin of fluorescence emission is thought to be the CD model, which consists of a hydrogenated carbon matrix with $sp^2$ domain inclusions generated by polycyclic aromatic hydrocarbons and derivatives [11].

### 8.12.2 Quantum confinement

Quantum confinement characterizes the state of the electrons that are enclosed in defined dimensions in the CQD structure, which adapt to the size of the blocks formed by the quantum confinement [43]. Also, the fluorescence emission made with PEG and PPEI-EI was highlighted [45].

## 8.13 The application of carbon quantum dots in bio-imaging

Under UV illumination, CDs emit a bright blue fluorescence [46]. Some of the major characteristics that allow them to be employed as fluorescent probes are their water dispersibility, photostability throughout a wide pH range, and low toxicity.

### 8.13.1 Detection of ionic components

The sensitivity of CQDs to certain metals would confer potential in the analysis of metal pollutants, and they are increasingly used in the design of biosensors that can detect metal ions [2, 6, 10, 18–49], pH variation [50, 51], hydrogen peroxide, and glucose [52, 53] in various environments. In addition, an innovative sensor based on spontaneous fluorescence resonance energy might be useful in cell imaging and vitamin B2 detection [2, 6, 10, 22–54]. The N-CD is biocompatible and may be used for cellular imaging and intracellular $Hg^{2+}$ detection [2, 6, 10, 25–55].

The quenching process for *Citrus sinensis* CQD and *Citrus limon* CQD fluorescent probes for cell bio-imaging and intracellular detection ranges from 0.01–1.0 M to 0.6–23.5 M, respectively, and the reduction in fluorescence intensity is linear. The content of tartrazine is linked to $Fe^{3+}$ [14]. Efficient sensors to determine the precise amount of $Fe^{3+}$ in body cells are critical for the early diagnosis of illnesses or disorders caused by an excess or deficit of iron [1–57].

In another work, exceptional carbon dots were generated by hydrothermal carbonization of *Fusobacterium nucleatum* (Fn-CDs), and can function as a biosensor for effective detection of intracellular $Fe^{3+}$ [2, 6, 10, 24–58]. The Fn-CD fluorescence was maintained over 24 h after entering the cells and was very sensitive to $Fe^{3+}$ ions even inside cells, indicating that it has a lot of potential for *in vivo* $Fe^{3+}$ ion detection [2, 6, 10, 23–57].

The detection selectivity was tested in blood serum and urine, with virtually no interference. It has been shown to be a multifunctional fluorescence detection system that can detect pH, Ag(I), and Fe(III) values in aqueous solutions, even though the fluorescence intensity is inversely proportional to pH [51–59]. Also, CDs have also been used in fungal bio-imaging, with promising results [60]. The synthesis, characterization, and application of nitrogen-doped red-emitting carbon dots for dual sensing of indium ($In^{3+}$) and palladium ($Pd^{2+}$) in water has been reported [61].

Human body, after adding $Cu^{2+}$ ion ions, the photoluminescence frequency of C-dots might decline, and after adding thiamine, it will preferentially rise. There is a linear relationship between the increase in fluorescence emission intensity at the point C of $Cu^{2+}$ ion-change CDs and a thiamine concentration range from 10 to 50 M, with a detection limit of 280 nM [60]. Organelles such as lysosomes and endoplasmic reticulum are important for normal cell activities and related to pH values; hence they were effectively used to visualize structural changes in cells [62]. Using the differential quenching capacity of iron ions towards the fluorescence of N-CDs, sensitive detection of $H_2O_2$ was produced, and cell toxicity tests revealed that it maintained minimal toxicity effects despite the addition of several nitrogen atoms [59].

Bio-imaging using an appropriate microscope makes it possible to track the various changes in the $H_2S$ content of the cell simply due to the change in the luminescence of the CQDs and to predict certain elements in biological cells. $H_2S$ administered to a laboratory mouse can turn the luminescence from blue to green [2–63].

### 8.13.2 Carbon quantum dots in cancer bio-imaging

CDs possess great potential for rapid and bright employment in cancer diagnostics [63]. For example, only 3 min was required for CDs to pass through hepatoma cells, and they stayed inside the cell up to 24 h [63].

Folate receptor-targeted photoluminescent CDs revealed almost minimal cytotoxicity (cell survival more than 80%) and a high affinity to the membranes that highly expressed folate receptors in Hela cells, and lit the cytoplasm as a fluorescent marker in *in vitro* tests [34]. These CDs exposed to UV (from 365 nm with maximal emission lengths of 530, 429, and 612 nm) was luminous for 24 h, and it displayed extended luminous (green, blue, and red luminescence) stability with a PL intensity greater than 90% [33]. These carbon nanoparticles were ingested into targeted cervical carcinoma cancer and the fluorescence emission lasted 30 min [34].

CDs integrated with molecularly imprinted polymers can be employed to detect cancer biomarkers and as biocompatible optical imaging tools, for example glucuronic acid is a biomarker for some malignancies. The CDs can be taken up by C6 glioma cells in a time- and concentration-dependent manner, with endosomal and mitochondrial organelles [64]. For example, the produced CQDs emit blue–green fluorescence, with a quantum yield of 13%, and the fluorescence characteristics were stable and resistant to photobleaching and temperature fluctuations (figure 8.5) [65].

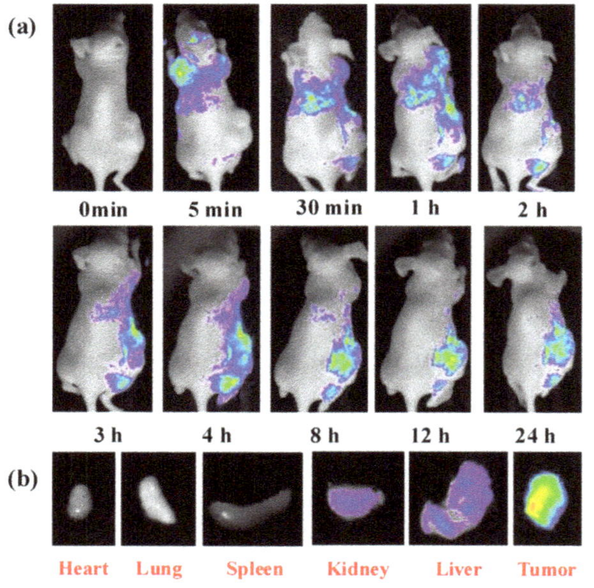


**Figure 8.5.** (a) *In vivo* fluorescence imaging of nude mice following intravenous injection of CQD-WS solution. (b) Typical fluorescence of dissected mouse organs after intravenous injection of CQD-WS solution for 24 h in nude mice. (Adapted from [65].)

Synthesized CQDs can be utilized for cell and tumor imaging in the brain, indicating that they have potential uses in bio-imaging and related sectors including phototherapy and imaging of gliomas *in vivo* [65]. Furthermore, the distribution of fluorescence in tissue slices revealed that they spread with a high intensity fluorescence in glioma [64, 65].

### 8.13.3 Fullerene based nanosystems for cancer imaging

A variety of innovative fullerene nanoparticles have been utilized in tumor treatment, and the distribution and metabolism of these agents in living beings has gained significant interest. For cancer diagnostics, functionalized fullerenes with particular physical and chemical characteristics have been produced [39, 66].

For example, C60 malonyldiserine alcohol amide (C60-ser) derived fullerene is highly water-soluble, non-ionic, and non-cytotoxic. It is studied as a potential nanocarrier for delivering biological and anticancer drugs across biological barriers via modified tumor vasculature and the blood–brain barrier [67].

Metal fullerenes are a novel form of magnetic resonance imagining (MRI) contrast enhancer that might have clinical diagnostic uses in positron emission tomography (PET) to measure the biodistribution of materials in the body [66, 68] and in photoacoustic imaging [39]. It was discovered that fluorophore-conjugated C60-Ser nanoparticles link to serum proteins in live cancer cells via distinct energy-dependent mechanisms and then escape the endocytic vesicles. Using a nuclear pore complex may localize the nanoparticles and cause them to accumulate in the nucleus [67].

This method combines thiolene chemistry with folic acid to synthesize fluorescent C60 nanoparticles, which demonstrated bright blue fluorescence and a quantum yield of 26%, which was greater than any previous fluorescent fullerene nanomaterial [66]. Fluorescent C60 nanoparticles demonstrated excellent photobleaching resistance after 5 h of continuous UV light exposure and excellent light stability after a few months under an aqueous medium. Fluorescent fullerene nanoparticles can also penetrate cancer cells with high levels of folate receptor expression utilizing the FA component, allowing for targeted cell imaging [66].

### 8.13.4 Microbial bio-imaging

Carbon dots have been investigated as a suitable alternative to chemical organic dyes in emission spectrum tissue labeling studies; they are non-toxic, enter cells rapidly, and have good bio-conjugation with bacterial cells [69]. Furthermore, novel *E. coli* derived CDs (WT-CDs) may be used for both living and deceased cell microbe image analysis. The WT-CDs may be employed as a biosensor for the extremely sensitive and fast observation of p-nitrophenol (p-NP), with a detection limit of 11 nM, which is the lowest limit recorded to date. As a result, p-NP sensing has become more feasible, contributing to the development of bio-imaging in numerous mouse organs, in addition to its ability to cross the blood–brain connection and its efficient excretion through the intestines [2, 6, 10, 16–51].

Carbonized polymer dots with a high-quality factor (QY), long-wavelength emission, and diverse functionalities are in great demand for developing fluorescent materials. They have been used to distinguish *Porphyromonas gingivalis*, *Streptococcus mutants*, *Escherichia coli*, and *Staphylococcus aureus* in dental plaque biofilms by measuring emission intensities and peaks under various pH conditions.

These findings might pave the way for a novel approach to carbonized polymer product development based on multi-emission center design [63]. A novel form of nitrogen and genipin co-doped carbon dots was made using hydrothermal techniques (N-CDs-GP). Furthermore, the N-CDs-GP provided high biocompatibility and light stability across a wide range of bacteria. It is also a fast method of detection for selective imaging of gram-positive bacteria [70]. The development of multiple CDs with the goal of identifying bacterial pathogens, as well as their mechanisms, are described here. Antimicrobial nanomaterials are being studied to replace traditional antibiotics by inhibiting bacterial growth and destroying cells [70, 71].

Carbon dots, for example, may be utilized to make a fluorescent chemosensor for $Fe^{3+}$ in aqueous solutions that can be used as a sensor to see cytoplasmic $Fe^{3+}$ in the fungal cell wall. Additionally, in fungal cells cultured in Tris–HCl solution containing $Fe^{3+}$, a rise in intracellular $Fe^{3+}$ concentration might cause fluorescence quenching of carbon dots [70]. Since the CD-stained fungal cells were intensely lit up on a confocal microscope stimulated at 405 nm, carbon dots can be utilized as a fluorescent agent for fungal bio-imaging [8].

## 8.14 Advantages and disadvantages

Fluorescent carbon dots offer huge potential in biological imaging because of their minimal photobleaching, absence of optical flashing, and good biocompatibility, although their *in vivo* imaging applications are limited because of their narrow excitation/emission wavelengths [64]. The fluorescence quenching mechanism comes from the electron transfer between the CQDs and metal ions that creates non-radiative recombination, resulting in fluorescence quenching. The visual decrease in blue fluorescence as a function related to variation in the concentration of inserted metal ions is noticeable. Fluorescence is significantly quenched when iron-containing solutions are used. Each nanoparticle has a 350 nm excitation wavelength at 0.005 M ion concentration [10]. This may be due to the increased tendency of $Fe^{3+}$ ions to take part in electron transfer with phenolic hydroxyl groups on the CQD surface. This electron transfer generates more emission interference, while decreasing fluorescence [10].

This phenomenon is not limited to copper, various other metal ions such as $Fe^{3+}$, $Zn^{2+}$, $Ca^{2+}$, $Mn^{2+}$, and $Mg^{2+}$ also have an effect, but the CQDs are significantly more sensitive for $Cu^{2+}$ and $Fe^{3+}$ ion detection.

## 8.15 Toxicity

Due to their low toxicity, higher water solubility, excellent biocompatibility, and sustained photoluminescence, CDs are widely used by researchers [2, 6, 10, 17–64, 70, 72, 73]. CDs at doses ranging from 0.1 to 1.0 mg $mL^{-1}$ cause a physiological response in the mung bean but have no influence on its growth. *In vivo* viewing of CDs presents the prospect of their effective utilization as delivery vectors in plants through the vascular system via the apoplastic pathway [74].

CDs used for bio-imaging and labeling have demonstrated their use in a variety of settings [6–37]. Not only can the CDs' precursor and entire synthesis technique be naturally occurring, but they also have a number of benefits and characteristics, including high luminescence, strong stability, chemical inertness, and sufficient resilience [2, 6–75].

N-CDs have almost low cytotoxicity and can be employed as an appropriate nanosensor, for cell cultures authentic scanning, and other biomedical applications [9,76].

## 8.16 Conclusion and future prospects

Fluorescent carbon dots (CDs) have low photobleaching, no optical scintillation, and excellent biocompatibility, making them suitable for bio-imaging. However, their applicability in *in vivo* bio-imaging is limited because of their relatively short excitation/emission wavelengths. Using plants as a precursor for carbon quantum dots demonstrates both an ecologically acceptable approach as novel source for carbon quantum dot synthesis and a means to manage natural waste. It was discovered that the characteristics of CQDs are affected not only by the upstream reagents but also by the synthesis process parameters (pH, reaction time, and temperature). Spectroscopic studies have revealed that these are made up of carboxylic acid chains as well as nitrogen and phosphorus impurities that can function as doping heteroatoms in carbon quantum dots. These natural dopants can provide high-quality luminescence without the need for any extra additives. Finally, such carbon quantum dots used for metal detection in solution are sensitive to changes in the concentration of $Cu^{2+}$ and $Fe^{3+}$ ions. Bio-conjugated carbon quantum dots as theranostic agents for targeted medication delivery and therapeutic applications require more research.

## Acknowledgments

No financial support was provided for this chapter. The author wishes to thank Dr Ravindra Pratap Singh and Mr Kshitij RB Singh for their suggestions on preparing the manuscript.

## References

[1] Li L, Shi L, Jia J, Chang D, Dong C and Shuang S 2020 $Fe^{3+}$ detection, bioimaging, and patterning based on bright blue-fluorescent N-doped carbon dots *Analyst* **145** 5450–7

[2] Yuan Y H *et al* 2016 Synthesis of nitrogen-doping carbon dots with different photoluminescence properties by controlling the surface states *Nanoscale* **8** 6770–6

[3] Yu P, Wen X, Toh Y-R and Tang J 2012 Temperature-dependent fluorescence in carbon dots *J. Phys. Chem.* C **116** 25552–7

[4] Demir B *et al* 2018 Tracking hyaluronan: molecularly imprinted polymer coated carbon dots for cancer cell targeting and imaging *ACS Appl. Mater. Interfaces* **10** 3305–13

[5] Ariga K and Shrestha L K 2020 Fullerene nanoarchitectonics with shape-shifting *Materials* **13** 2280

[6] Dugam S, Nangare S, Patil P and Jadhav N 2021 Carbon dots: a novel trend in pharmaceutical applications *Ann. Pharm. Fr.* **79** 335–45

[7] Yang X, Zhuo Y, Zhu S, Luo Y, Feng Y and Dou Y 2014 Novel and green synthesis of high-fluorescent carbon dots originated from honey for sensing and imaging *Biosens. Bioelectron.* **60** 292–8

[8] Chen Y, Sun X, Pan W, Yu G and Wang J 2020 $Fe^{3+}$-sensitive carbon dots for detection of $Fe^{3+}$ in aqueous solution and intracellular imaging of $Fe^{3+}$ inside fungal cells *Front. Chem.* **7** 911

[9] Gu D, Hong L, Zhang L, Liu H and Shang S 2018 Nitrogen and sulfur co-doped highly luminescent carbon dots for sensitive detection of Cd (II) ions and living cell imaging applications *J. Photochem. Photobiol.* B **186** 144–51

[10] Hong G, Diao S, Antaris A L and Dai H 2015 Carbon nanomaterials for biological imaging and nanomedicinal therapy *Chem. Rev.* **115** 10816–906

[11] Stepanidenko E A *et al* 2020 Influence of the solvent environment on luminescent centers within carbon dots *Nanoscale* **12** 602–9

[12] Jani M, Arcos-Pareja J A and Ni M 2020 Engineered zero-dimensional fullerene/carbon dots-polymer based nanocomposite membranes for wastewater treatment *Molecules* **25** 4934

[13] Zhao X J, Zhang W L and Zhou Z Q 2014 Sodium hydroxide-mediated hydrogel of citrus pectin for preparation of fluorescent carbon dots for bioimaging *Colloids Surf.* B **123** 493–7

[14] Chatzimitakos T, Kasouni A, Sygellou L, Avgeropoulos A, Troganis A and Stalikas C 2017 Two of a kind but different: luminescent carbon quantum dots from *Citrus* peels for iron and tartrazine sensing and cell imaging *Talanta* **175** 305–12
[15] Fu X, Fu X, Li W, Chen Y and Cai Z 2019 Ovalbumin as a precursor for green synthesis of highly fluorescent carbon dots for cell imaging *J. Biomed. Nanotechnol.* **15** 1232–40
[16] Qin K *et al* 2020 Applications of hydrothermal synthesis of *Escherichia coli* derived carbon dots *in vitro* and *in vivo* imaging and p-nitrophenol detection *Analyst* **145** 177–83
[17] Humaera N A, Fahri A N, Armynah B and Tahir D 2021 Natural source of carbon dots from part of a plant and its applications: a review *Luminescence* **36** 1354–64
[18] Dong Y *et al* 2012 Blue luminescent graphene quantum dots and graphene oxide prepared by tuning the carbonization degree of citric acid *Carbon* **50** 4738–43
[19] Sachdev A and Gopinath P 2015 Green synthesis of multifunctional carbon dots from coriander leaves and their potential application as antioxidants, sensors and bioimaging agents *Analyst* **140** 4260–9
[20] Kuo T-R, Sung S-Y, Hsu C-W, Chang C-J, Chiu T-C and Hu C-C 2015 One-pot green hydrothermal synthesis of fluorescent nitrogen-doped carbon nanodots for *in vivo* bioimaging *Anal. Bioanal.Chem.* **408** 77–82
[21] Chu H-W *et al* 2017 Pulse laser-induced fragmentation of carbon quantum dots: a structural analysis *Nanoscale* **9** 18359–67
[22] Kundu A, Nandi S, Das P and Nandi A K 2016 Facile and green approach to prepare fluorescent carbon dots: emergent nanomaterial for cell imaging and detection of vitamin B2 *J. Colloid Interface Sci.* **468** 276–83
[23] Liu L *et al* 2020 Bifunctional carbon dots derived from an anaerobic bacterium of *Porphyromonas gingivalis* for selective detection of $Fe^{3+}$ and bioimaging *Photochem. Photobiol.* **97** 574–81
[24] Liu L *et al* 2021 Carbon dots derived from *Fusobacterium nucleatum* for intracellular determination of $Fe^{3+}$ and bioimaging both *in vitro* and *in vivo Anal. Methods* **13** 1121–31
[25] Tan A, Yang G and Wan X 2021 Ultra-high quantum yield nitrogen-doped carbon quantum dots and their versatile application in fluorescence sensing, bioimaging and anti-counterfeiting *Spectrochim. Acta* A **253** 119583
[26] Iohara D 2019 Preparation and evaluation of fullerene based nanomedicine *Yakugaku Zasshi* **139** 1539–46
[27] Jeong J, Cho M, Lim Y T, Song N W and Chung B H 2009 Synthesis and characterization of a photoluminescent nanoparticle based on fullerene-silica hybridization *Angew. Chem. Int. Ed.* **48** 5296–9
[28] Liu N *et al* 2014 One-step catalase controllable degradation of $C_3N_4$ for N-doped carbon dot green fabrication and their bioimaging applications *J. Mater. Chem.* B **2** 5768
[29] Tan L *et al* 2016 Water-soluble photoluminescent fullerene capped mesoporous silica for pH-responsive drug delivery and bioimaging *Nanotechnology* **27** 315104
[30] Tang Q *et al* 2018 A fullerene-rhodamine B photosensitizer with pH-activated visible-light absorbance/fluorescence/photodynamic therapy *J. Mater. Chem.* B **6** 2778–84
[31] Liu J-H *et al* 2010 Fullerene-conjugated doxorubicin in cells *ACS Appl. Mater. Interfaces* **2** 1384–9
[32] Wang W, Lu Y-C, Huang H, Feng J-J, Chen J-R and Wang A-J 2014 Facile synthesis of water-soluble and biocompatible fluorescent nitrogen-doped carbon dots for cell imaging *Analyst* **139** 1692–6

[33] Bankoti K, Rameshbabu A P, Datta S, Das B, Mitra A and Dhara S 2017 Onion derived carbon nanodots for live cell imaging and accelerated skin wound healing *J. Mater. Chem.* B **5** 6579–92

[34] Li L and Dong T 2018 Photoluminescence tuning in carbon dots: surface passivation or/and functionalization, heteroatom doping *J. Mater. Chem.* C **6** 7944–70

[35] Sahu S, Behera B, Maiti T K and Mohapatra S 2012 Simple one-step synthesis of highly luminescent carbon dots from orange juice: application as excellent bio-imaging agents *Chem. Commun.* **48** 8835

[36] Edison T N J I, Atchudan R, Sethuraman M G, Shim J-J and Lee Y R 2016 Microwave assisted green synthesis of fluorescent N-doped carbon dots: cytotoxicity and bio-imaging applications *J. Photochem. Photobiol.* B **161** 154–61

[37] Wang C, Sun D, Zhuo K, Zhang H and Wang J 2014 Simple and green synthesis of nitrogen-, sulfur-, and phosphorus-co-doped carbon dots with tunable luminescence properties and sensing application *RSC Adv.* **4** 54060–5

[38] Krysmann M J, Kelarakis A, Dallas P and Giannelis E P 2011 Formation mechanism of carbogenic nanoparticles with dual photoluminescence emission *J. Am. Chem. Soc.* **134** 747–50

[39] Wang J *et al* 2018 Co-reductive fabrication of carbon nanodots with high quantum yield for bioimaging of bacteria *Beilstein J. Nanotechnol.* **9** 137–45

[40] Zhu Z *et al* 2019 Synthesis of diverse green carbon nanomaterials through fully utilizing biomass carbon source assisted by KOH *ACS Appl. Mater. Interfaces* **11** 24205–11

[41] Miao X *et al* 2017 Synthesis of carbon dots with multiple color emission by controlled graphitization and surface functionalization *Adv. Mater.* **30** 1704740

[42] Kong W, Wu H, Ye Z, Li R, Xu T and Zhang B 2014 Optical properties of pH-sensitive carbon-dots with different modifications *J. Lumin.* **148** 238–42

[43] Bourlinos A B, Stassinopoulos A, Anglos D, Zboril R, Karakassides M and Giannelis E P 2008 Surface functionalized carbogenic quantum dots *Small* **4** 455–8

[44] Demchenko A P and Dekaliuk M O 2016 The origin of emissive states of carbon nanoparticles derived from ensemble-averaged and single-molecular studies *Nanoscale.* **8** 14057–69

[45] Sun Y-P *et al* 2006 Quantum-sized carbon dots for bright and colorful photoluminescence *J. Am. Chem. Soc.* **128** 7756–7

[46] Han C *et al* 2016 Highly fluorescent carbon dots as selective and sensitive 'on–off–on' probes for iron(III) ion and apoferritin detection and imaging in living cells *Biosens. Bioelectron.* **83** 229–36

[47] Zhang Z *et al* 2014 Quinoline derivative-functionalized carbon dots as a fluorescent nanosensor for sensing and intracellular imaging of $Zn^{2+}$ *J. Mater. Chem.* B **2** 5020–7

[48] Qu K, Wang J, Ren J and Qu X 2013 Carbon dots prepared by hydrothermal treatment of dopamine as an effective fluorescent sensing platform for the label-free detection of iron(III) ions and dopamine *Chem.—A Eur. J.* **19** 7243–9

[49] Cui F *et al* 2019 Ultrasensitive fluorometric determination of iron(iii) and inositol hexaphosphate in cancerous and bacterial cells by using carbon dots with bright yellow fluorescence *Analyst* **144** 5010–21

[50] Nie H *et al* 2014 Carbon dots with continuously tunable full-color emission and their application in ratiometric pH sensing *Chem. Mater.* **26** 3104–12

[51] Ding Y-Y *et al* 2018 Facile preparation of bright orange fluorescent carbon dots and the constructed biosensing platform for the detection of pH in living cells *Talanta* **189** 8–15

[52] Shan X, Chai L, Ma J, Qian Z, Chen J and Feng H 2014 B-doped carbon quantum dots as a sensitive fluorescence probe for hydrogen peroxide and glucose detection *Analyst* **139** 2322–5

[53] Wang H, Yi J, Velado D, Yu Y and Zhou S 2015 Immobilization of carbon dots in molecularly imprinted microgels for optical sensing of glucose at physiological pH *ACS Appl. Mater. Interfaces* **7** 15735–45

[54] Du F *et al* 2020 A label-free multifunctional nanosensor based on N-doped carbon nanodots for vitamin B12 and $Co^{2+}$ detection, and bioimaging in living cells and zebrafish *J. Mater. Chem.* B **8** 5089–95

[55] Liu Y, Liao M, He X, Liu X, Kou X and Xiao D 2015 One-step synthesis of highly luminescent nitrogen-doped carbon dots for selective and sensitive detection of mercury(II) Ions and cellular imaging *Anal. Sci.* **31** 971–7

[56] Zhai Y, Zhu Z, Zhu C, Ren J, Wang E and Dong S 2014 Multifunctional water-soluble luminescent carbon dots for imaging and $Hg^{2+}$ sensing *J. Mater. Chem.* B **2** 6995–9

[57] Kalaiyarasan G, Joseph J and Kumar P 2020 Phosphorus-doped carbon quantum dots as fluorometric probes for iron detection *ACS Omega.* **5** 22278–88

[58] Pu J, Liu C, Wang B, Liu P, Jin Y and Chen J 2021 Orange red-emitting carbon dots for enhanced colorimetric detection of $Fe^{3+}$ *Analyst* **146** 1032–9

[59] Qian Z, Ma J, Shan X, Feng H, Shao L and Chen J 2014 Highly luminescent N-doped carbon quantum dots as an effective multifunctional fluorescence sensing platform *Chem.—A Eur. J.* **20** 2254–63

[60] Purbia R and Paria S 2016 A simple turn on fluorescent sensor for the selective detection of thiamine using coconut water derived luminescent carbon dots *Biosens. Bioelectron.* **79** 467–75

[61] Pawar S, Kaja S and Nag A 2020 Red-emitting carbon dots as a dual sensor for $In^{3+}$ and $Pd^{2+}$ in water *ACS Omega* **5** 8362–72

[62] Shuang E, Mao Q-X, Yuan X-L, Kong X-L, Chen X-W and Wang J-H 2018 Targeted imaging of the lysosome and endoplasmic reticulum and their pH monitoring with surface regulated carbon dots *Nanoscale* **10** 12788–96

[63] Duan Q *et al* 2019 Rapid cancer diagnosis by highly fluorescent carbon nanodots-based imaging *Anal. Bioanal. Chem.* **411** 967–72

[64] Ruan S *et al* 2014 A simple one-step method to prepare fluorescent carbon dots and their potential application in non-invasive glioma imaging *Nanoscale* **6** 10040

[65] Huang C, Dong H, Su Y, Wu Y, Narron R and Yong Q 2019 Synthesis of carbon quantum dot nanoparticles derived from byproducts in bio-refinery process for cell imaging and *in vivo* bioimaging *Nanomaterials* **9** 387

[66] Ma Y, Fu S, Tan Y and Zhang A 2021 Design and synthesis of highly fluorescent and stable fullerene nanoparticles as probes for folic acid detection and targeted cancer cell imaging *Nanotechnology* **32** 195501

[67] Raoof M, Mackeyev Y, Cheney M A, Wilson L J and Curley S A 2012 Internalization of C60 fullerenes into cancer cells with accumulation in the nucleus via the nuclear pore complex *Biomaterials* **33** 2952–60

[68] Zaibaq N G, Pollard A C, Collins M J, Pisaneschi F, Pagel M D and Wilson L J 2020 Evaluation of the biodistribution of serinolamide-derivatized C60 fullerene *Nanomaterials* **10** 143

[69] Das P *et al* 2017 Green approach to photoluminescent carbon dots for imaging of gram-negative bacteria *Escherichia coli Nanotechnology* **28** 195501
[70] Chu X *et al* 2020 Genipin cross-linked carbon dots for antimicrobial, bioimaging and bacterial discrimination *Colloids Surf.* B **190** 110930
[71] Anand A, Manavalan G, Mandal R P, Chang H-T, Chiou Y-R and Huang C-C 2020 Carbon dots for bacterial detection and antibacterial applications-a minireview *Curr. Pharm. Design.* **25** 4848–60
[72] Jiang D *et al* 2017 Investigation of luminescent mechanism: N-rich carbon dots as luminescence centers in fluorescent hydroxyapatite prepared using a typical hydrothermal process *J. Mater. Chem.* B **5** 3749–57
[73] Zhang Y, Shen Y, Teng X, Yan M, Bi H and Morais P C 2015 Mitochondria-targeting nanoplatform with fluorescent carbon dots for long time imaging and magnetic field-enhanced cellular uptake *ACS Appl. Mater. Interfaces* **7** 10201–12
[74] Li W *et al* 2016 Phytotoxicity, uptake, and translocation of fluorescent carbon dots in mung bean plants *ACS Appl. Mater. Interfaces* **8** 19939–45
[75] Ostadhossein F and Pan D 2016 Functional carbon nanodots for multiscale imaging and therapy *WIREs Nanomed. Nanobiotechnol.* **9** e1436
[76] Lu S *et al* 2016 Hydrothermal synthesis of nitrogen-doped carbon dots with real-time live-cell imaging and blood–brain barrier penetration capabilities *Int. J. Nanomed.* **11** 6325–36

# Chapter 9

# Zero-dimensional carbon nanomaterials for central nervous system diseases

**Tyciane de Souza Nascimento, Stephen Rathinaraj Benjamin, Albert Layo Costa de Assis, Jessica Rabelo Bezerra, Jessica Maria Pessoa Gomes, Geanne Matos de Andrade and Reinaldo Barreto Oriá**

The complexity of the central nervous system (CNS), its inadequate self-repair capacity, and the inefficient transmission to the brain of most CNS medications exacerbate the progressive and irreversible aspects of many neurological disorders, and the intensity of their outcomes has made CNS medications the largest field in the development of new diagnostic and therapeutic technologies. In recent years, zero-dimensional carbon nanomaterials have been investigated as a potential strategy to diagnose, treat, and monitor CNS diseases, for potential therapeutic, biosensor, and imagery applications, based on their antioxidant properties, inherent photoluminescence, and ability to cross the blood–brain barrier, providing enormous potential.

## 9.1 Introduction

### 9.1.1 Overview of zero-dimensional carbon nanomaterials

Carbon, the most dynamic element in the periodic table, serves as the foundation for all forms of life on the planet. Carbon exhibits sophisticated allotropy depending on the complexity of its hybridization and has served humans in several different ways, including as coal, which is the world's most important source of energy. When its dimensions reach the nanoscale (1–100 nm), carbon nanomaterials and related colloids showed promising characteristics for thermal, chemical, mechanical, and electrical applications [1]. On this scale, carbon appears in three different dimensions, including zero-dimensional (carbon quantum dots (CQDs) or carbon dots (CDs), graphene quantum dots (GQDs) and fullerene, as seen in figure 9.1), one-dimensional (carbon nanotubes (CNTs)), and two-dimensional (graphene and graphene oxide), and each different structure shows extraordinary properties [2]. Zero-dimensional nanomaterials are materials in the three-dimensional space of the

doi:10.1088/978-0-7503-4048-9ch9  

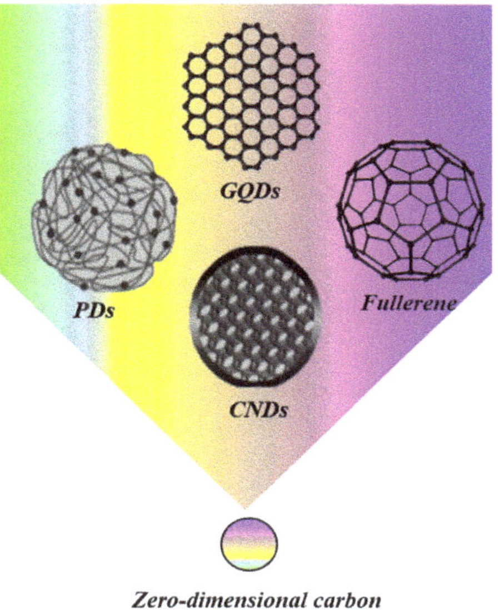


**Figure 9.1.** Types of zero-dimensional carbon: graphene quantum dots (GQDs), carbon nanodots (CNDs), polymer dots (PDs), and fullerene.

nanometer scale, including certain nanoparticles and atomic and quantum clusters, typically made up of a tiny number of atoms and molecules. CDs are usually described as tiny carbon nanoparticles. They are monodisperse carbon-based spherical nanoparticles with a high number of oxygen containing groups on their surface, renowned for their small size and fluorescence properties [3].

In addition to their excellent functional properties compared to semiconductor quantum dots, they also overcome the inadequacies of classical materials in terms of cellular and environmental toxicity, biohazards, water solubility, stability, and photobleaching tolerance, as well as ease of surface modification and large-scale preparation, as seen in figure 9.2 [4, 5]. CDs are mostly used as nonmetal quantum dots and shown great promise in the advancement of bioimaging and another clinical research due to their high biocompatibility.

Since the discovery of graphene, its characteristics and usage have been studied and developed extensively. However, it has a lot of shortcomings, including a zero bandgap and poor absorptivity. Subsequently, structural modifications of graphene were explored to address its weaknesses. In 2008 Ponomarenko and Geim developed GQDs based on previous work by Xu and colleagues on CDs. Unlike CDs, however, GQDs are distinguished by having atom-thin graphitized planes, typically one or two layers thick, with a thickness of less than 2 nm and transverse dimensions of less than 10 nm [6–8]. GQDs feature a variety of novel characteristics, including their distinct fluorescent abilities. Numerous investigations have shown that the photoluminescence excitation and emission wavelengths may be altered by changing the size, shape, or dopant concentration of the photoluminescent materials. Furthermore, there are a large number of oxygen-rich functional groups at their edges, which adds to their

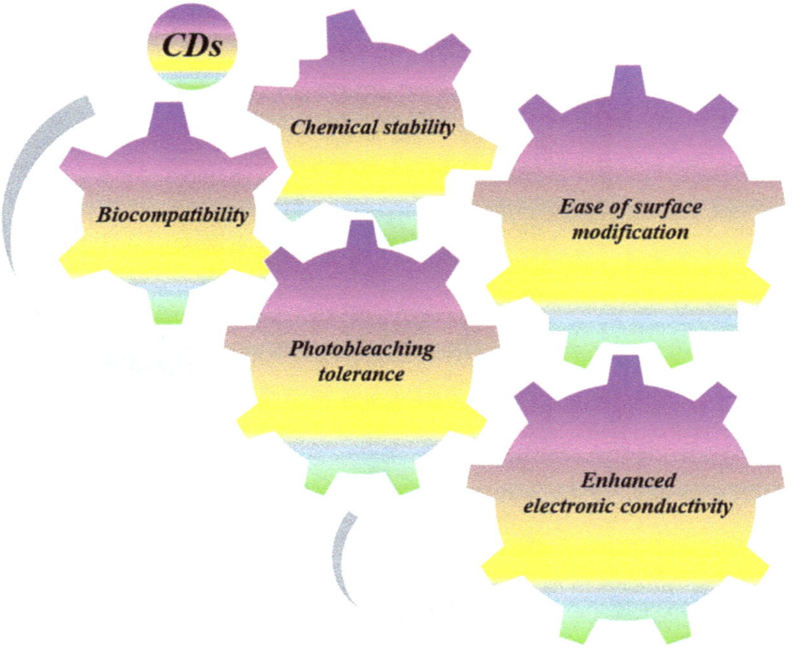


**Figure 9.2.** Interesting properties of CDs for nanomedicine.

excellent hydrophilic nature and biocompatibility, allowing them to be used in biomarker analysis for the detection of serious diseases [9].

## 9.2 Central nervous system (CNS) diseases

### 9.2.1 An ageing population and neurodegenerative diseases

Neurological disorders are a frequent and increasing cause of mortality and intellectual disability worldwide, in particular in the aged population, and represent the biggest health and social care issue of the twenty-first century [10]. In recent years, the fast increase of elderly populations across the globe, particularly in industrialized nations, has resulted in a substantial increase in the incidence of different age-related disorders, notably neurodegenerative diseases linked with cognitive impairment. In 2050, one in six people on the globe will be over the age of 65, increasing from one in eleven in 2019 [11], and ageing will be the main risk factor for most neurological illnesses, such as stroke, Alzheimer's disease (AD), and Parkinson's disease (PD), as the prevalence of these disorders grows rapidly with age [12].

The progressive degeneration of preferentially susceptible groups of neurons is related to many basic aspects involved with progressing neuronal dysfunction and mortality in neurodegenerative disorders, including proteotoxic stress and related abnormalities of the ubiquitin–proteasomic and autophagosomal/lysosomal systems, oxidative stress, programmed cell mortality, and neuroinflammation [13]. Individual neurodegenerative diseases differ in their clinical presentations and underlying physiology, despite the fact that they often share characteristics, therefore accurate diagnosis is essential, since it allows for more accurate prognosis and often directs particular therapy and care [14].

### 9.2.2 Challenges in the diagnosis of CNS diseases

The CNS diseases have a clinical presentation through characteristic symptoms and behaviors, from neuroinfections to neurocognitive disorders. However, these syndromic manifestations often occur a considerable period after the onset of the disease, mainly in neurodegenerative disorders such as PD and AD [15, 16]. Research has provided important advances in the therapeutic approaches adopted to treat these diseases but much still needs to be done. In part this is due to the challenge of developing specific diagnostic tools for CNS diseases that provide early diagnosis, detection of specificities, continuous monitoring of disease evolution, and early and appropriate therapeutic interventions for each clinical condition [15, 17]. The lack of precise biomarkers for early detection and the diversity of CNS disorders provides an opportunity to develop these tools [16, 17]. To face the challenges of crossing the blood–brain barrier (BBB) and other limits on the transportation of drugs into the brain, nanoparticles are being investigated as promising vehicles for transporting drugs which act on CNS disorders. These particles appear to provide a strategy for conventional and systemic drug administration systems, such as the oral and gastrointestinal routes, which currently have compromised efficacy and considerable side effects for these diseases [17, 18]. Early diagnosis and treatment may substantially enhance the quality of care of affected patients with neurodegenerative disorders. However, the diagnostic tools used are still not sensitive and specific enough for early action to be taken, meaning the medical therapies adopted are focused only on minimizing symptoms and in addition have many side effects. Thus nanotechnology can help because it is a non-invasive tool, which can be personalized and directed to the specific area that the drug aims to reach [18, 19].

### 9.2.3 Challenges of drug delivery in the CNS

CNS neurons need control so that they can perform a specific function. This control is performed by the BBB created by the endothelial cells that line the cerebral blood vessels. The endothelium regulates molecular entry and outflow via the BBB, as seen in figure 9.3. The BBB protects the brain but hinders drug delivery. [20]. Many medicines do not cross the BBB and therefore compromise the therapy of CNS disorders such as brain tumors, strokes, and neuropathy such as PD and AD [20, 21].

The permeability of the BBB has been studied widely, as it can provide important insights into the mechanisms by which different treatments and therapeutic approaches cross this barrier and the main challenges that exist in this traffic [20, 22]. However, even with many studies having been carried out, we still do not know what the best method is for a given therapy to permeate the BBB. In practice, the treatment of certain diseases of the CNS has its effectiveness compromised by oral or systemic administrations due to the difficulty of penetration of these therapies through the BBB [22, 23].

Faced with the problem of BBB permeability in relation to the efficacy of pharmacotherapies used to treat disorders of the CNS, studies with nanoparticles are

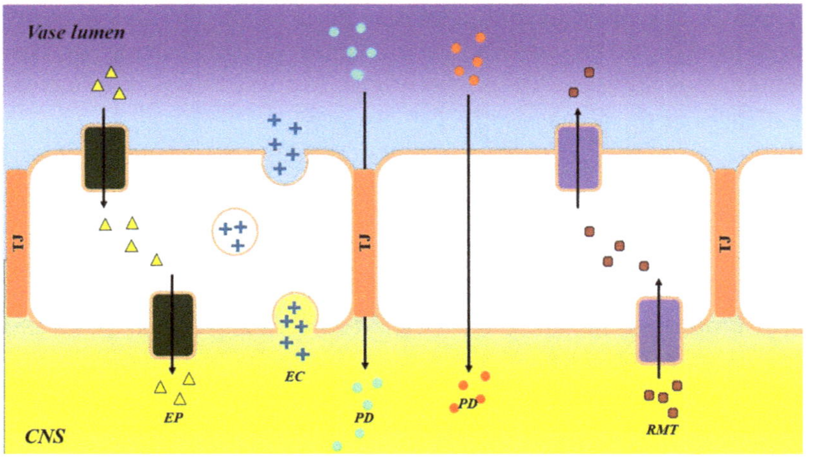


**Figure 9.3.** The passage of molecules to the CNS tissue is tightly regulated by the multiple transport systems present in the BBB, such as efflux pumps (EP), endocytosis mechanisms (EC), passive diffusion (PD) of small molecules, and receptor-mediated transports (RMT).

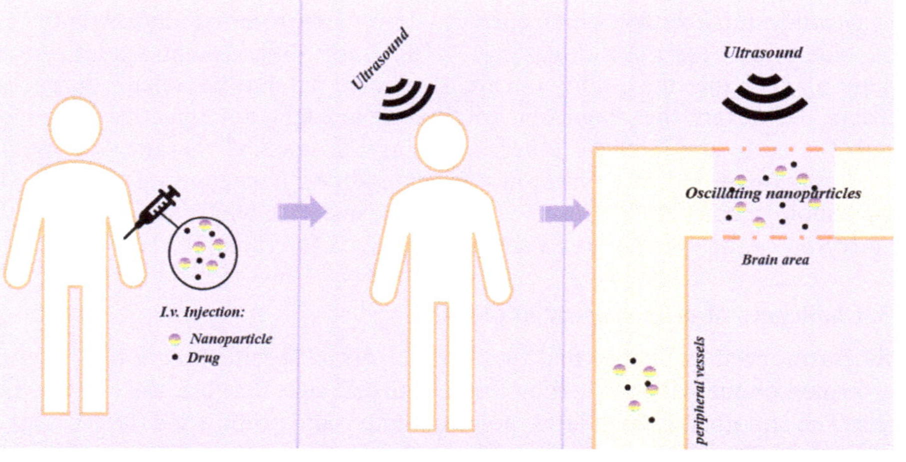


**Figure 9.4.** Co-administration of drugs with nanoparticles is an effective way to delivery substances to the brain, once it can cross the BBB and reach the target area, making the drug or substance carried by the nanoparticle able to reach the desired site of action. These proprieties of nanoparticles can be effectively used in the improvement of therapy delivery into the brain or maximizing the effectiveness and sensitivity of diagnostic imaging methods for CNS, like ultrasound, or even replacing invasive techniques diagnostic methods.

presented as a promising strategy to solve this issue, as they have been shown to help in the effectiveness of CNS therapies [23, 24]. The advantages of using nanoparticles are many [25]. They are exceedingly small particles, capable of crossing BBB capillaries and reaching the target cell, making the drug, carried by the nanoparticle, able to reach the desired site of action [25, 26], as seen in figure 9.4. Furthermore, nanoparticles are biocompatible, can be biodegradable, and can achieve therapeutic efficacy with fewer side effects [26, 27]. However, it is important to highlight that nanotechnology still requires more robust research to prove its safety before its widespread use [26–28].

## 9.3 Zero-dimensional carbon nanomaterials and diagnosis of CNS diseases

### 9.3.1 Zero-dimensional carbon nanomaterials and diagnosis in Parkinson's disease

Parkinson's disease is a complex, multifactorial and chronic neurodegenerative disease whose neuropathologic hallmarks are the progressive dopaminergic degeneration of substantia nigra leading to dopamine (DA) depletion, and aggregates of non-soluble intraneural proteins, such as $\alpha$-synuclein and ubiquitin, called Lewy bodies [29, 30]. The pathophysiology of PD also involves several molecular and cellular changes that include oxidative stress, excitotoxicity, neuroinflammation, pro-apoptotic mechanisms, ubiquitin–proteasome system impairment, and mitochondrial dysfunction [31–33]. However, the exact pathogenesis of PD is not yet fully known.

PD is the second most common age-related degenerative neurological disorder in humans, second only to AD, affecting 1% of the global population above the age of 60 years and 5% above the age of 85 years [34]. Its rate ranges from 5 to 35 cases per 100 000 annually [35]. The Global Burden of Disease Study estimated that the burden of PD more than doubled between 1990 and 2017; in 2017 there were more than seven million individuals with PD worldwide [36], and the expectation is that this number will increase dramatically.

Clinically PD is characterized by a broad spectrum of progressive motor symptoms, such as resting tremors, akinesia, muscle rigidity, instability, postural abnormalities, and acinesia [37], and non-motor symptoms, such as constipation, hyposmia, REM (rapid eye movement) sleep disorders, anxiety, depression, fatigue, hypotension, sexual dysfunction, attention and memory disorders, dementia, and mood disorders [38–40]. The clinical diagnosis of PD is primarily based on motor features, however, these manifest only after the loss of about 70%–80% of the nervous system (NS) dopaminergic neurons [41]. The motor symptoms of PD are preceded by a prodromal phase characterized by extensive systemic changes in which neurodegeneration spreads through the central nervous system, and by nonspecific symptoms that can precede the motor diagnostic by 10–20 years, although which are insufficient to be used as a framework for a diagnosis [42]. However, these changes indicate fertile ground for the study of biomarkers that can help identify early high-risk individuals and help in an accurate diagnosis. Many experienced and skilled neurologists and clinicians still dispute the early diagnosis of PD. Currently PD's diagnostic is based on a history of prodromal features associated with examination of the presence of symptoms such as motor abnormalities, psychological or cognitive problems, and single-photon emission computed tomography (SPECT) for DA transporters, but due to its heterogeneity it still remains challenging, slowing the time and accuracy of PD diagnosis [43].

Furthermore, in order to diagnose PD, a distinct molecular biomarker of PD pathogenesis to enhance its evaluation, early identification, and administration is essential [44]. In recent decades, nanotechnology has evolved quickly and may offer significant potential benefits for PD diagnosis and treatment, such as developing methods for the detection of closely linked PD pathophysiology, such as $\alpha$-synuclein abnormalities, and DA and mitochondrial failure [18].

With this purpose gold (Au) nanoparticles encapsulated in G4-polyamidoamine dendrimer were used in the development of a electrochemical immunosensor used for $\alpha$-synuclein, with high sensitivity and stability with a threshold of detection of 14.6 pg $mL^{-1}$ [45]. Au-doped titanium dioxide ($TiO_2$) nanotube arrays were also used in the development of a immunosensor designed for photoelectrochemical detection of $\alpha$-synuclein with a high-sensitivity limit of detection of 34 pg $mL^{-1}$ [46]. Also to identify aggregated $\alpha$-synuclein in cerebrospinal fluid (CSF), Kobayashi, Sawamura, and Sohgawa built a sensor based on liposomes immobilized in a cantilever system associated with a real-time quaking-induced conversion technique with sensitivity up to 7 pM [47].

Regarding the detection of DA changes, an arrangements of vertically aligned zinc oxide prepared from graphene lather nanowire arrays is a promising method with high sensitivity and specificity to detect changes in DA levels in the serum of PD patients with a detection limit of 1 nM [48]. Likewise, a carbon paste electrode associated with silver nanoparticles was developed for the quantification of DA levels with high sensitivity and selectivity, with detection limits of $3.0 \times 10^{-7}$ M [49]. Pandey *et al* also developed a portable and flexible electronic CD@graphite paper-based sensor for electrochemical sensing of DA levels with a detection threshold of 0.015 $\mu$M, providing a simple tool for the determination of DA levels with sensitivity, selectivity, operational stability, and robustness [50]. Additionally, Varela *et al* developed a single molecule quantum dot-based tracking platform to label and track DA receptors in rat brain slices, and was able to reveal the behavior and activity of these receptors, opening doors for future nanoscale strategies for accurate assessment of the dopaminergic receptor system [51].

Aiming to identify signs of mitochondrial failure, a complex using cadmium (Cd), selenium (Se), zinc (Zn), and sulfur (S) quantum dots (QDs) (CdSe/ZnS@QDs) was developed for the detection of complex I levels of the mitochondrial electron transport chain, which is affected in PD patients. The QDs were extremely sensitive to changes in electron transfer and their fluorescence intensity change reflected the levels of complex I with a detection range of 0.02 U to 4.9 U [52]. Likewise, a new proposal for early diagnosis of PD was tested in an innovative clinical trial, where Feinberg *et al* used an electronic sensor array system to detect volatile molecules in exhaled breath. This was a 40 cross-reactive sensor based on single walled CNT, used to test samples of exhaled breath before medication in association with investigation of other average diagnostic tests for PD, such as ultrasound, smell tests, and the detection of non-motor symptoms, with results of 79% sensitivity, 84% specificity, and 81% accuracy [53]. The simplicity of the method associated with early diagnosis is an example of what is expected as the next step in the field of diagnosis and the main objective of using nanotechnology for PD.

### 9.3.2 Zero-dimensional carbon nanomaterials and diagnosis in Alzheimer's disease

AD is a predominant neuronal degeneration illness and a major cause of dementia; the population of AD sufferers is projected to double about every 20 years [54]. Globally, according to the World Alzheimer's Disease Survey 2019, the number of those suffering with AD is expected to reach 152 million by the year 2050 [55]. The pathophysiology of AD can be distinguished primarily by the depletion of amyloid $\beta$ (A$\beta$) protein in intracellular aggregates in the cerebrum tissue, which is accompanied by neuronal loss and enhanced phosphorylation and secretion of tau, which further occurs long before the onset of clinical symptoms [56, 57]. Intraneuronal neurofibrillary tangles (NFTs) are formed as a result of abnormal tau metabolism, increasing the likelihood of degeneration [58]. Memory and spatial attention impairments, disability in speech, difficulty solving problems, and a loss in people's understanding are all typical signs of AD. Other traditional symptoms of the disease include depression, hallucination, delusion, anomic speech, acalculia, and apathy, among others [59–61]. Sporadic AD is associated with aging and is characterized as one of the most widely accepted forms of the disease, defined by a cumulative complicated interaction between A$\beta$ accumulation, vessel modification, microglia and astrocyte stimulation, other pathologies (neural $\alpha$-synuclein and transactive DNA-bound protein 43 (TDP-43) inclusions), genetic susceptibility of the $\varepsilon$4 allele of the apolipoprotein E genetic factor (APOE4), and conditions that affect brain safety during growth [62–64].

Mutation in three genes was also recognized to be a major contributor to AD, such as chromosomal abnormality 21 and the amyloid precursor protein (APP) genetic factor, present in Down's syndrome, which result in the aggregation of A$\beta$, the development of NFTs, and the early onset of AD [65]. Studies in individuals with autosomal recessive gene modifications in the APP or presenilin 1 or 2 genetic factors have also revealed that pathophysiological changes occurred more than 20 years before the start of illness symptoms [66]. Reduced levels of 42 residue A$\beta$ isoform (A$\beta$42) and a greater concentration of CSF is also shown to be correlated with dementia symptoms after 3–4 years, as high A$\beta$42 in the blood and low A$\beta$42 in the CSF lead to amyloid aggregation in the brain and an increased chance of AD [67, 68].

Research has revealed that A$\beta$42, the total of tau protein and the abnormal phosphorylation of tau protein may serve as AD biomarkers in the CSF [69–71]. Nuclear medical imaging and brain nuclear magnetic resonance imaging, together with certain molecular specimens, represent a promising method for determining the levels of A$\beta$42 and tau protein [72–75]. The use of nanotechnology in the production of samples may enhance the resolution and imaging sensitivity of MRI and PET methods, and improve the detecting specificity of certain disorders, for example AD diagnosis. Polymorphic gadolinium (AGuIX) nanoparticles $< 5$ nm in diameter, were utilized to enhance MIR sensitivity [76]. AguIX nanoparticles in combination with Pittsburgh Compound B are safe and are able to transport A$\beta$ add-ons inside the BBB, enabling a good imaging of A$\beta$ aggregates an *in vivo* protocol for AD, without adverse effects [76].

A sensitive and distinct biomarker would be extremely effective in the development of the diagnosis of AD. Interactions among carbon nanomaterials and A$\beta$ may be utilized as novel AD sensors, thus early diagnosis may improve the success of medication therapy because no effective treatment is currently available [77]. In the study developed by Bordur *et al* in 2021 a new acetylcholine biosensor amperometric neurotransmitter was developed using a CD modified carbon paste electrode, where carbon paste based on nanodots-(3-aminopropyl) triethoxysilane (CDs-APTES) was prepared using the enzymes acetylcholine esterase (AChE) and choline acetyltransferase [78]. The acetylcholine biosensor developed in this study permits the determination of acetylcholine at low and multiple concentrations due to the high operating scope and low boundary of sensing. The acetylcholine biosensor is repeatable and is easy to produce in the laboratory [78].

The research by Huang *et al* from 2017 suggested a technique to determine the quantity of amyloid peptide (A) monomers in solution by using the fluorescence properties of GQDs. The comparable A$\beta$ aggregation dynamical observation utilizing GQDs and thioflavin T has shown that the GQD-based identification procedure is feasible without a co-incubation using soluble amyloid peptide monomers. GQDs have been demonstrated for use as a novel diagnostic for monitoring amyloid monomers in the identification and pathologic diagnosis of degenerative and other structural conditions [79].

### 9.3.3 Zero-dimensional carbon nanomaterials and diagnosis in stroke

Stroke is one of the world's major causes of mortality and morbidity and accounts for 13.5% of the world's mortality of people over 70 years, with a projection of a 27% of rise in stroke between 2017 and 2047 [80, 81]. A rapid reduction in brain tissue blood flow may lead to symptoms such as weakness in the face, legs, and arms, disorientation, and aphasia. The reduced brain blood flow signs and symptoms result in a prognosis that can vary from the spectrum of full patient recovery to serious neurological deficits and death [82]. Two methods of neuroimaging are used regularly to diagnose and monitor a patient's susceptible to acute ischemic stroke: x-ray computed tomography (CT) and MRI [83]. Despite CT being widely accessible and cheaper, it is not responsive to ischemic stroke and new ischemic events. Less than one-third of cerebral ischemia patients display CT features 3 h after the onset of symptoms [84].

The presence of stroke-like indications, particularly in different non-vascular disorders, including epileptic seizures and migraine, provides the possibility of misdiagnosis and incorrect application of thrombolytic therapy [85]. In this respect, significant advantages in the diagnosis and treatment of stroke may be offered by designing adapted nanoparticles that might enhance current imaging methods with improved efficacy, selectivity, and temporal and spatial resolution [86]. The unique characteristics of nanoparticles are also being integrated into future techniques of contrast agents and diagnostic equipment for stroke [87].

Nanoparticles have been shown to be effective imaging agents with no negative side effects. Winter *et al* showed that a fibrin-targeted paramagnetic nanoparticle formulation demonstrates improvement in fibrin-specific molecular imaging compared to previous techniques, improving the capacity for the detection and quantification of microthrombus [88]. Likewise, iron oxide nanoparticles have demonstrated a high efficiency in the MRI of inflammation in both experimental ischemia and human stroke [89]. Regarding the diagnosis of ischemic stroke, nanomaterials can be used in the construction of devices with high sensitivity and selectivity, some reaching a 1 $\mu$M detection limit, and have the potential to serve as a screening tool or as key components in the construction of sensors for a range of biomolecules (from reactive oxygen species (ROS) to neurotransmitters), that can be used for the investigation of the pathophysiological mechanisms of stroke and follow-up of post-stroke patients for pathophysiological improvements [86, 90]. There has also been much interest in designing nanoparticles capable of aiding diagnosis through damaged and normal BBBs and targeting ischemia tissue as well as the penumbra area of the affected tissue [91, 92].

## 9.4 Zero-dimensional carbon nanomaterials in drug delivery in CNS diseases

### 9.4.1 Zero-dimensional carbon nanomaterials in drug delivery in Parkinson's disease

Current therapies for PD aim only at symptoms control, increasing DA levels in the brain or mimicking its effects, such as levodopa (L-DOPA), which provides only a temporary solution, since progressive neuronal degeneration persists and complications resulting from the dopamine based treatment emerge, such as clinical fluctuations, shortening or inconsistency of its effects, and the development of L-DOPA-induced dyskinesia about 5 years after the start of treatment [93–96]. Surgical interventions, such as deep brain stimulation (DBS), has still been an option for advanced cases and those resistant to pharmacological treatment, however, these interventions have a high cost and limited time of effectiveness [97, 98]. Accordingly, numerous efforts have been made by several research centers in an attempt to discover new neuroprotective strategies that can be used to inhibit the advance of PD, but most novel drugs developed have a low success rate due to unsatisfactory drug metabolism, nonspecific targeting, and the possibility of increased adverse effects [18]. Nanotechnology can also be applied in PD treatment improvement, since nanomedicine tools present the ability to cross the BBB, and increase specificity, bioavailability, efficacy, and safety [99]. Due to these much desired properties, it has been widely tested.

In a computational study using classical molecular dynamics and density functional theory, Frazão *et al* demonstrated that $C_{60}$ fullerene seems to be a good delivery system for the L-DOPA state with $COO^-$ and $NH_3^+$ protonated (LDOPAc), in an association called LDOPAc@$C_{60}$, for oral and transdermal drug delivery for the treatment of PD, and may improve its bioavailability at the local of action, prolonging DA stimulation and decreasing side effects [100].

Corona-Morales *et al* showed that zero-dimensional carbon nanomaterials can be used to prevent the neurotoxicity generated by L-DOPA treatment, using an association between fullerene $C_{60}$ and ascorbic acid to protect adrenal chromaffin cells from L-DOPA toxicity, preventing cell death and apoptosis [101]. This suggests a new perspective of strategies for L-DOPA clinical use, where its association with compounds such as fullerene and other antioxidant compounds can prevent its toxicity from long term clinical use, such as dyskinesia.

The efficient delivery of neuroprotective agents is another use of nanomedicine in the search for new treatments for PD. A study using a combination of a vascular endothelial growth factor and a neurotrophic factor in lactic-co-glycolic acid nanospheres showed a significant protective effect from dopaminergic degeneration *in vitro* and *in vivo*, [102]. Similarly Dugan *et al* showed that the use of carboxy-fullerenes effectively inhibited lipid peroxidation by the elimination of superoxide anions and $H_2O_2$, protected cortical cell cultures against excitotoxicity and apoptosis, and reversed mesencephalic dopaminergic neuron degeneration induced from both 1-methyl-4-phenylpyridinium (MPP1) and 6-hydroxydopamine [103]. Likewise, Cai *et al* revealed that polyhydroxylated fullerene protected derivative human neuroblastoma cells from death, prevented mitochondrial dysfunction by protecting complex I and II activity, reduced oxidative damage, decreased ROS levels, and increased the activity and expression of glutathione and g-glutamyl cysteine ligase levels in a MPP1-induced PD model [104]. Additionally, Jin *et al* showed that fullerenols protected primary neuron cultures from glutamate-induced neurotoxicity by about 80% through the inhibition of glutamate receptors and the reduction of $Ca^{2+}$ levels [105].

Efforts in employment of nanotechnology on targeting one of the most significant neuropathologic hallmarks of PD, $\alpha$-synuclein aggregates, has also been made. Kim *et al* used GQDs to attack aggregated $\alpha$-synuclein that interacted directly with mature fibrils, triggering their disaggregation and preventing its fibrilization in *in vitro* and *in vivo* assays, and protecting animals from dopaminergic degeneration, behavioral deficits and Lewy body formation in a mouse model of $\alpha$-synuclein preformed fibril [106].

### 9.4.2 Zero-dimensional carbon nanomaterials in drug delivery in Alzheimer's disease

Known for being a slowly progressing neurodegenerative disease and the accumulation of amyloid fibers, AD has no known effective treatment [107]. Existing approaches for the treatment of AD include therapy with the noncompetitive *N*-methyl-D-aspartate (NMDA) receptor antagonist memantine, increasing the acetylcholine concentration in the cerebrum with acetylcholinesterase inhibitors (i.e. donepezil, galantamine, and rivastigmine), which promotes a greater amount of the neurotransmitter acetylcholine in the synapses, non-steroidal anti-inflammatory factors, antioxidants, $A\beta$ fibrillation inhibitors, and hormone replacement therapy [108, 109].

Additionally, modified nanomaterials have the potential to be used during monitoring and diagnosis of neurodegenerative illnesses such as AD [57]. The intervention of carbon nanomaterials with $A\beta$ organization was evidenced by recent observations in cellular and animal models, as well as molecular modeling studies, with the main results being their inhibition, the defeat of A aggregation, and the subsequent formation of amyloid fibrils when combined with appropriate radical and peptide scavenging frameworks [77, 110–112]. These findings give rise to the possibility that, in the future, the connection between carbon nanomaterials and A$\beta$ peptides might be utilized to develop intelligent nanomaterials to recognize and eliminate the causes of Alzheimer's at various stages, thus substantially and directly affecting the quality of life of patients.

GQDs are a relatively new addition to the collection of carbon-based nanomaterials that were able to show benefits in the progress of AD treatment [113]. A study utilizing hen egg white lysozyme as a pattern protein found that interfacial charge delivery with graphene after exfoliation reduced fibril production in amyloid lysozyme. This was shown to be due to both charge transport and stacking interactions involving graphene and aromatic residues [114].

Polymerized o-phenylenediamine-derived CTs were able to solubilize Cu(II) ions, prevent amyloid accumulation, photo-oxygenate amyloid peptides, and also prevent the oxidizing deterioration of neurons [115]. Chrysin-loaded PEGylated magnetic silica nanospheres, owing to their antioxidant capabilities in a cellular amyloidogenous milieu, enhanced protection against the oxidant stresses produced by accumulation [116]. Recent studies also indicate that carbon nanomaterials' redox reactions in cellular and animal models are proven to be helpful as scavengers for ROS problems in ROS-associated degenerative diseases, as seen in AD [117].

A recent study showed that CDs produced from brown coal were able to suppress islet amyloid protein and A$\beta$ amyloidogenesis, indicating a possible therapeutic function for type 2 diabetes and AD [118]. Guerrero *et al* [119] studied the capacity to use Na-citrate-derived CDs in an amyloid fibril phase transformation using hen lysozyme, obtained from egg white, as an amyloidogenic protein pattern and the results showed that the Na-citrate CDs could be used with a safe content of 2.5 mg $ml^{-1}$ and that hen egg white lysozyme amyloidogenesis could be prevented by CDs (if not up to 5 mg mL).

Chung *et al* reported recently that photoluminescent carbon nanodots (CNDs) can eventually prevent the self-assembly of Alzheimer's A$\beta$ by disintegrating A$\beta$ aggregates and reducing neurotoxicity using biocompatible, visible-light active CNDs with ammonium citrate as a source of carbon and by passivating them into branched polyethylene as a $\beta$-sheet deaggregator [120]. Chung *et al* demonstrated that red-light-responsive, A$\beta$-targeted CNDs could be engineered to inhibit A$\beta$ aggregation and neurotoxicity in a spatiotemporal pattern and to lower the total brain A$\beta$ load. Apta@CNDs produced by the attachment of A$\beta$ aptamers (Apta) to CNDs shown an excellent capacity to target amyloid species, indeed in complicated neurophysiological conditions. Apta@CNDs below red light showed photomodulating actions over A$\beta$ aggregation and oxidation of contiguous amyloid species, in addition to successfully inhibiting the development of neurotoxicity $\beta$ leaf abundant amyloid amassment. The anti-amyloidogenic functions of Apta@CNDs in a living organism are also shown by employing the photomodulation technique on five familial AD mutation mouse brains. Moreover, A$\beta$-oriented Apta@CNDs have shown their therapeutic potential *in vivo* as an efficient nano-agent for aberrant amyloid loads [121].

### 9.4.3 Zero-dimensional carbon nanomaterials in drug delivery in stroke

Stroke therapy is still mainly focused on reperfusion, restoring blood flow to the brain, using intravenous thrombolysis through recombinant tissue plasminogen activator, or/and endovascular rehabilitation with mechanical thrombectomy based on the hours since stroke symptom onset and the size and level of artery occlusion, yet often with a lack of success which continues to stimulate the search for more effective treatments [86, 122].

The use of carbon nanoparticles and related technology has also been shown to be an improvement in stroke treatment. The efficacy of such nanoparticles was shown in providing the tissue plasminogen stimulator to the thrombin with the absence of adverse side effects, including the intracerebral bleeding associated with exogenous plasminogen activators [123]. Its benefits include its preferred position in thrombi development, efficient thrombolysis, and improved safety due to a significant decrease of the fibrinolytic medication dose and a downstream reduction of side effects [88, 123–125]. The transport of neuroprotective medicines, antioxidant enzymes, or anti-inflammatory medications to the infarction zone is another potential use for nanoparticles [87]. Vani *et al* demonstrated that the administration of fullerene nanoparticles is able to reduce the infarct area by 80%, behaving as a scavenger of ROS, defending the cells in the brain against ischemia/reperfusion injury, and inhibiting the brain damage caused by oxidative stress in a rodent model of stroke [126]. Likewise, Fluri *et al* combined the antioxidant power of fullerene with the anti-inflammatory properties of glucosamine in a glucosamine-fullerene conjugate (GlcN-F), which has been shown to be able to reduce the infarction area, pro-inflammatory cytokine profile, and neurological performance in an animal model of stroke associated with hypertension [127]. Yang *et al* found similar results with the chronic administration of hexasulfonated $C_{60}$ (FC4S), indicating that these nanocompounds can possibly be used as a treatment for stroke and protect the brain against cerebral ischemia [128].

## 9.5 The use of zero-dimensional carbon nanomaterials for CNS diseases: perspectives

Zero-dimensional nanoparticles are characterized by a small volume and by being biocompatible [9]. In this way, they can act as biosensors and transport vehicles for the pharmacotherapy of CNS diseases, improving the effectiveness of medical therapies and aiding in the detection and diagnosis of these diseases. Even with the large number of recent studies on these materials, research with nanoparticles needs to go a long way to produce more robust and safer clinical implications [9, 129]. Carbon-based nanoparticles are an interesting alternative to CNS, as carbon is a good electrical conductor and has good interaction with biosystems, in particular with CNS, due to its ability to modulate neuronal electrical activity [129]. In addition, it has antiviral properties, further attracting the focus of research to nanotechnology [130]. However, the health risks for carbon nanoparticles to be used in clinical practice need to be clarified [130, 131]. Despite the success of these materials, in particular in the diagnosis of CNS diseases, their toxicity needs to be studied more widely. The growing complexity of animal models and the small amount of research that evaluates the biotoxicity and biodegradation related to these particles makes their use more challenging in clinical practice [129, 131]. With the immense potential of applying these types of nanoparticles, it is essential to develop research that allows for the expansion of nanotechnology for clinical research and the subsequent practical application of these tools [131].

## 9.6 Conclusion

Zero-dimensional carbon nanomaterials present immense possibilities in the field of new diagnostic and therapeutic technologies to diagnose, treat, and monitor CNS diseases, as seen in the applications for Parkinson's disease, Alzheimer's disease, and stroke. Their ability to cross the BBB, antioxidant properties, inherent photoluminescence, ability to improve the effectiveness of medical therapies, and early and precise detection of pathophysiology markers of those diseases highlight this potential. There is a need for more investment in this research field, in order to learn more about the potential of zero-dimensional nanomaterials for brain medicine, and medicine in general, as well as the health risks and environmental implications for its use in clinical practice. We believe in the future growth prospects in the field of 0D nanomaterials, and their major role in biosensing and point-of-care development

## Acknowledgments

We would like to thank everyone involved in carrying out, organizing, and writing this chapter. We also thank our professors Dr Geanne Matos de Andrade and Dr Reinaldo Barreto Oriá for their precious teachings. We are very grateful to CAPES and CNPq, our main Brazilian research funding agencies, for their financial support and for believing in the importance of science for the country's development, given the current obscurantism present in our country; it is our goal to persist in spreading the light of science of our country.

## References

[1] Han X, Li S, Peng Z and Al-yuobi A O 2016 Interactions between carbon nanomaterials and biomolecules *J. Oleo Sci.* **6** 1–7

[2] Henna T K and Pramod K 2020 Graphene quantum dots redefine nanobiomedicine *Mater. Sci. Eng.* C **110** 2–21

[3] Semeniuk M, Yi Z, Poursorkhabi V, Tjong J, Lu Z and Sain M. 2019 Future perspectives and review on organic carbon dots in electronic applications *ACS Nano.* **13** 6224–55

[4] Wang X, Feng Y, Dong P and Huang J. 2019 A mini review on carbon quantum dots: preparation, properties, and electrocatalytic application *Front. Chem.* **7** 671

[5] Yang S *et al* 2014 Large-scale fabrication of heavy doped carbon quantum dots with tunable-photoluminescence and sensitive fluorescence detection *J. Mater. Chem.* A **2** 8660–7

[6] Ponomarenko L A *et al* 2008 Chaotic Dirac billiard in graphene quantum dots *Science* **320** 356–8

[7] Xu X *et al* 2004 Electrophoretic analysis and purification of fluorescent single-walled carbon nanotube fragments *J. Am. Chem. Soc.* **126** 12736–7

[8] Tian P, Tang L, Teng K S and Lau S P 2018 Graphene quantum dots from chemistry to applications *Mater. Today Chem.* **10** 221–58

[9] Wang Z, Hu T, Liang R and Wei M 2020 Application of zero-dimensional nanomaterials in biosensing *Front. Chem.* **8** 320

[10] Si Z, Sun L and Wang X 2021 Evidence and perspectives of cell senescence in neurodegenerative diseases *Biomed. Pharmacother.* **137** 2–11

[11] UN Department of Social and Economic Affairs 2019 *World Population Ageing 2019: Highlights* (New York: United Nations) p 46

[12] Hou Y *et al* 2019 Ageing as a risk factor for neurodegenerative disease *Nat. Rev. Neurol.* **15** 565–81

[13] Dugger B N and Dickson D W 2017 Pathology of neurodegenerative diseases *Cold Spring Harb. Perspect. Biol.* **9** a028035

[14] Erkkinen M G, Kim M and Geschwind M D 2018 Clinical neurology and epidemiology of the major neurodegenerative diseases *Cold Spring Harb. Perspect. Biol.* **10** a033118

[15] Kaushik A, Jayant R D, Bhardwaj V and Nair M 2019 Personalized nanomedicine for CNS diseases *Drug Discov. Today* **23** 1007–15

[16] Brambilla D *et al* 2011 Nanotechnologies for Alzheimer's disease : diagnosis, therapy, and safety issues *Nanomed. Nanotechnol. Biol. Med.* **7** 521–40

[17] Zhang X, Zhou J, Gu Z, Zhang H, Gong Q and Luo K 2020 Advances in nanomedicines for diagnosis of central nervous system disorders *Biomaterials* **269** 120492

[18] Zhang W, Wang W, Yu D X, Xiao Z and He Z 2018 Application of nanodiagnostics and nanotherapy to CNS diseases *Nanomedicine* **13** 2341–71

[19] Shaw L M, Korecka M, Clark C M and Trojanowski J Q 2007 Biomarkers of neurodegeneration for diagnosis and monitoring therapeutics *Nat. Rev. Drug Discov.* **6** 295–303

[20] Abbott N J 2013 Blood–brain barrier structure and function and the challenges for CNS drug delivery *J. Inherit. Metab. Dis.* **36** 437–49

[21] Bors L A and Erd F 2019 Overcoming the blood–brain barrier. Challenges and tricks for CNS drug delivery *Sci. Pharm.* **87** 6

[22] Serwer L P and James C D 2012 Challenges in drug delivery to tumors of the central nervous system: an overview of pharmacological and surgical considerations *Adv. Drug Deliv. Rev.* **64** 590–7

[23] Barchet T M and Amiji M M 2020 Challenges and opportunities in CNS delivery of therapeutics for neurodegenerative diseases *Expert Opin. Drug Deliv.* **6** 211–25

[24] Akhtar A, Andleeb A, Waris T S, Moradi A, Awan N R and Yar M 2020 Neurodegenerative diseases and effective drug delivery: a review of challenges and novel therapeutics *J. Control. Release* **20** 1152–67

[25] Patel T, Zhou J, Piepmeier J M and Saltzman W M 2013 Polymeric nanoparticles for drug delivery to the central nervous system *Adv. Drug Deliv. Rev.* **64** 701–5

[26] Stockwell J, Abdi N, Lu X, Maheshwari O and Taghibiglou C 2014 Novel central nervous system drug delivery systems *Chem. Biol. Drug Des.* **2014** 507–20

[27] Karami Z, Reza M, Zanjani S and Hamidi M 2019 Nanoemulsions in CNS drug delivery: recent developments, impacts and challenges *Drug Discov. Today.* **24** 1104–15

[28] Vashist A *et al* 2019 Nanogels as potential drug nanocarriers for CNS drug delivery *Drug Discov. Today.* **23** 1436–43

[29] Wirdefeldt K, Adami H-O, Cole P, Trichopoulos D and Mandel J 2011 Epidemiology and etiology of Parkinson's disease: a review of the evidence *Eur. J. Epidemiol.* **26** 1–58

[30] Spillantini M G, Schmidt M L, Lee V M, Trojanowski J Q, Jakes R and Goedert M 1997 Alpha-synuclein in Lewy bodies *Nature* **388** 839–40
[31] Jost W H and Reichmann H 2017 'An essay on the shaking palsy' 200 years old *J. Neural. Transm.* **124** 899–900
[32] Pereira D and Garrett C 2010 Risk factors for Parkinson disease: an epidemiologic study *Acta Med. Port.* **23** 15–24
[33] Dextera D T and Jenner P 2013 Parkinson disease: from pathology to molecular disease mechanisms *Free Radic. Biol. Med.* **62** 132–44
[34] Dumurgier J and Tzourio C 2020 Epidemiology of neurological diseases in older adults *Rev. Neurol.* **176** 642–8
[35] Simon D K, Tanner C M and Brundin P 2020 Parkinson disease epidemiology, pathology, genetics, and pathophysiology *Clin. Geriatr. Med.* **36** 1–12
[36] Deuschl G *et al* 2020 The burden of neurological diseases in Europe: an analysis for the Global Burden of Disease Study 2017 *Lancet Public Heal.* **5** e551–67
[37] Obeso J A *et al* 2017 Past, present, and future of Parkinson's disease: a special essay on the 200th Anniversary of the Shaking Palsy *Mov. Disord.* **32** 1264–310
[38] Schrag A, Horsfall L, Walters K, Noyce A and Petersen I 2015 Prediagnostic presentations of Parkinson's disease in primary care: a case-control study *Lancet Neurol.* **14** 57–64
[39] Postuma R B and Berg D 2019 Prodromal Parkinson's disease: the decade past, the decade to come *Mov. Disord.* **34** 665–75
[40] Schapira A H V, Chaudhuri K R and Jenner P 2017 Non-motor features of Parkinson disease *Nat. Rev. Neurosci.* **18** 435–50
[41] Fereshtehnejad S M, Yao C, Pelletier A, Montplaisir J Y, Gagnon J F and Postuma R B 2019 Evolution of prodromal Parkinson's disease and dementia with Lewy bodies: a prospective study *Brain* **142** 2051–67
[42] Lotankar S, Prabhavalkar K S and Bhatt L K 2017 Biomarkers for Parkinson's disease: recent advancement *Neurosci. Bull.* **33** 585–97
[43] Armstrong M J and Okun M S 2020 Diagnosis and treatment of Parkinson disease: a review *JAMA—J. Am. Med. Assoc.* **323** 548–60
[44] Ghazy E, Rahdar A, Barani M and Kyzas G Z 2021 Nanomaterials for Parkinson disease: recent progress *J. Mol. Struct.* **1231** 129698
[45] An Y *et al* 2012 Sensitive electrochemical immunosensor for $\alpha$-synuclein based on dual signal amplification using PAMAM dendrimer-encapsulated Au and enhanced gold nanoparticle labels *Biosens. Bioelectron.* **32** 224–30
[46] An Y *et al* 2010 A photoelectrochemical immunosensor based on au-doped $TiO_2$ nanotube arrays for the detection of $\alpha$-synuclein *Chem—A. Eur. J.* **16** 14439–46
[47] Kobayashi R, Noda M, Sawamura M, Yamakado H and Sohgawa M 2019 A novel detection of biomarker molecule of $\alpha$-synuclein for Parkinson disease by phospholipid liposome-immobilized cantilever biosensor using real-time quaking-induced conversion method *Proc. IEEE Sensors (Montreal, 27–30 Oct. 2019)* pp 1–3
[48] Yue H Y *et al* 2014 ZnO nanowire arrays on 3D hierachical graphene foam: biomarker detection of Parkinson's disease *ACS Nano* **8** 1639–46
[49] Tashkhourian J, Nezhad M R H, Khodavesi J and Javadi S 2009 Silver nanoparticles modified carbon nanotube paste electrode for simultaneous determination of dopamine and ascorbic acid *J. Electroanal. Chem.* **633** 85–91

[50] Pandey I, Tiwari J D, Furukawa H, Khosla A and Sekhar P K 2019 Flexible Prussian blue/carbon dots nanocomposite modified exfoliated graphite paper based sensor for simultaneous monitoring of hypertension and Parkinson disease *Microsyst. Technol.* **2** 109–19

[51] Varela J A, Dupuis J P, Etchepare L, Espana A, Cognet L and Groc L 2016 Targeting neurotransmitter receptors with nanoparticles *in vivo* allows single-molecule tracking in acute brain slices *Nat. Commun.* **7** 10947

[52] Ma W *et al* 2013 Ubiquinone-quantum dot bioconjugates for *in vitro* and intracellular complex I sensing *Sci. Rep.* **3** 1537

[53] Finberg J P M *et al* 2018 Sensor array for detection of early stage Parkinson's disease before medication *ACS Chem. Neurosci.* **9** 2548–53

[54] Winters T (ed) 2020 Alzheimer's disease and related dementias: experience and caregiving *Epidemiology, and Models of Care* 1st edn (Washington, DC: National Academies Press ) p 12

[55] Lynch C 2019 World Alzheimer Report 2019: Attitudes to Dementia *Alzheim. Dementia* **16** e038255

[56] DeTure M A and Dickson D W 2019 The neuropathological diagnosis of Alzheimer's disease *Mol. Neurodegener.* **14** 32

[57] Blennow K, Hampel H, Weiner M and Zetterberg H 2010 Cerebrospinal fluid and plasma biomarkers in Alzheimer disease *Nat. Rev. Neurol.* **6** 131–44

[58] Selkoe B D J 2018 Early network dysfunction in Alzheimer's disease *Science* **365** 540

[59] Braak E, Griffing K, Arai K, Bohl J, Bratzke H and Braak H 1999 Neuropathology of Alzheimer's disease: what is new since A Alzheimer ? *Eur. Arch. Psych. Clin. Neurosci.* **249** 14–22

[60] Du X, Wang X and Geng M 2018 Alzheimer's disease hypothesis and related therapies *Transl. Neurodegener.* **7** 2

[61] Sharma P, Srivastava P, Seth A, Tripathi P N and Banerjee A G 2019 Comprehensive review of mechanisms of pathogenesis involved in Alzheimer's disease and potential therapeutic strategies *Prog. Neurobiol.* **174** 53–89

[62] De Strooper B and Karran E 2016 The cellular phase of Alzheimer's disease *Cell* **164** 603–15

[63] Frigerio C S *et al* 2019 The major risk factors for Alzheimer's disease: age, sex, and genes modulate the microglia response to A$\beta$ plaques resource *Cell Rep.* **27** 1293–306

[64] Kivipelto M, Mangialasche F and Ngandu T 2018 Lifestyle interventions to prevent cognitive impairment, dementia and Alzheimer disease *Nat. Rev. Neurol.* **14** 653–66

[65] Moncaster J A *et al* 2010 Alzheimer's disease amyloid-$\beta$ links lens and brain pathology in down syndrome *PLoS One* **5** e10659

[66] Storandt M, Head D, Fagan A M, Holtzman D M and Morris J C 2012 Toward a multifactorial model of Alzheimer disease *Neurobiol. Aging* **33** 2262–71

[67] Goate A *et al* 2012 Clinical and biomarker changes in dominantly inherited Alzheimer's disease *N. Engl. J. Med.* **367** 795–804

[68] Palmqvist S *et al* 2019 Cerebrospinal fluid and plasma biomarker trajectories with increasing amyloid deposition in Alzheimer's disease *EMBO Mol. Med.* **11** e11170

[69] Blennow K and Zetterberg H 2018 Biomarkers for Alzheimer's disease: current status and prospects for the future *J. Intern. Med.* **284** 643–63

[70] Jagust W 2018 Imaging the evolution and pathophysiology of Alzheimer disease *Nat. Rev. Neurosci.* **19** 687–700

[71] Song C, Deng P and Que L 2018 Rapid multiplexed detection of beta-amyloid and total-tau as biomarkers for Alzheimer's disease in cerebrospinal fluid *Nanomed. Nanotechnol. Biol. Med.* **14** 1845–52

[72] Kincses Z T, Király A, Veréb D and Vécsei L 2015 Structural magnetic resonance imaging markers of Alzheimer's disease and its retranslation to rodent models *J. Alzheimer's Dis.* **47** 277–90

[73] Zhang X Y, Yang Z L, Lu G M, Yang G F and Zhang L J 2017 PET/MR imaging: new frontier in Alzheimer's disease and other dementias *Front. Mol. Neurosci.* **10** 343

[74] Yilmaz U 2015 Alzheimer-demenz *Radiologe* **55** 386–8

[75] Li F, Takechi H, Saito R, Ayaki T and Kokuryu A 2018 A comparative study: visual rating scores and the voxel-based specific regional analysis system for Alzheimer's disease on magnetic resonance imaging among subjects with Alzheimer's disease, mild cognitive impairment, and normal cognition *Psychogeriatrics* **19** 95–104

[76] Pansieri J *et al* 2018 Magnetic nanoparticles applications for amyloidosis study and detection: a review *Nanomaterials* **8** 740

[77] Mohajeri M, Behnam B, Barreto G E and Sahebkar A 2019 Carbon nanomaterials and amyloid-beta interactions: potentials for the detection and treatment of Alzheimer's disease? *Pharmacol. Res.* **143** 186–203

[78] Bodur O C, Dinc S, Ozmen M and Arslan F 2020 A sensitive amperometric detection of neurotransmitter acetylcholine using carbon dot-modified carbon paste electrode *Biotechnol. Appl. Biochem.* **68** 20–9

[79] Huang H *et al* 2017 Graphene quantum dots for detecting monomeric amyloid peptides *Nanoscale* **9** 5044–8

[80] Mattiuzzi C and Lippi G 2020 Worldwide disease epidemiology in the older persons *Eur. Geriatr. Med.* **11** 147–53

[81] Wafa H A, Wolfe C D A, Emmett E, Roth G A, Johnson C O and Wang Y 2020 Burden of stroke in Europe *Stroke* **51** 2418–27

[82] Kyle S and Saha S 2014 Nanotechnology for the detection and therapy of stroke *Adv. Healthc. Mater.* **3** 1703–20

[83] Wang J *et al* 2016 High-performance upconversion nanoprobes for multimodal MR imaging of acute ischemic stroke *Small* **12** 3591–600

[84] Chalela J A *et al* 2007 Magnetic resonance imaging and computed tomography in emergency assessment of patients with suspected acute stroke: a prospective comparison *Lancet* **369** 293–8

[85] Mouhieddine T H *et al* 2015 Nanotheragnostic applications for ischemic and hemorrhagic strokes: improved delivery for a better prognosis *Curr. Neurol. Neurosci. Rep.* **15** 505

[86] Fernandes L F, Bruch G E, Massensini A R and Frézard F 2018 Recent advances in the therapeutic and diagnostic use of liposomes and carbon nanomaterials in ischemic stroke *Front. Neurosci.* **12** 453

[87] Shcharbina N, Shcharbin D and Bryszewska M 2013 Nanomaterials in stroke treatment: perspectives *Stroke* **44** 2351–5

[88] Winter P M *et al* 2003 Improved molecular imaging contrast agent for detection of human thrombus *Magn. Reson. Med.* **50** 411–6

[89] Jander S, Schroeter M and Saleh A 2007 Imaging inflammation in acute brain ischemia *Stroke* **38** 642–5

[90] Kruss S, Hilmer A J, Zhang J, Reuel N F, Mu B and Strano M S 2013 Carbon nanotubes as optical biomedical sensors *Adv. Drug Deliv. Rev.* **65** 1933–50

[91] Mendonça M C P *et al* 2015 Reduced graphene oxide induces transient blood–brain barrier opening: an *in vivo* study *J. Nanobiotechnol.* **13** 78

[92] Kafa H *et al* 2015 The interaction of carbon nanotubes with an *in vitro* blood–brain barrier model and mouse brain *in vivo* *Biomaterials* **53** 437–52

[93] Torres-Ortega P V, Saludas L, Hanafy A S, Garbayo E and Blanco-Prieto M J 2019 Micro- and nanotechnology approaches to improve Parkinson's disease therapy *J. Control. Release* **295** 201–13

[94] Clarke C E, Vieregge P, Khoshnood B and Deschaseaux-voinet C 2005 Effects of dyskinesias in Parkinson's disease on quality of life and health-related costs: a prospective European study *Eur. J. Neurol.* **12** 956–63

[95] Katzenschlager R and Lees A J 2002 Treatment of Parkinson's disease: levodopa as the first choice *J. Neurol.* **249** ii19–ii24

[96] Stocchi F, Tagliati M and Olanow C W 2008 Treatment of levodopa-induced motor complications *Mov. Disord.* **23** S599–612

[97] Fargel M, Grobe B, Oesterle E, Hastedt C and Rupp M 2007 Treatment of Parkinson's disease: a survey of patients and neurologists *Clin. Drug Investig.* **27** 207–18

[98] Benabid A L, Chabardes S, Mitrofanis J and Pollak P 2009 Deep brain stimulation of the subthalamic nucleus for the treatment of Parkinson's disease *Lancet Neurol.* **8** 67–81

[99] Silindir Gunay M, Yekta Ozer A and Chalon S 2015 Drug delivery systems for imaging and therapy of Parkinson's disease *Curr. Neuropharmacol.* **14** 376–91

[100] Frazão N F *et al* 2012 Four-level levodopa adsorption on C60 fullerene for transdermal and oral administration: a computational study *RSC Adv.* **2** 8306–22

[101] Corona-morales A A and Escobar A 2003 Fullerene C60 and ascorbic acid protect cultured chromaffin cells against levodopa toxicity *J. Neurosci. Res.* **126** 121–6

[102] Ruiz-ortega J A, Ugedo L and Pedraz J L 2014 Increased antiparkinson efficacy of the combined administration of VEGF- and GDNF-loaded nanospheres in a partial lesion model of Parkinson's disease *Int. J. Nanomed.* **9** 2677–87

[103] Dugan L L, Lovett E G, Quick K L, Lotharius J, Lin T T and Malley K L O 2001 Fullerene-based antioxidants and neurodegenerative disorders *Parkinsonism Oth. Disord.* **7** 243–6

[104] Cai X *et al* 2008 Polyhydroxylated fullerene derivative C60 (OH) 24 prevents mitochondrial dysfunction and oxidative damage in an MPP 1-induced cellular model of Parkinson's disease *J. Neurosci. Res.* **3634** 3622–34

[105] Jin H *et al* 2000 Polyhydroxylated C60, fullerenols, as glutamate receptor antagonists and neuroprotective agents *J. Neurosci. Res.* **607** 600–7

[106] Kim D *et al* 2018 Graphene quantum dots prevent $\alpha$-synucleinopathy in Parkinson's disease *Nat. Nanotechnol.* **62** 600–7

[107] Stefani M 2010 Protein misfolding and cellular stress in disease and aging *Protein Misfolding and Cellular Stress in Disease and Aging: Concepts and ProtocolsMethods in Molecular Biology* vol 648 ed P Bross and N Gregersen (Totowa, NJ: Humana Press) p 41

[108] Wilson B 2011 Therapeutic compliance of nanomedicine in Alzheimer's disease *Nanomedicine* **6** 1137–9

[109] Yiannopoulou K G and Papageorgiou S G 2020 Current and future treatments in Alzheimer disease : an update *J. Cent. Nerv. Syst. Dis.* **12** 1–12

[110] Lee C, Huang S, Huang S, Lin H, Tsai H and Wu J C 2011 60 fullerene-pentoxifylline dyad nanoparticles enhance autophagy to avoid cytotoxic effects caused by the $\beta$-amyloid peptide *Nanomed. Nanotechnol. Biol. Med.* **7** 107–14

[111] Li H, Luo Y, Derreumaux P and Wei G 2011 Carbon nanotube inhibits the formation of $\beta$-sheet-rich oligomers of the Alzheimer's amyloid-$\beta$(16–22) peptide *Biophys. J.* **101** 2267–76

[112] Mahmoudi M, Akhavan O, Ghavami M, Rezaee F and Ghiasi S M A 2012 Graphene oxide strongly inhibits amyloid beta fibrillation *Nanoscale* **4** 7322

[113] Henna T K and Pramod K 2020 Graphene quantum dots redefine nanobiomedicine *Mater. Sci. Eng.* C **110** 110651

[114] Gregory W E, Sharma B, Hu L, Raghavendra A J and Podila R 2020 Interfacial charge transfer with exfoliated graphene inhibits fibril formation in lysozyme amyloid *Biointerphases* **15** 031010

[115] Chung Y J, Lee B I and Park C B 2019 Multifunctional carbon dots as a therapeutic nanoagent for modulating Cu(II)-mediated $\beta$-amyloid aggregation *Nanoscale* **11** 6297–306

[116] Nday Christiane M, Eleftheriadou D and Jackson G 2019 Magnetic chrysin silica nanomaterials behavior in an amyloidogenic environment *Hell. J. Nucl. Med.* **22** 42–50

[117] Akhtar M J, Ahamed M and Alhadlaq H A 2017 Mechanism of ROS scavenging and antioxidant signalling by redox metallic and fullerene nanomaterials: potential implications in ROS associated degenerative disorders *BBA—Gen. Subj.* **1861** 802–13

[118] Koppel K *et al* 2020 Elevated amyloidoses of human IAPP and amyloid beta by lipopolysaccharide and their mitigation by carbon quantum dots *Nanoscale* **12** 12317–28

[119] Guerrero E D, Lopez-Velazquez A M, Ahlawat J and Narayan M 2021 Carbon quantum dots for treatment of amyloid disorders *ACS Appl. Nano Mater.* **4** 2423–33

[120] Chung Y J, Kim K, Lee B I L and Park C B 2017 Carbon nanodot-sensitized modulation of Alzheimer's β-amyloid self-assembly, disassembly, and toxicity *Small* **13** 1700983

[121] Chung Y J, Lee C H, Lim J, Jang J, Kang H and Park C B 2020 Photomodulating carbon dots for spatiotemporal suppression of Alzheimer's $\beta$-amyloid aggregation *ACS Nano* **14** 16973–83

[122] Rabinstein A A 2020 Update on treatment of acute ischemic stroke *Continuum* **26** 268–86

[123] Erdem S S, Sazonova I Y, Hara T, Jaffer F A and McCarthy J 2012 Detection and treatment of intravascular thrombi with magnetofluorescent nanoparticles *Methods Enzymol.* **508** 209

[124] Chen J P, Yang P C, Ma Y H, Tu S J and Lu Y J 2012 Targeted delivery of tissue plasminogen activator by binding to silica-coated magnetic nanoparticle *Int. J. Nanomed.* **7** 5137–49

[125] Marsh J N *et al* 2011 A fibrin-specific thrombolytic nanomedicine approach to acute ischemic stroke *Nanomedicine* **6** 605–15

[126] Rasouli Vani J, Mohammadi M T, Sarami Foroshani M and Jafari M 2016 Polyhydroxylated fullerene nanoparticles attenuate brain infarction and oxidative stress in rat model of ischemic stroke *EXCLI J.* **15** 378–90

[127] Fluri F *et al* 2015 Fullerenols and glucosamine fullerenes reduce infarct volume and cerebral inflammation after ischemic stroke in normotensive and hypertensive rats *Exp. Neurol.* **265** 142–51

[128] Yang D Y, Wang M F, Chen I L, Chan Y C, Lee M S and Cheng F C 2001 Systemic administration of a water-soluble hexasulfonated C60 (FC4S) reduces cerebral ischemia-induced infarct volume in gerbils *Neurosci. Lett.* **311** 121–4

[129] Rauti R, Musto M, Bosi S, Prato M and Ballerini L 2018 Properties and behavior of carbon nanomaterials when interfacing neuronal cells: how far have we come? *Carbon* **143** 430–46

[130] Teleanu D M, Chircov C, Grumezescu A M, Volceanov A and Teleanu R I 2018 Impact of nanoparticles on brain health: an up to date overview *J. Clin. Med.* **7** 490

[131] Henna T K, Raphey V R, Sankar R, Shirin V K A and Gangadharappa H V 2020 Carbon nanostructures: the drug and the delivery system for brain disorders *Int. J. Pharm.* **587** 119701

**IOP** Publishing

Zero-dimensional Carbon Nanomaterials
Fundamentals and applications
**Pratima R Solanki and Ravindra Pratap Singh**

# Chapter 10

# Zero-dimensional carbon nanomaterials for targeted drug delivery and photothermal therapy applications

**Charles Oluwaseun Adetunji, John Tsado Mathew, Kshitij RB Singh, Abel Inobeme, Vanya Nayak, Olugbemi T Olaniyan, Jay Singh and Ravindra Pratap Singh**

Zero-dimensional (0D) nanomaterials are considered suitable nanomaterials for several biomedical applications. Typical examples of such nanomaterials include graphene quantum dots (GQDs), carbon quantum dots (CQDs), fullerenes, inorganic quantum dots (QDs), and polymer dots (P-dots). These 0D nanomaterials possess several applications that include biomolecular recognition, ion detection, illness diagnosis, quantum confinement effects, pathogen detection, etc, which are possible due to their ultra-small size, superior chemical and physical properties, and strong biomedical applicability. This chapter aims to enhance the reader's knowledge on the various utilities of zero-dimensional carbon nanomaterials for photothermal therapy and targeted drug delivery.

## 10.1 Introduction

The growing interest in carbon-based nanomaterials (CBNs) has boosted their use in the biomedical domain; however, their applications are still restricted due to a lack of proper knowledge of their side-effects on human health and the environment. For researchers, making an informed decision on the suitability of a flexible carbon-based nanocarrier device and its future uses can be difficult. Surface modification, enzymatic degradation, and biological interactions aid in the production of CBNs for the delivery of drugs. Nanotechnology combines chemistry, physics, biology, and material science, therefore, it is considered one of the most promising fields for creating cancer-treating drugs with a higher healing index [1]. Several theoretical and practical investigations have been conducted over the last few years to develop carbon-related nanomaterials, carbon being among the most fascinating elements used in nanoscale materials. The convergence of nanostructure with biomedicine has

doi:10.1088/978-0-7503-4048-9ch10  

been a critical issue in recent decades, thanks to the rapid growth of nanotechnology. Several nanostructures, particularly those manufactured through self-assembly, tend to improve the concentration dramatically, decrease the cytotoxicity, and advance therapeutic effectiveness due to unique and superior features. Nanostructures have been applied to transport various drugs, small compounds, nucleic acids, proteins, and peptides, with great effectiveness. Moreover, polymers, liposomes, metals, ceramics, nucleic acids, peptides, and even medication molecules are used to construct a wide range of nanostructures from different types of materials [2].

Zero-dimensional (0D) nanomaterials, such as graphene quantum dots (GQDs), carbon quantum dots (CQDs), fullerenes, inorganic quantum dots (QDs), and polymer dots (P-dots), have recieved widespread attention in the biosensing domain in recent years. 0D nanomaterials have shown tremendous promise in biomolecular recognition, ion detection, illness diagnosis, quantum confinement effects, and pathogen detection due to their ultra-small size and superior chemical and physical properties [3]. Nanoparticles, nanocrystals, quantum dots, and nanoclusters are all 0D nanomaterials with different names. Moreover, 0D nanomaterials are considered the basic building blocks for planning and synthesizing 1D, 2D, and complicated 3D nanomaterials [4]. Many works in the literature have mentioned novel uses of various bionanomaterials in drug delivery [5,6,7]. Carbon nanotubes (CNTs) also exhibit a variety of properties, but their most appealing feature is their distinctive photoluminescence. However, carbon dots are a rapidly increasing family of carbonaceous nanomaterials that have attracted a lot of research interest, as they exhibit low toxicity, cost-effectiveness, fluorescence, luminescence, environmental friendliness, ease of synthesis, and interface passivation. Carbon dots have been used widely in bioimaging, biosensing, catalysis, and nanomedicine due to their unique features [8]. A two-dimensional layer of $sp^2$ hybridized carbon is known as graphene. Its honeycomb structure, which can be built to produce 3D graphite, stretched to generate 1D carbon nanotubes, and twisted to produce 0D carbon nanotubes, is the primary component of the other key carbon allotropes. Hence, graphene is the 'mother' of all graphitic carbon-based nanomaterials. This chapter describes the numerous characteristics of graphene, and its many uses, including cancer therapeutics [9]. Graphene and carbon nanotubes are both low-dimensional $sp^2$ carbon nanotechnologies with various fascinating chemical and physical properties that are helpful in an assortment of uses, particularly in nanomedicine. In the same vein, carbon nanotubes have been studied intensively as a medium to deliver drugs, chemotherapeutic medicines, and with genes, and for intracellular transfer of proteins. Several groups have demonstrated *in vivo* delivery of chemotherapeutic drugs as a pathway for cancer therapy by utilizing carbon nanotubes in animal tests [10]. Moreover, 0D materials have piqued researchers' attention worldwide thanks to their chemical and physical properties and large surface area, which make them ideal organic compounds. Antimicrobial activity, bioimaging, cancer diagnostics, drug administration, targeting, tissue engineering, biosensors, skincare, and dental care are just a few of their potential biological uses. Photosensitizers (PSs) are those materials that can absorb light from light sources, generate oxidative stress in cellular settings, and are commonly used in chemotherapy. Various kinds of

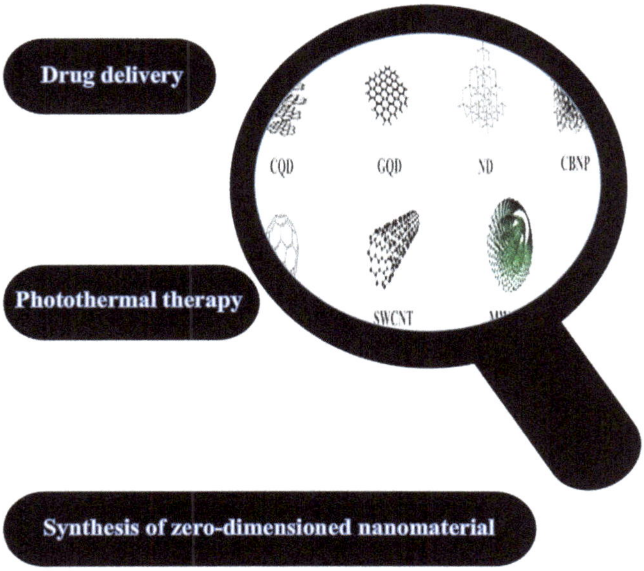


**Figure 10.1.** A schematic illustration of the details covered in this chapter.

innocuous PSs have been fabricated on CBNs and have exhibited several benefits in immunotherapy. The prospective role and possible influence of CBNs and phototherapy are discussed in this chapter, including the functionalization and mechanism of PSs on nanoparticles and their promised benefits in drug development [11].

The purpose of this chapter is to present review-based literature on some of the most intriguing and essential advancements in 0D carbon nanomaterials synthesis, and applications in the biomedical domain. Hence, this chapter aims to impart a comprehensive discussion of the utilities of 0D carbon nanomaterials for photothermal therapy and targeted drug delivery, as represented in figure 10.1.

## 10.2 Synthesis of zero-dimensional carbon nanomaterials

Carbon is one of the fundamental elements of life, and researchers are largely focused on 0D and 1D carbon nanoparticles even though they have a highly porous structure. These 0D carbon nanomaterials have attracted great attention due to their small size and unique properties. The 0D carbon nanomaterials can be synthesized through physical, chemical, and biological routes. However, toxicity still remains a significant concern when using carbon nanomaterials for medicinal applications; therefore, researchers have focused their attention on the biological pathways for synthesizing 0D carbon nanomaterials. Since biogenic synthesis utilizes biological media such as plants, bacteria, fungi, eggs, honey, etc, it eliminates the use of harmful chemicals, enhances the biocompatibility, and decreases the bio-toxicity of 0D carbon nanomaterials. Moreover, biological routes are more cost-effective than chemical and physical routes. For example, using the leaves of *Lawsonia inermis,* highly fluorescent carbon dots were synthesized and, further doped with $Fe^{2+}$ to study the effect of doping on the fluorescence emission [12].

Similarly, among the fluorescent nanomaterials, GQDs have gained widespread attention due to their extremely small size that helps them exhibit unique physicochemical properties. GQDs exhibit high biocompatibility and high photoluminescence properties, making them suitable therapeutic and diagnostic agents. GQDs can be synthesized using strong acidic oxidation, electrochemical exfoliation, pyrolysis, or solvothermal processes. For example, using dimethylfuran (DMF) as a solvent, GQDs were synthesized through a one-step solvothermal method. However, the excessive use of toxic reagents such as DMF, sulfuric acid, and nitric acids decreases the biocompatibility of the GQDs and increases their cytotoxicity. In contrast, biogenic routes to synthesize GQDs increase their bio-photoluminescence properties and decreased their cytotoxicity and cost. By using glucose and water, GQDs were synthesized through a one-pot hydrothermal method. These synthesized GQDs ranged from 1.5–3.0 nm and exhibited excellent water solubility and high photoluminescent emission. Further, their *in vitro* compatibility and *in vivo* toxicity analysis showed that the synthesized GQDs showed no apparent cytotoxicity and damage to any organ, which gives the surety that these GQDs can be efficiently utilized in biological applications [13]. Although various routes have been applied to synthesize 0D carbon nanomaterials, significant attention has been directed toward the biogenic pathways discussed in chapter 2 of this book.

## 10.3 Photothermal therapy

Photothermal therapy (PTT) has received much attention as a potential cancer treatment option with fewer side-effects than standard radiotherapy, chemotherapy, or surgical resection, because it does not show any adverse effects. PTT is based on the photoconversion capability of photothermal agents (PTAs), and a wide range of nanostructures have been used as PTAs for cancer treatment. PTAs are locally or systematically delivered to improve elimination effectiveness and become concentrated in cancer cells [14]. The unique properties of carbon dots, such as high biocompatibility, low toxicity, and high photostability, make them suitable imaging agents for optical bioimaging. Many carbon dots have been reported to be activated by ultraviolet/blue light or visible/first near-infrared (NIR) emissions that are shorter than 820 nm, which hampers their imaging applications. Therefore, there is an immediate need to develop a carbon dot-based material that exhibits second near-infrared (NIR-II) emission at 1000−1700 nm. By keeping this in mind, a novel theranostic NIR-II-emitting CD-based nanoprobe was synthesized using watermelon juice, which was activated at a wavelength of 808 nm, and showed NIR-II emission at 900–1200 nm with high quantum yield and biocompatibility; therefore, this method can be used to synthesize probes for *in vivo* NIR-II bioimaging. Moreover, these CDs were found to be suitable for photothermal therapy as they showed a high photothermal efficiency of 30.6%, which is assumed to break the barriers *in vivo* and allow highly sensitive optical imaging. Further, it was also observed that 65% of carbon dots were excreted out of the mouse's body within 6 h.

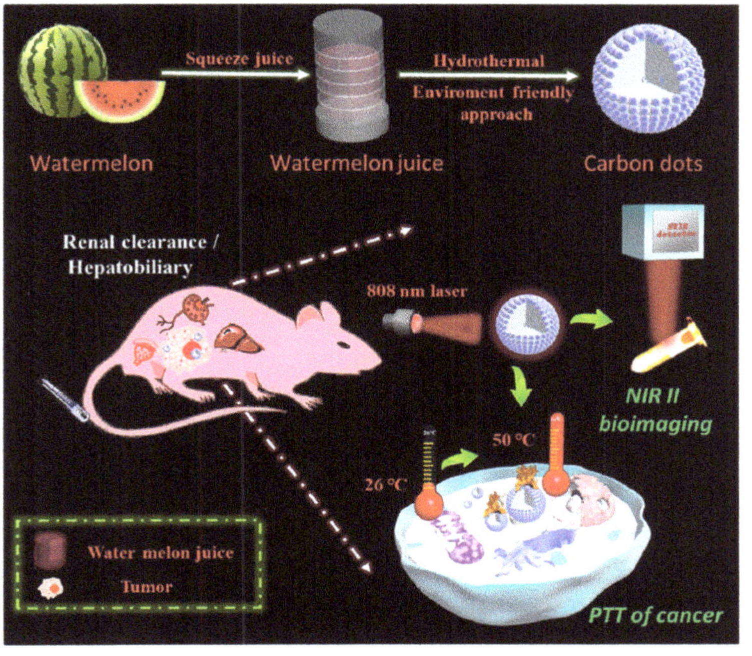


**Figure 10.2.** A schematic illustration of the synthesis of NIR-II emitting carbon dots for rapid renal clearance, NIR-II bioimaging, and PTT of cancer. (Reproduced with permission from [15]. Copyright 2019 American Chemical Society.)

Figure 10.2 exhibits a stepwise process for synthesizing NIR-II emitting carbon dots for rapid renal clearance in NIR-II bioimaging and PTT of cancer [15].

Similarly, an experiment was conducted in which a mixture of carbon dots with gold nanorods (CD@GNRs) was used to release doxorubicin (DOX) under ideal physiological conditions. These CD@GNRs exhibited high drug loading capacity due to the high porosity of the carbon dots. The carbon dots had a high fluorescence property and biocompatible nature, and gold nanorods are widely recognized for their excellent photothermal activity. When combined in the CD@GNRs, they exhibited excellent photothermal properties at NIR of 808 nm in tumorous cells, as illustrated in figure 10.3 [16]. However, research on the utilization of 0D carbon nanomaterials for PTT is still in its early stages. Many limitations, such as cytotoxicity, route of elimination, biosafety, etc, are still to be addressed, restricting their commercial use.

## 10.4 Drug delivery

Owing to the unique structural properties of CNTs, their applications have been identified to differ from other carbon allotropes, as they enable an appropriate passage for the transfer of numerous molecules of drugs, most significantly into living cells. The drug molecules are joined to the sidewalls of functionalized CNTs

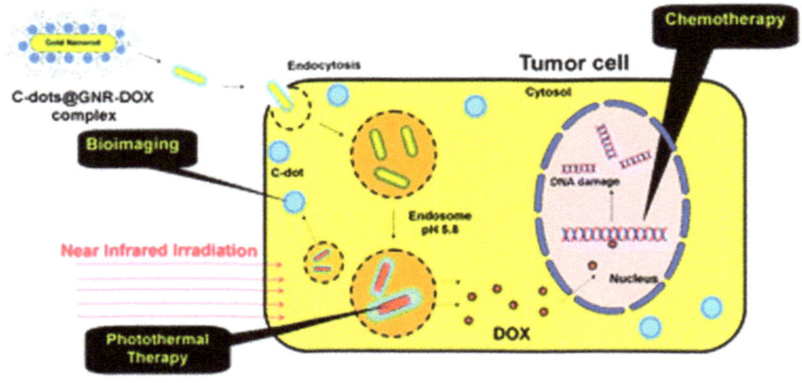


**Figure 10.3.** An illustration of a prepared C-dots@GNR-DOX complex used for bioimaging, photothermal therapy, and chemotherapy. (Reproduced with permission from [16]. Copyright 2013 Royal Society of Chemistry.)

through non-covalent and covalent bonding. In 2015 Shi *et al* applied an electric field to release ibuprofen from a hybrid hydrogel made up of bacterial cellulose, sodium alginate, and multi-walled carbon nanotubes (MWCNTs) [17]. Moreover, several drugs have been loaded onto CNTs, which include oxaliplatin [18], doxorubicin [19], and docetaxel [20], to establish their effectiveness for *in vivo* and *in vitro* cancer management. The utilization of graphene as a potential drug carrier could be linked to the presence of $\pi$ electrons and an exceptionally large surface area. For example, it was observed that when an enormous amount of doxorubicin was applied on phospholipid monolayer coated graphene, better doxorubicin release in the presence of acidic pH compared to basic pH was observed [24, 25]. Further, it was also observed that doxorubicin could be applied on graphene sheets through physisorption followed by surface modification by O-(2-aminoethyl)polyethylene glycol (PEG-$NH_2$) to improve their compatibility and stability in a biological medium [21]. Similarly, paclitaxel (PTX) was applied onto graphene oxide-polyethylene(glycol) (GO-PEG) using hydrophobic interactions and $\pi$–$\pi$ stacking, whose loading effectiveness was observed at 11.2 wt% [22].

It has been shown that GQDs have a single atomic layer that includes an oxygen-rich surface and minimal lateral size, making them suitable for the sustainable application of drugs and improving their stability, particularly in the physiological media. Also, the fluorescent features of GQDs makes them appropriate platforms for adequate delivery of drugs into cancer cells [23]. A GQD-conjugated gemcitabine-loaded HSA nanoformulation was developed to transfer drugs adequately. Also, it was observed that the presence of albumin assists in the adequate delivery of gemcitabine to tumor cells through the gp60 pathway [24]. However, many limitations, such as dose limiting side-effects, restrict the potential of GQDs as nanocarrier. Therefore, to overcome these limitations water soluble GQDs were synthesized using purified pristine MWCNTs through acidic oxidation. These synthesized GQDs were covalently bonded to the tumor-targeting module biotin

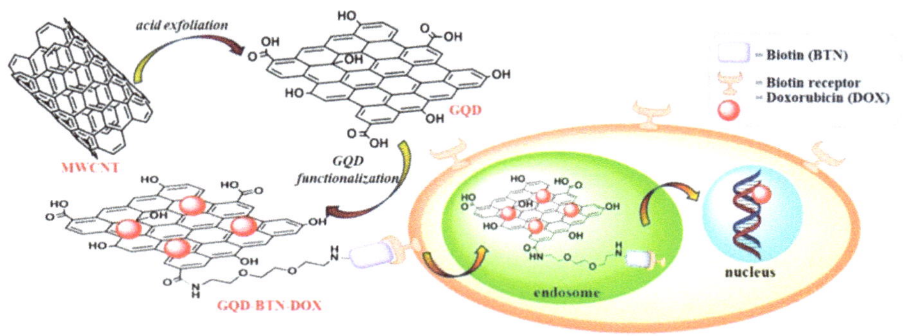


**Figure 10.4.** An illustration of biotin-conjugated DOX-loaded GQDs utilized for targeted drug delivery in cancer therapy. (Reproduced with permission from [25]. Copyright 2017 Elsevier).

(GQD-BTN) through a strategically designed cleavable linker that can be specifically activated inside cells. These GQD-BTNs were used to detect over-expressed biotin receptors on cancer cells. Further, with the help of $\pi$–$\pi$ interaction, DOX was loaded on the surface of synthesized GQDs. It was reported that the intrinsic fluorescence properties helped in tracing of the release of the drug, as illustrated in figure 10.4. Due to the acidic environment of cancer cells, the drug was detached from the nanosystem, therefore a delayed nuclear internalization occurred [25]. Several nanomaterials have been identified for their suitability in cancer treatment, which might be due to certain factors such as surface functional groups, low toxicity, intrinsic fluorescence properties, unique physicochemical features, and good hydrophilicity. These features enable them to function for radioisotopes and chemotherapeutics and also allows them to be utilized in drug delivery for cancer treatment [25]. GQD capped magnetic mesoporous silica nanoparticles were synthesized to generate heat under NIR irradiation and an alternating magnetic field (AMF). *In vitro* studies showed that the material exhibited various effective magnetic hyperthermia and chemo-photothermal therapy effects [26].

Further, graphene oxide has also been recognized as a sustainable nanomaterial in certain applications, because it exhibits multipurpose surface properties, inherent size- and shape-dependent optical properties, and a huge surface-to-volume ratio, that enable them to serve as nanomaterials for cancer therapy [27]. Moreover, the Chen group synthesized a DOX-loaded-reduced graphene oxide (RGO) gold nanorod vehicle that involved combined chemotherapy and photothermal treatment. It was observed that a high amount of DOX was liberated due to the acidic nature and NIR photothermal heating influence of the tumor microenvironment [28]. Another unique material that could include dual chemotherapeutics, a loaded sponge-like carbon material that can hold and fit into graphene nanosheets (GNSs), was developed. It was modified with tumor-targeting proteins so that it could assist through the lipid bilayers (lipo-GNS). It was observed that these well fabricated ultra-small lipo-GNSs of size 40 nm produced a noteworthy build-up in the tumor site, and an efficacious destruction of the xenograft tumors in 16 days was observed

[29]. Therefore, it can be concluded that the 0D carbon nanomaterials can act as a suitable drug delivery agent and hold the potential to bring about a revolution in disease treatment. However, a lack of proper studies of their route of elimination, cytotoxicity, biocompatibility, and prolonged side-effects limits their uses and market potential. However, it is believed that 0D carbon nanomaterials will be the most popular theragnostic agent within a few decades.

## 10.5 Conclusion

This chapter has provided detailed information on the properties of 0D carbon nanomaterials and their utilization in photothermal therapy and drug delivery. Furthermore, this chapter aimed to provide a description of different CBNs in biomedical applications, such as carbon nanotubes for drug delivery, graphene oxide for drug delivery, graphene oxide for cancer therapy, GQDs for drug delivery, and cancer therapy.

## References

[1] Debnath S K and Srivastava R 2021 Drug delivery with carbon-based nanomaterials as versatile nanocarriers: progress and prospects *Front. Nanotechnol.* **3** 644564

[2] Lu H, Wang J, Wang T, Zhong J, Bao Y and Hao H 2016 Recent progress on nanostructures for drug delivery applications *J. Nanomater.* **2016** 5762431

[3] Wang Z, Hu T, Liang R and Wei M 2020 Application of zero-dimensional nanomaterials in biosensing *Front. Chem.* **8** 320

[4] Cao H 2017 Synthesis, characterization, and applications of zero-dimensional (0D) nanostructures *Synthesis and Applications of Inorganic Nanostructures* (Weinheim: Wiley) pp 21–146

[5] Singh K R, Nayak V, Singh J, Singh A K and Singh R P 2021 Potentialities of bioinspired metal and metal oxide nanoparticles in biomedical sciences *RSC Adv.* **11** 24722–46

[6] Singh K R, Solanki P R, Malhotra B D, Pandey A C and Singh R P 2021 Introduction to nanomaterials *Nanomaterials in Bionanotechnology* (Boca Raton, FL: CRC Press) pp 1–35

[7] Singh R P and Choi J-W 2010 Bio-nanomaterials for versatile bio-molecules detection technology *Adv. Mater. Lett.* **1** 83–4

[8] Xu Q, Li W, Ding L, Yang W, Xiao H and Ong W-J 2019 Function-driven engineering of 1D carbon nanotubes and 0D carbon dots: mechanism, properties and applications *Nanoscale* **11** 1475–504

[9] Sur U K 2013 Graphene: the two-dimensional carbon nanomaterial *Nano Sci. Nano Technol.* **7** 156–61

[10] Liu Z, Robinson J T, Tabakman S M, Yang K and Dai H 2011 Carbon materials for drug delivery and cancer therapy *Mater. Today* **14** 316–23

[11] Sundaram P and Abrahamse H 2020 Phototherapy combined with carbon nanomaterials (1D and 2D) and their applications in cancer therapy *Materials* **13** 4830

[12] Mary Alex A, Kiran M D, Hari G, Krishnan A, Jayan J S and Saritha A 2020 Carbon dots: a green synthesis from *Lawsonia inermis* leaves *Mater. Today Proc.* **26** 716–9

[13] Yan C *et al* 2020 Highly biocompatible graphene quantum dots: green synthesis, toxicity comparison and fluorescence imaging *J. Mater. Sci.* **55** 1198–215

[14] Sun R, Chen H, Sutrisno L, Kawazoe N and Chen G 2021 Nanomaterials and their composite scaffolds for photothermal therapy and tissue engineering applications *Sci. Technol. Adv. Mater.* **22** 404–28

[15] Li Y, Bai G, Zeng S and Hao J 2019 Theranostic carbon dots with innovative NIR-II emission for *in vivo* renal-excreted optical imaging and photothermal therapy *ACS Appl. Mater. Interfaces* **11** 4737–44

[16] Pandey S, Thakur M, Mewada A, Anjarlekar D, Mishra N and Sharon M 2013 Carbon dots functionalized gold nanorod mediated delivery of doxorubicin: tri-functional nano-worms for drug delivery, photothermal therapy and bioimaging *J. Mater. Chem.* B **1** 4972

[17] Shi X *et al* 2015 Dual stimulus responsive drug release under the interaction of pH value and pulsatile electric field for a bacterial cellulose/sodium alginate/multi-walled carbon nanotube hybrid hydrogel *RSC Adv.* **5** 41820–9

[18] Lee P-C, Lin C-Y, Peng C-L and Shieh M-J 2016 Development of a controlled-release drug delivery system by encapsulating oxaliplatin into SPIO/MWNT nanoparticles for effective colon cancer therapy and magnetic resonance imaging *Biomater. Sci.* **4** 1742–53

[19] Huang H, Yuan Q, Shah J S and Misra R D K 2011 A new family of folate-decorated and carbon nanotube-mediated drug delivery system: synthesis and drug delivery response *Adv. Drug. Deliv. Rev.* **63** 1332–9

[20] Jain S, Spandana G, Agrawal A K, Kushwah V and Thanki K 2015 Enhanced antitumor efficacy and reduced toxicity of docetaxel loaded estradiol functionalized stealth polymeric nanoparticles *Mol. Pharm.* **12** 3871–84

[21] Zhang H, Grüner G and Zhao Y 2013 Recent advancements of graphene in biomedicine *J. Mater. Chem.* B **1** 2542

[22] Xu Z, Wang S, Li Y, Wang M, Shi P and Huang X 2014 Covalent functionalization of graphene oxide with biocompatible poly(ethylene glycol) for delivery of paclitaxel *ACS Appl. Mater. Interfaces* **6** 17268–76

[23] Srivastava A, Yadav T, Sharma S, Nayak A, Akanksha Kumari A and Mishra N 2016 Polymers in drug delivery *J. Biosci. Med.* **4** 69–84

[24] Nigam P, Waghmode S, Louis M, Wangnoo S, Chavan P and Sarkar D 2014 Graphene quantum dots conjugated albumin nanoparticles for targeted drug delivery and imaging of pancreatic cancer *J. Mater. Chem.* B **2** 3190–5

[25] Iannazzo D *et al* 2017 Graphene quantum dots for cancer targeted drug delivery *Int. J. Pharm.* **518** 185–92

[26] Yao X *et al* 2017 Graphene quantum dots-capped magnetic mesoporous silica nanoparticles as a multifunctional platform for controlled drug delivery, magnetic hyperthermia, and photothermal therapy *Small* **13** 1602225

[27] Nejabat M, Charbgoo F and Ramezani M 2017 Graphene as multifunctional delivery platform in cancer therapy *J. Biomed. Mater. Res. Part* A **105** 2355–67

[28] Song L, Shi J, Lu J and Lu C 2015 Structure observation of graphene quantum dots by single-layered formation in layered confinement space *Chem. Sci.* **6** 4846–50

[29] Yang K, Zhang S, Zhang G, Sun X, Lee S-T and Liu Z 2010 Graphene in mice: ultrahigh *in vivo* tumor uptake and efficient photothermal therapy *Nano Lett.* **10** 3318–23

**IOP** Publishing

Zero-dimensional Carbon Nanomaterials
Fundamentals and applications
**Pratima R Solanki and Ravindra Pratap Singh**

# Chapter 11

## Zero-dimensional carbon nanomaterials for tissue engineering applications

**Monika Vats, Mansi Swami, Awadesh Kumar Dwivedi and Kushagra Rajendra**

The field of tissue engineering is about replacing damaged/diseased tissues with biological replacements, such as in organ transplantation and regenerative medicine. The field continuously needs new tools/techniques to monitor the growth of engineered tissues. Carbon nanoparticles (CNPs), in particular zero-dimensional materials, possess remarkable chemical, physical, and electrical properties. Thus they can be used as contrasting agents for the evaluation and monitoring of tissue formation. However, the zero-dimensional CNPs have a certain associated cytotoxicity. Addressing this concern may lead them to be promising materials in the field of tissue engineering. This chapter provides an overview of zero-dimensional carbon nanomaterials and their applications in tissue engineering.

### 11.1 Introduction

Nanotechnology has become a rapidly growing field over the past few decades. Nanoparticles are characterized by their nanoscale size, which allows them to develop unique physicochemical properties, thus making them advantageous for a wide range of applications. The carbon-based nanomaterials have received much attention due to their remarkable and novel physical, chemical, and biological properties. These exclusive properties make carbon allotropes, such as carbon nanotubes [1], nano-fullerenes [2], and graphene [3], suitable for use in a wide range of applications. These carbon-based materials are of great interest and have aroused significant research attention in the fields of drug delivery, bio-sensing, bio-imaging, etc. Additionally, carbon-based nanomaterials may be used to reinforce two- and three-dimensional scaffolds (e.g. polymers) to enhance their mechanical properties [4]. Polymeric nano-architectures have already been known to have importance in the field of tissue engineering and regenerative medicine [5]. The polymeric and carbon-based materials doped/composited with nanoparticles can be used for

doi:10.1088/978-0-7503-4048-9ch11  

cell labeling for active tracking of artificial tissues, as well as for enhancing the mechanical properties of scaffolds [6].

The carbon-based nanomaterials can be categorized based upon their physical dimensions as zero-dimensional (e.g. carbon dots), one-dimensional (e.g. single-walled carbon nanotubes), two-dimensional (e.g. graphene), and three-dimensional (e.g. diamond). The zero-dimensional nanomaterials include graphene, quantum dots, carbon quantum dots, fullerenes, inorganic quantum dots, magnetic nanoparticles, and polymer dots [7]. Zero-dimensional carbon nanomaterials have also started making inroads into the field of tissue engineering [8]. Tissue engineering is a rapidly developing method of replacing and repairing damaged and defective tissues in the human body. Figure 11.1 describes the importance and feasibility of tissue engineering in different parts of the human body. This process requires porous scaffolds as a matrix to support tissue growth. The scaffolds form the space required for cells to grow in the appropriate anatomical form. Depending on the design of the nanostructures the applications may vary, for example scaffolds with good electrical conductivity can be used for the development of nerve tissue scaffolds, while for preparing scaffolds for bone and cartilage cells the prime requirement is improved mechanical properties. Tissue engineering has had a great impact on the modern society. It is expected to continue to improve the quality of human life and reduce the social and economic costs associated with medical care and improve life expectancy.

The main requirement for these scaffolds is biocompatibility, to avoid inhibiting cell growth and for better coordination with the body. Every year, many people need bone replacement to cure bone defects that cannot heal on their own. Various advanced biocompatible materials have been used to form scaffolds that can facilitates new bone growth [9]. Scaffolds based on carbon-based nanomaterials are key examples, with the advantages of biocompatibility, mechanical stability, and commercial availability. The materials have the potential to provide early detection of pathological conditions,

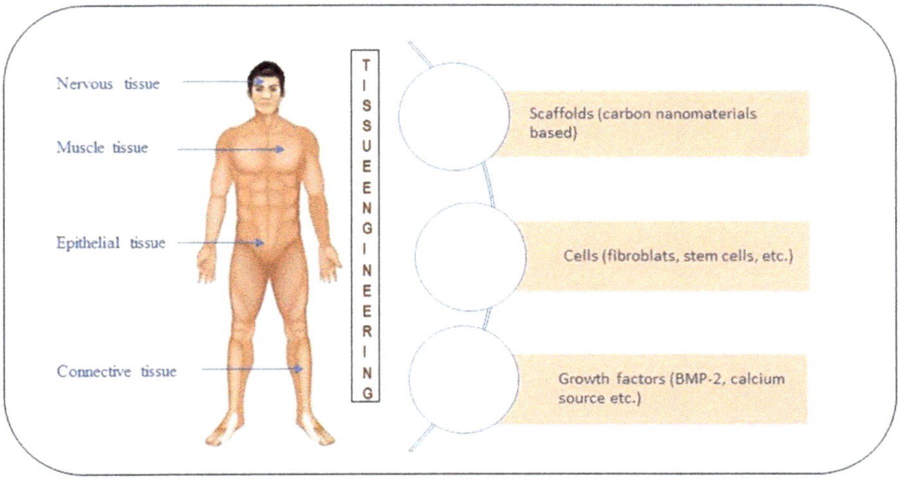


**Figure 11.1.** Tissue engineering in the human body.

reduce the severity of treatment, and bring about better clinical results for patients. These biological properties and advantages make the carbon-based nanomaterials, in particular zero-dimensional carbon nanomaterials, useful in the field of cell/tissue regeneration or replacement in preparing the devices which are needed to monitor tissue growth, cell imaging, and other tissue engineering tools.

In this chapter we present a contemporary overview of carbon nanomaterials, and the usability and advantages of zero-dimensional carbon nanomaterials for tissue engineering and other biomedical applications, briefly cover their synthesis and characterization, and address the limitations and concerns associated with zero-dimensional carbon nanomaterials in tissue engineering. The intrinsic physiochemical and biomedical properties of zero-dimensional carbon nanomaterials may be examined further and prove to be a suitable alternative for tissue repair and regeneration, and overcome the current obstacles in tissue engineering.

## 11.2 Types of carbon nanomaterials

Carbon has excellent polymerizing properties at the atomic level and can form several long chain combinations. The carbon atom is quaternary and can form bonds via single, double, or triple covalent bonds, or with other elements due to its atomic configurations and size. These properties of carbon atoms are possible due to their small size. Carbon can exist in a wide range of different molecular forms called allotropes, which are composed of the same type of atom but, due to different structures, possess different properties. Along with the three types of naturally occurring carbon allotropes that were known previously, i.e. amorphous carbon, diamond, and graphite, several new forms have been described, including carbon nanomaterials [10]. Figure 11.2 shows the different types of carbon nanomaterials. The structure and different types of carbon nanomaterials are tabulated and summarized based on contemporary research findings in table 11.1.

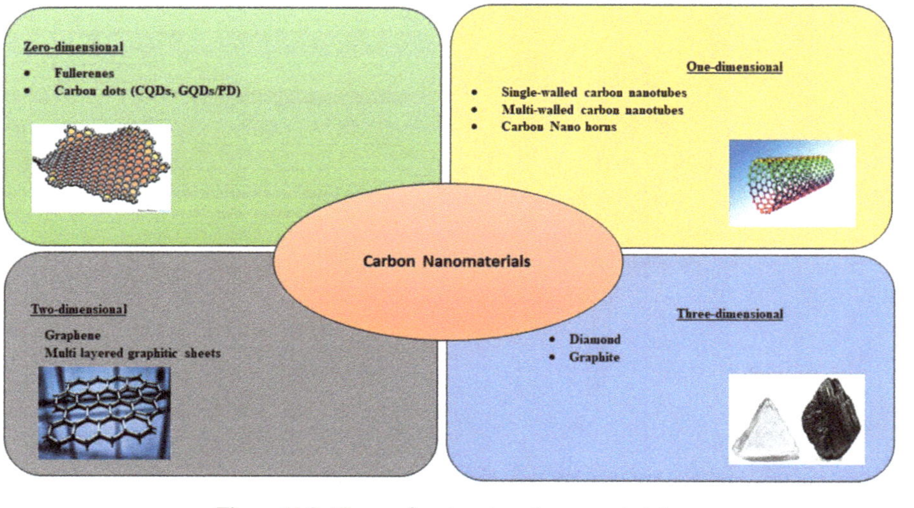


**Figure 11.2.** Types of carbon-based nanomaterials.

**Table 11.1.** Different types of carbon allotropes in nanoform.

| S. No. | Nanomaterial | Dimensions | Structure | Reference |
|---|---|---|---|---|
| 1 | Graphene quantum dots | Zero-dimensional | 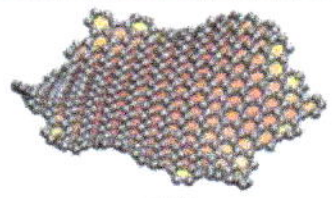 | [12] |
| 2 | Fullerene | Zero-dimensional |  | [13] |
| 3 | Carbon quantum dots | Zero-dimensional |  | [14] |
| 4 | Carbon nanotubes | One-dimensional | 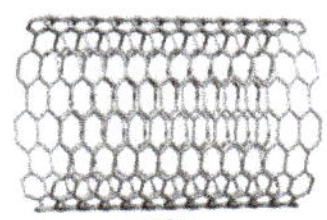 | [13] |
| 5 | Carbon nanohorns | One-dimensional | 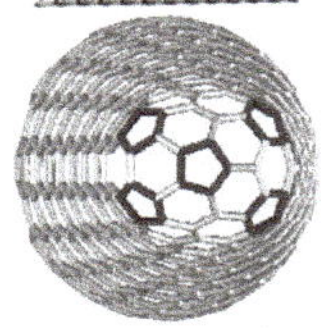 | [15] |
| 6 | Nano-graphene | Two-dimensional | 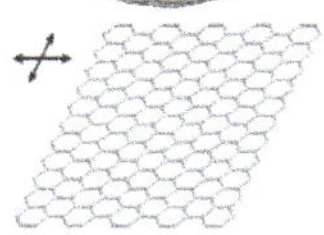 | [13] |
| 7 | Multi-layered graphitic sheets | Two-dimensional | 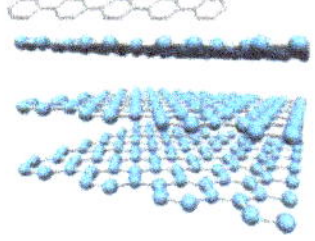 | [16] |
| 8 | Nano-diamond | Three-dimensional | 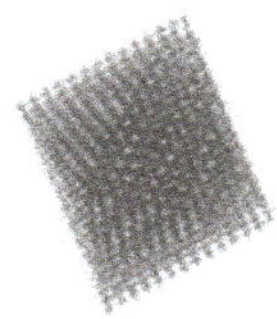 | [17] |
| 9 | Nano-graphite | Three-dimensional | 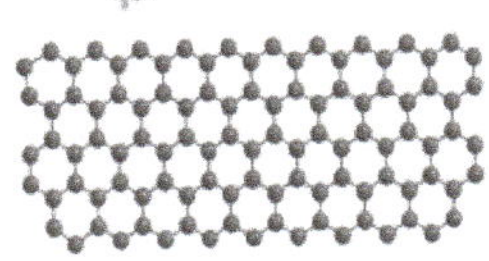 | [18] |

These newly discovered materials, such as carbon nanotubes (CNTs), graphene oxide (GO), graphene quantum dots (GQDs), and fullerene, are well explored in terms of their specific properties and applications [11]. Due their unique characteristics, such as their optical, electrical, physical, mechanical, chemical, and biological properties, these materials have shown the potential for different application [4].

## 11.3 The importance of zero-dimensional (0D) carbon nanomaterials

One of the main characteristics that distinguishes various types of nanostructures is their dimensionality. Zero-dimensional nanostructures are the materials in which all the dimensions are within the nanoscale (no dimension is greater than 100 nm) [19–21]. Carbon-based nanomaterials are well explored in the area of nanotechnology [22]. Examples of carbon-based 0D nanomaterials are fullerenes and carbon dots (CDs). Based on a core structure of carbon and attached surface groups, CDs can be further characterized as (i) carbon quantum dots, (ii) graphene/graphene oxide quantum dots, and (iii) carbonized polymer dots [23–26]. Among all the carbon nanomaterials, zero-dimensional carbon nanomaterials exhibit unique electrical and optical properties, and lower toxicity with high quantum yield. These materials are highly suited for use in preparing high performance micro-sensors with low power consumption [27, 28].

Zero-dimensional nanoparticles have been used in a variety of tissue engineering functions, ranging from improving the biomedical, physical, electrical, and mechanical properties of scaffolds, to DNA transfection, drug delivery, molecular growth of tissues, cell imaging, and bio-sensing applications [29].

### 11.3.1 Fullerenes

Fullerenes have a cage-like spherical structure where carbon is placed at the corners of the pentagons and hexagons of the polyhedral structure. The spherical fullerene looks like a football, a so called ‘bucky ball’. Fullerenes are named after the designer of geodesic domes similar to fullerenes, architect Richard Buckminster Fuller. In fact, researchers at Rice University accidentally discovered fullerene in a laser spectroscopy experiment in 1985 [11]. The diameter of fullerenes has been reported to be in the range 1.40 nm to 1.6 nm $\pm$ 0.05 nm [30]. These are inert and hollow, and can be modified indefinitely. They have been the center of attention in various scientific fields since their discovery. The unique carbon cage structure and huge derivatization range make fullerenes a potential therapeutic agent [31]. Their applications range from molecular electronics, environmental remediation, and water purifications, to biomedicine [24, 32, 33]. Fullerene has excellent anti-oxidant properties and hence shows applications in dermatological and cosmetic products [34]. Some other applications are shown in figure 11.3.

### 11.3.2 Carbon dots

Carbon dots (CDs) are zero-dimensional carbon-based nanomaterials with a size range smaller than 10 nm [35]. They can be synthesized by breaking down graphite or graphene oxide. These are a new addition to the carbon family and have attracted the attention of researchers due to their low toxicity, small size, excellent photoluminescence, low-cost sources, and important applications in the field of biomedicine, optoelectronic devices, and many more [25, 36, 37]. CDs can be of different shape or nature, namely quasi-spherical, crystalline or amorphous, and contain $sp^2$

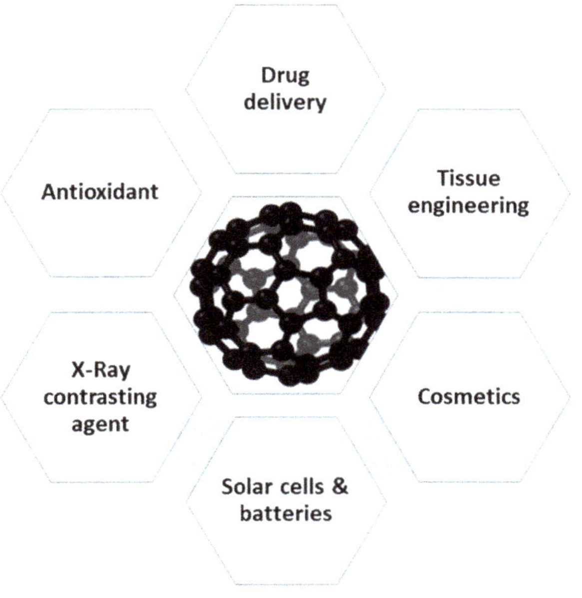


**Figure 11.3.** Applications of fullerenes.

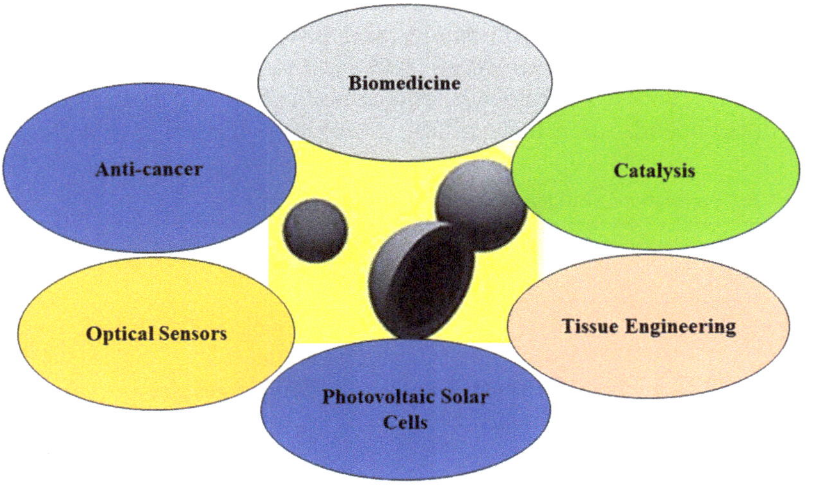


**Figure 11.4.** Applications of carbon dots.

and $sp^3$ hybridized carbon [26, 38]. The specific physical and optical properties of CDs have been explored for biomedical applications [39]. Their unique cage-like carbon-based core structure, low chemical activity, remarkable optical properties, economy, and better biological properties make CDs a suitable alternative to semiconductor quantum dots [40]. Some other applications of CDs are listed in figure 11.4. The diverse range of carbon dots are classified into (i) carbon quantum dots (CQDs), (ii) graphene quantum dots (GQDs) and graphene oxide quantum dots (GOQDs), and (iii) carbonized polymer dots (CPDs).

### *11.3.2.1 Carbon quantum dots*

Carbon quantum dots (CQDs) have improved emission quantum yield, which is a prerequisite for better imaging of cells. In addition to passivation of the surface, high quantum efficiency can be achieved by doping of heteroatoms such as nitrogen, sulfur, and even magnesium ions, which may provide a quantum yield of about 80%–83%. In addition to creating new functions, metal ion doping improves optical properties as well. There is, however, a higher chance of increased toxicity after metal ion doping, a downside for biomedical applications [41].

Yang *et al* explored the probability of PEGylated CQDs as a fluorescent contrasting agent [42]. The combination of medicine and bio-imaging shows great prospects for studying the drug distribution pattern and monitoring [43]. Molaei revived the applicability of DCQs in therapeutics and biomedical uses [44]. Mehta *et al* synthesized CQDs from *Saccharum officinarum* juice and showed its application for imaging cells of different bacterial groups [45]. CQDs have been utilized in preparing biosensors for monitoring blood glucose [46]. The relevant studies and established applications of CQDs are given in table 11.2.

### *11.3.2.2 Graphene and graphene oxide quantum dots*

Graphene quantum dots (GQDs) and graphene oxide quantum dots (GOQDs) are 0D nanomaterials of the carbon family and have shown promising biomedical applications due to their ultra-small size, non-toxicity, biocompatibility, light weight, stability, tunable fluorescence, and solubility [56, 57]. The structure of these quantum dots depends upon the synthesis conditions. Basically, they have a honey

**Table 11.2.** A few applications of CQDs.

| S. No. | Type of CQD | Application | References |
|---|---|---|---|
| 1. | Hollow CQDs | Doxorubicin (DOX) drug delivery | [47] |
| 2. | CQDs doped with inorganic salts (e.g. ZnS) | *In vivo* bio-imaging | [48] |
| 3. | CQDs prepared from phenylenediamine isomers | Functionalized nano-probes | [49] |
| 4. | CQDs passivated by poly(propionyl ethylenimine-*co*-ethylenimine) | Bio-imaging | [50] |
| 5. | Nitrogen doped CQDs | Imaging of HeLa and HepG2 cells | [51] |
| 6. | Human hair and pig skin derived CQDs | Imaging probes | [52] |
| 7. | CQDs coated with hydrophilic polymer coating | Targeting at the tumor site for better image | [53] |
| 8 | Photo luminescent cationic CQDs | Plasmid SOX9 delivery | [54] |
| 9 | Alginate derived cationic CQDs | Gene delivery | [55] |

comb crystal lattice of carbon atoms, which are arranged in a ring of six atoms. The p–p bonding in the atomic plane gives GQDs superior thermal and electrical conductivity and low toxicity as compared to conventional semiconductor quantum dots [58]. They show suitable applications in drug delivery, bio-imaging, bio-sensing, and antibacterial applications as well [59–62]. Figure 11.5 shows a few applications of GQDs and GOQDs.

*11.3.2.3 Polymer/carbonized polymer dots (PPDs)*

Polymer dots are unique and useful quantum dots, designed and applicable for advanced hybrid nanomaterial uses. These dots combine with other quantum dots and nanoparticles to form advanced hybrid nanomaterials. Polymer dot nanomaterials have appeared in many applications, such as solar cells, super-capacitors, electronics, probes, gas sensors, biosensors, bio-imaging, and drug delivery [24, 63, 64].

Semiconductor polymer dots prepared via reprecipitation followed by immobilizing on an indium tin oxide glass electrode showed their application for PEC sensing of L-cysteine. The polymer dots displayed PEC activity, a rapid response, enhanced stability, and selectivity [65]. A co-reactant-free two-fold amplified electrochemical luminescence (ECL) sensing device conjugated with luminophore polymer dots was designed by Luo *et al.* This polymer-dot-based device presented ultra-sensitive miRNA detection and recognition. These conjugated polymer dots provide a good platform for constructing ECL biosensors without co-reactants, and extended applications in clinical and medical analysis [66]. Multifunctional carbonized polymer dots (CPDs) doped with nitrogen have been reported, which were found to be potent scavengers as well as detectors for Alzheimer's $\beta$-amyloid plaques [67]. Indole based carbonized polymer dots (CPDs) have been observed to be

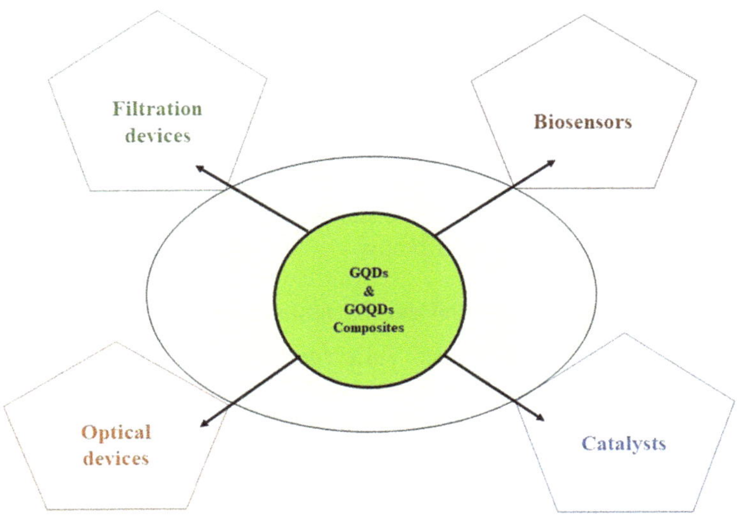


**Figure 11.5.** Applications of GQDs and GOQD composites.

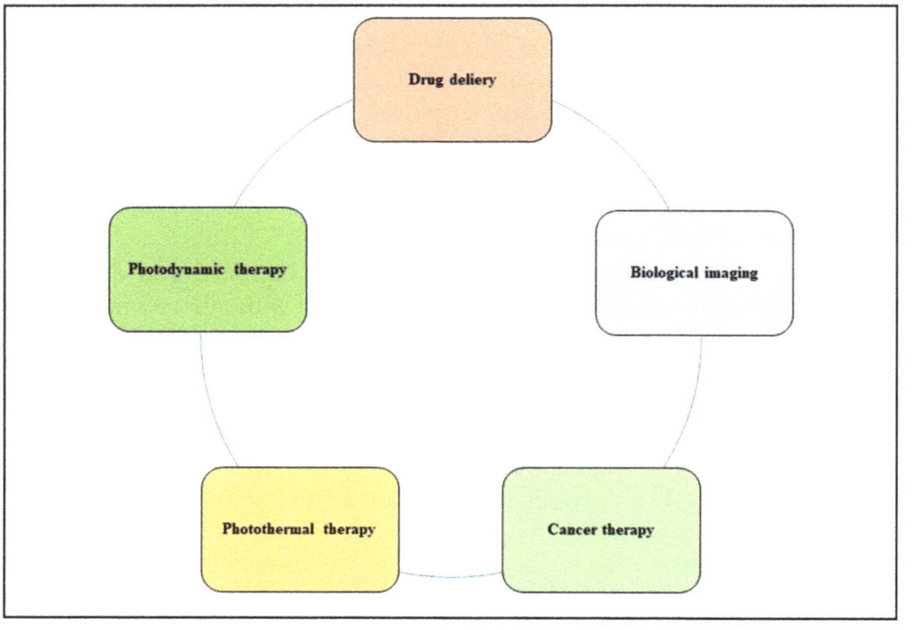


**Figure 11.6.** A few applications of polymer dots.

potential imaging tools for the future [68]. Less toxic carbonized polymer dots have been synthesized via a hydrothermal method. They have shown high fluorescence quantum yield, which is useful as a pH indicator. The various applications of polymer dots and carbonized polymer dots are described in figure 11.6.

## 11.4 Synthesis and characterization of zero-dimensional carbon nanomaterials

Zero-dimensional nanostructures, including nanoparticles and quantum dots, have attracted the attention of researchers over the past few years due to their remarkable physical properties, such as their electronic, optical, and magnetic properties. Therefore, their synthesis and characterization has become significant. A few of the synthesis methods are discussed below.

### 11.4.1 Synthesis of zero-dimensional carbon nanomaterials

Kratschmer and Hoffman produced fullerenes by passing a graphite electrode through helium atmosphere and later on modification of the reactor was performed by setting an arc between the two charges of graphite electrode. This method is known as the arch discharge method [10]. This synthesis pulse arc discharge method operates at high temperatures. The components of the device are a furnace for heating purposes, a quartz tube, a carbon electrode, a water cooling trap, and high voltage pulse power input. For the carbon group annealing process, the buffer gas passes through the quartz tube. The temperature of the tube is between 25 °C and 1000 °C. In this way, the electrode of the negative electrode is consumed, and the condensed carbon is quenched to form fullerenes. The fullerene and other carbon particles are then collected in a water-cooled trap. The analysis of fullerene has been carried out by high performance liquid chromatography (HPLC). This HPLC system can detect low concentrations of fullerenes. The fullerene concentration, high voltage pulse duration, and temperature dependence of the buffer gas type were measured [69]. An arc discharge apparatus is as shown in figure 11.7.

Another method for synthesizing fullerenes is the vaporization method. In this method a carbon source is vaporized via laser irradiation, resulting in the production of C60 and C70 fullerenes, however, this method is less efficient for higher fullerenes. The vaporization method now uses pyrolysis techniques, arc discharge-plasma, or radio-frequency-plasma for vaporization of graphite or other carbon sources [70].

A method for the preparation of fullerenes where fusing two similar hemispherical hydrocarbons was proposed was found to be unsuccessful [71]. A new method of synthesis of corannulene, which used vacuum pyrolysis in ring-closure step(s), was developed and improved [72, 73].

Carbon dots are synthesized chiefly by two methods: (a) the top-down approach and (b) bottom-up approach. The top-down method means the decomposition of larger carbon structural forms by either chemical oxidation, various ultrasonic methods, and electrochemical oxidation, whereas the bottom-up approach means converting smaller carbon structures into a CD of the anticipated shape and properties. This method includes thermal decomposition, ultrasonic treatment, microwave synthesis, and carbonization. Basically, in the top-down approach, energy is applied to convert larger carbon structures into ultra-small fragments,

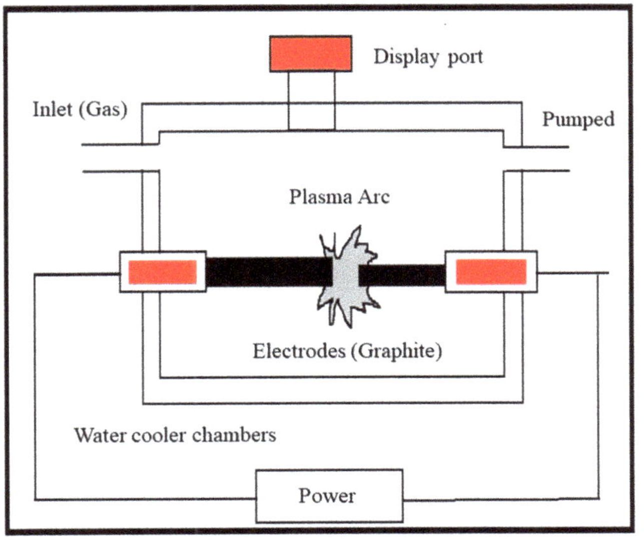


**Figure 11.7.** Arc discharge apparatus.

and, in the case of the bottom-up approach, carbon dots are synthesized from smaller carbon units through the application of energy, and the molecules ionize, dissociate, evaporate, and then condense to form carbon dots [74]. An illustration of both methods is shown in figure 11.8.

The bottom-up method requires complex reaction steps, so it is difficult to optimize the conditions. Hence, the top-down method is preferred over the bottom-up method. The former involves the cutting of large pieces of carbon material into smaller pieces (figure 11.9). The raw materials required by this method are abundant carbon materials, which are cheap and easily available, and the method is relatively simple and easy to replicate. It involves various methods, such as the chemical oxidation method, pulsed laser ablation technique, and chemical vapor deposition method [75].

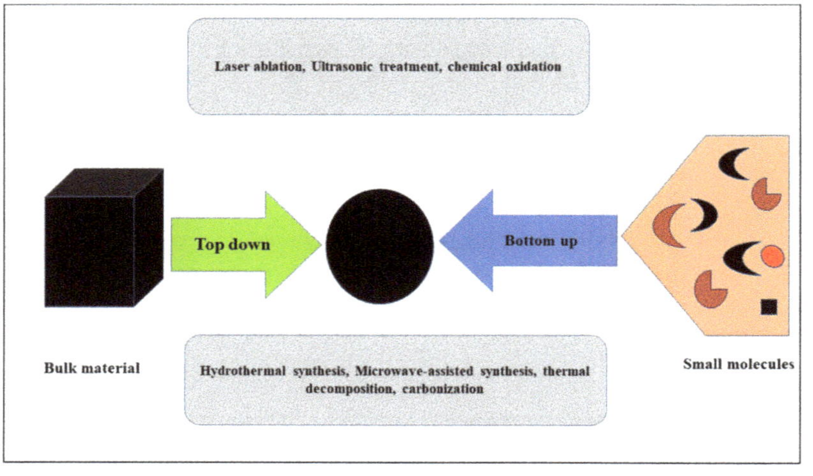


**Figure 11.8.** Synthesis of carbon dots.

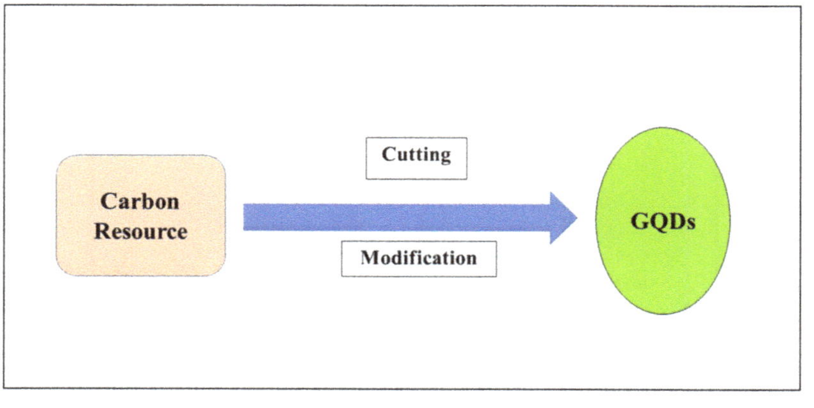


**Figure 11.9.** The top-down method.

Various methods have been reported for the synthesis of polymer dots. One of the simplest methods is microwave-assisted synthesis. The aqueous-dispersible as well as fluorescent polymer dots have been synthesized from hyper-branched poly-ethylenimine and glycerol by a microwave-assisted technique [76]. Tang *et al* synthesized PDs through an amidation reaction between hyper-branched poly-ethylenimine and 5-aminosalicylic acid, which is an example of a self-assembly method. The preparation of core–shell structured carbon polymer dots by hydrothermal synthesis has also been reported [77]. Red-emissive carbon polymer dots have been also synthesized via of oxidative polymerization from *p*-phenylenediamine [78].

### 11.4.2 The characterization of zero-dimensional carbon nanomaterials

The most popular tools for studying the particle size and morphology of carbon nanomaterials are microscopy techniques, i.e. atomic force microscopy (AFM), ultra-violet visible spectroscopy (UV–vis), x-ray photoelectron spectroscopy (XPS), transmission electron microscopy (TEM), scanning electron microscopy (SEM), x-ray diffraction (XRD), Raman spectroscopy, photoluminescence (PL), scanning tunneling microscopy (STM), zeta potential analyzer, dynamic light scattering analysis (DLS), thermogravimetric analysis (TGA), Fourier transform infrared spectroscopy (FTIR), and x-ray photoelectron energy spectroscopy (XPS) [41, 79–83].

One of the most prominent physical and chemical properties of fullerenes has been that they tend to aggregate to form clusters of various shapes and sizes. Understanding the degree of aggregation of fullerenes, and their size distribution and shape is essential for assessing their toxicity and tracking their transportation and fate in the environment [84]. Raman spectroscopy helps in revealing the structural properties and characteristics of carbon atoms in a carbon dot. Typically, two broad peaks are observed in the CD Raman spectrum, located at 1300 $cm^{-1}$ and 1580 $cm^{-1}$, corresponding to the D band (disordered $sp^3$ carbon) and the G band, respectively. The crystal band G is related to the vibration in the plane of the $sp^2$ carbon. The relative strength ($I_D/I_G$) of these bands can be used to determine the structural characteristics of the carbon skeleton, in particular the crystallinity and relative abundance inside the carbon dot [85].

## 11.5 Application of zero-dimensional carbon nanomaterials in tissue engineering

In recent years, a lot of theoretical and experimental research has been conducted that resulted in the development of the most fascinating feature of nanotechnology, known as carbon-based nanomaterials. Various synthetic techniques have been employed to prepare carbon nanomaterials with unique structural shapes and features. Among these the zero-dimensional carbon nanomaterials have emerged as chemically stable, less toxic, and biocompatible materials with decent water dispersibility, that display a wide range of applications such as bio-sensing, gene delivery, cell imaging, etc [86]. Utilizing new approaches to formulate scaffolds for tissue engineering is important, such as the fusion of nanotechnology and tissue engineering [87]. Carbon nanomaterials appear to be a promising substitute for the conventional cell regeneration and repair approaches.

Non-natural scaffolds that are used for tissue regeneration have to be analogous to the natural extracellular matrix in terms of composition, structure, and strength [88, 89]. Carbon nanomaterials have been found to have physical analogy to extra cellular matrix (ECM) components. The ease of function, high conductivity, and remarkable mechanical properties of carbon-based nanomaterials further support the utilization of these materials in artificial scaffolds [90, 91]. Zero-dimensional carbon nanomaterials also started making inroads in the area of tissue engineering [8]. However, the use of these nanostructures in tissue engineering is still in its early stages and is mired in few challenges. The zero-dimensional carbon nanomaterials are not only found to be useful in the field of cell/tissue regeneration or replacement, but also in preparing the devices which are needed to monitor tissue growth, cell imaging, and tissue engineering.

Fullerenes have attractive physical and chemical properties that can be used in medical chemistry. These materials have been reported recently for their application in bone tissue engineering. Fullerenes have a spherical molecular structure which facilitates free radical scavenging, leading to its application in biomedicine. In recent years, due to the observation of increased cell adhesion to fullerene-based bio-nanomaterials, their application in bone tissue engineering has attracted the attention of many researchers. Carbon nanofiber coated fullerene layers have been developed which can improve the adhesion of MG-63 osteoblasts and enhances cell proliferation [92]. The morphology of the fullerenes is layered, which also plays a crucial role in cell growth. Additionally, fullerene-based biomaterials limit cell growth to 1 mm height. Krishnan *et al* provided a new method, which is to fabricate and deposit fullerene nano-whiskers on a scaffold to stimulate the growth of MG-63 osteoblasts. This is achieved by the rotating flow of a solution containing fullerene nano-shells, allowing the matrix deposition to be evenly arranged on the glass substrate. It was found that the regular arrangement of fullerene nano-whisker deposition shows better performance than a random deposition. The distance between the glass substrate and the center of the vortex played a key role in the arrangement. The sample prepared at the edge of the vortex solution had a more regular morphology than the sample from the center of the vortex [93].

The design of an appropriate three-dimensional nanostructure with attractive tissue engineering applications is a multidisciplinary task. The cells and scaffolds are the basic building blocks necessary for tissue constructs, which combine signals and therapeutic agents. The scaffold serves as a template which is important for the transformation of cells into tissues and to regulate cell functions, namely (i) cell adhesion, (ii) cell growth, (iii) cell proliferation, and (iv) cell differentiation. At the same time the scaffold also controls the deposition of the extracellular matrix. Additionally, carbon dot incorporated polymer-based scaffolds have been found to be useful in tissue engineering. Silk fibroin PLA nanofiber scaffold preparation mediated via CQDs (functionalized *p*-phenylenediamine) has been reported by Ren *et al.* The prepared scaffolds exhibited improved mechanical properties with expanded use in cardiac tissue engineering [94]. CQDs have also been reported to limit bacterial growth at a wound infection site, thus accelerating wound healing.

The antibacterial activity of any CQD has been found to be related to its surface and size. Additionally, CQDs induce ROS and prevent bacterial growth [95]. CQDs have been reported for bactericidal activity and encouraging fibroblast proliferation in a mouse model [86]. Recently, a CD incorporated fluorescent scaffold was found to be useful for tissue engineering applications [94].

Researchers fabricated an antibacterial system that combined graphene quantum dots (GQD) with hydrogen peroxide, $H_2O_2$. A high concentration of $H_2O_2$ is harmful to normal tissues and can also delay wound healing. GQD-$H_2O_2$ can even kill bacteria at a lower concentration. The GQDs show activity like peroxide and catalyze the $H_2O_2$ decomposition to yield hydroxyl radicals which shows high antibacterial activity [96].

As a zero-dimensional members of the carbon family, GQDs and CQDs are expected to actively replace these conventional fluorophores due to their strong and adjustable PL, photo-stability, excellent biocompatibility, and effective renal clearance, providing unprecedented opportunities for biological imaging. CQDs and GQDs have shown their application in two-photon cellular/nucleus deep and high-resolution bio-imaging [97, 98] and, therefore, may be used for observing tissue proliferation and cell regeneration. Thus, these materials can be designed to make tools and probes that can be used in image guided tissue engineering [94].

Polymer dots (PDs) have been reported for their usefulness in the preparation of fluorescence probes for bio-sensing and bio-imaging [64]. Huang *et al* prepared Coumarin-dye-embedded polymer dots that displayed fluoride ion sensing in an aqueous solution and as well as cellular bio-imaging [99].

Ren *et al* synthesized CQDs which were functionalized with *p*-phenylenediamine and prepared nano-fibrous silk scaffolds. The scaffolds had enhanced mechanical properties and were found to be useful in cardiac tissue engineering [100].

Pal *et al* prepared a polycaprolactone-gelatin matrix and carbon nanodot containing fluorescent nano-fibrous scaffold. They showed its application as a non-invasive imaging tool in *in vivo* skin tissue regeneration [101]. A few of the applications of zero dot carbon nanomaterials in tissue engineering are listed in table 11.3.

## 11.6 Future prospects and conclusion

To summarise, the carbon-based nanomaterials possess remarkable and novel physical, chemical, and biological properties, such as a high surface-to-volume ratio, quantum confinement effects, very small size, optical properties, biocompatibility, and ease of modification and functionalization. The unique properties of the carbon allotropes has led to a wide range of applications. The carbon-based nanostructures can be categorized as zero-, one-, two-, and three-dimensional nanomaterials. The 0D carbon nanomaterials include graphene quantum dots, carbon quantum dots, fullerenes, and polymer dots. The materials can be prepared via various synthetic routes (such as vaporization methods, ultrasonic treatments,

**Table 11.3.** Applications of carbon-based zero-dimensional (0D) nanomaterials in tissue engineering.

| S. No. | 0D carbon nanomaterial | Methodology | Properties | Applications | Reference |
|---|---|---|---|---|---|
| 1. | Chitosan-CDs | Hydrogel or solvent casting | Bactericidal, biocompatibility. | Skin tissue regeneration | [102] |
| 2. | PLA-CQDs | Scaffold/3D printing | Biocompatibility, scaffold imaging. | Tissue engineering | [103] |
| 3. | $C_{60}$NW | LLIP method | Promotes growth and differentiation of stem cell (osteoblasts and neural) | Tissue engineering and sensors | [104, 105] |
| 4. | GQDs/PVA | Electro-spinning | Formation of surface oxide traps for electrons. | Smart probe for bio-sensing | [106] |
| 5. | PDotsNH2-PEG | Fluorescence/*in vivo* | Large two-photon-absorption, good water dispersibility, high biocompatibility. | Brain imaging | [107] |
| 6. | SF-CQD-PLA | Electro-spun | Enhanced cell viability. | Tissue engineering | [100] |

and microwave synthesis) and can be characterized using XRD, TEM, SEM, PL, etc. The zero-dimensional nanomaterials have small size, high specific surface area, edge and good bio-compatibility. Due to their unique structure and performance, the zero-dimensional nanomaterials are expanding their application in the field of bio-sensing. These promising results suggest that 0D carbon nanomaterials present numerous physical, chemical, and biological properties that may be utilized in multidisciplinary research, such as bio-sensing, cell imaging, and tissue engineering.

The incredible rate of progress in nano-biotechnology has created great enthusiasm within industry and academia for the use of zero-dimensional carbon nanomaterials in tissue engineering applications, yet there remain several limiting issues. The investigation of all these nanomaterials in biosensors is still at an early stage. To make advancements in this field and encourage clinical applications, several challenges must be addressed. The major problem and concern about these materials is their toxicity, in particular for biological applications such as tissue engineering. These materials may accumulate in normal mammalian cells and at high concentrations may cause damage such as hemolysis, mutations, etc. Blending these materials with appropriate scaffolds to obtain biocompatibility may help in resolving this issue. Additionally, the use of new synthesis methods to functionalize and enhance hydrophilicity may help in reducing cytotoxicity.

The development of numerous synthetic methods can make it feasible to produce desirable multifaceted, tuned, and functionalized zero-dimensional nanomaterials, although detailed research on the the structure–activity relationship is still to be done. Additionally, due to non-standard preparation methods, the actual reaction mechanism, reaction conditions, nucleation mechanism, and effects of various precursors are still not known. Moreover, there is an absence of an efficient, fascile, and scalable synthetic method to yield high-quality 0D carbon nanomaterials with desirable structures and properties. Therefore, there is a need for an efficient production method that can produce carbon nanomaterials on a large scale. For real-time applications, a lot of systematic research has to be performed to develop an effective purification system for carbon nanomaterials on the basis of size, charge, and polarity. There is an immediate requirement for the development of an *in situ* technique and advanced technology for studying the reaction mechanism, surface groups, crosslinking, excited states, etc. This may help in controlling the synthesis of carbon nanostructures with homogeneous size, shape, and properties, promising a bright future for these materials.

Additionally, to deal with the existing limitations, continuous efforts and research must be ongoing to develop fluorescent 0D carbon nanostructure mediated scaffolds for non-invasive image-aided tissue engineering. Significant work has to be performed in preparing the next-generation red-to-NIR emissive 0D carbon nanomaterials that may result in increased efficiency, deeper penetration, more light scattering, and reduced loss of photons. The fabrication of tissue-compatible carbon nanomaterials may also help in achieving fewer clinical trial errors. The zero-dimensional nanomaterials have great scope for real-time applications in the area of tissue engineering in the near future.

## Acknowledgments

The authors would like to acknowledge the support provided under the DST-FIST Grant No. SR/FST/PS-I/2018/48 of the Government of India.

## References

[1] Wang Q H and Strano M S 2013 Carbon nanotubes: a bright future for defects *Nat. Chem.* **5** 812–3

[2] Hirsch A 2008 *The Chemistry of the Fullerenes* (Weinheim, New York: Wiley)

[3] Singh V, Joung D, Zhai L, Das S, Khondaker S I and Seal S 2011 Graphene based materials: past, present and future *Prog. Mater. Sci.* **56** 1178–271

[4] Harrison B S and Atala A 2007 Carbon nanotube applications for tissue engineering *Biomaterials* **28** 344–53

[5] Joshi V, Srivastava C M, Gupta A P and Vats M 2020 Electrospun nano-architectures for tissue engineering and regenerative medicine *Nanosci. Med.* **1** 213–48

[6] Walmsley G G *et al* 2015 Nanotechnology in bone tissue engineering *Nanomedicine* **11** 1253–63

[7] Kumar R, Kumar V B and Gedanken A 2020 Sonochemical synthesis of carbon dots, mechanism, effect of parameters, and catalytic, energy, biomedical and tissue engineering applications *Ultrason. Sonochem.* **64** 105009

[8] Zhao H, Ding R, Zhao X, Li Y, Qu L, Pei H, Yildirimer L, Wu Z and Zhang W 2017 Graphene-based nanomaterials for drug and/or gene delivery, bioimaging, and tissue engineering *Drug Discovery Today* **22** 1302–17

[9] Eivazzadeh-Keihan R, Maleki A, de la Guardia M, Bani M S, Chenab K K, Pashazadeh-Panahi P, Baradaran B, Mokhtarzadeh A and Hamblin M R 2019 Carbon based nanomaterials for tissue engineering of bone: building new bone on small black scaffolds: a review *J. Adv. Res.* **18** 185–201

[10] Zaytseva O and Neumann G 2016 Carbon nanomaterials: production, impact on plant development, agricultural and environmental applications *Chem. Biol. Technol. Agric.* **3** 17

[11] Mostofizadeh A, Li Y, Song B and Huang Y 2011 Synthesis, properties, and applications of low-dimensional carbon-related nanomaterials *J. Nanomater.* **2011** 685081

[12] Brotchie A 2016 Graphene quantum dots: it's all in the twist *Nat. Rev. Mater* **1** 16006

[13] Segawa Y, Ito H and Itami K 2016 Structurally uniform and atomically precise carbon nanostructures *Nat. Rev. Mater.* **1** 15002

[14] Nguyen T P, Nguyen D L T, Nguyen V H, Le T H, Vo D V N, Ly Q V, Kim S Y and Van Le Q 2019 Recent progress in carbon-based buffer layers for polymer solar cells *Polymers* **11** 1858

[15] Berber S, Kwon Y K and Tománek D 2000 Electronic and structural properties of carbon nanohorns *Phys. Rev.* B **62** R2291

[16] Lloyd-Hughes J and Jeon T I 2012 A review of the terahertz conductivity of bulk and nano-materials *J. Infrared, Millimeter, Terahertz Waves* **33** 871–925

[17] Komatsu N 2019 Various surfactants for 0–3 dimensional nanocarbons: separation, exfoliation and solubilization *Adv. Mater. Lett.* **10** 520–3

[18] Yuan X, Zhang X, Sun L, Wei Y and Wei X 2019 Cellular toxicity and immunological effects of carbon-based nanomaterials *Part. Fibre Toxicol.* **16** 18–44

[19] Liu J N, Bu W and Shi J 2017 Chemical design and synthesis of functionalized probes for imaging and treating tumor hypoxia *Chem. Rev.* **117** 6160–24

[20] Sondhi P, Maruf M H U and Stine K J 2020 Nanomaterials for biosensing lipopolysaccharide *Biosensors* **10** 2
[21] Chen J B, Yousefi H, Nemr C R, Gomis S, Atwal R, Labib M, Sargent E and Kelley S O 2020 Nanostructured architectures for biomolecular detection inside and outside the cell *Adv. Funct. Mater.* **30** 1–20
[22] Panwar N, Soehartono A M, Chan K K, Zeng S, Xu G, Qu J, Coquet P, Yong K T and Chen X 2019 Nanocarbons for biology and medicine: sensing, imaging, and drug delivery *Chem. Rev.* **119** 9559–656
[23] Zhu S, Song Y, Zhao X, Shao J, Zhang J and Yang B 2015 The photoluminescence mechanism in carbon dots (graphene quantum dots, carbon nanodots, and polymer dots): current state and future perspective *Nano Res.* **8** 355–81
[24] Wang Z, Hu T, Liang R and Wei M 2020 Application of zero-dimensional nanomaterials in biosensing *Front. Chem.* **8** 320
[25] Liu J, Li R and Yang B 2020 Carbon dots: a new type of carbon-based nanomaterial with wide applications *ACS Cent. Sci.* **6** 2179–95
[26] Xia C, Zhu S, Feng T, Yang M and Yang B 2019 Evolution and synthesis of carbon dots: from carbon dots to carbonized polymer dots *Adv. Sci.* **6** 1–23
[27] Shi X, Meng H, Sun Y, Qu L, Lin Y, Li Z and Du D 2019 Far-red to near-infrared carbon dots: preparation and applications in biotechnology *Small* **15** 1–17
[28] Zhou B, Guo Z, Lin Z, Zhang L, Jiang B P and Shen X C 2019 Recent insights into near-infrared light-responsive carbon dots for bioimaging and cancer phototherapy *Inorg. Chem. Front.* **6** 1116–28
[29] Hasan A, Morshed M, Memic A, Hassan S, Webster T J and Marei H E S 2018 Nanoparticles in tissue engineering: applications, challenges and prospects *Int. J. Nanomed.* **13** 5637–55
[30] Sloan J *et al* 2000 The size distribution, imaging and obstructing properties of C60 and higher fullerenes formed within arc-grown single walled carbon nanotubes *Chem. Phys. Lett.* **316** 191–8
[31] Yadav B C and Kumar R 2008 Structure, properties and applications of fullerenes *Int. J. Nanotechnol. Appl.* **2** 15–24
[32] Bhakta P and Barthunia B 2020 Fullerene and its applications: a review *J. Indian Acad. Oral Med. Radiol.* **32** 159–63
[33] Chae S R, Hotze E M and Wiesner M R 2014 Possible applications of fullerene nanomaterials in water treatment and reuse *Nanotechnology Applications for Clean Water: Solutions for Improving Water Quality* 2nd edn (Amsterdam: Elsevier) pp 329–38
[34] Chen Z, Mao R and Liu Y 2012 Fullerenes for cancer diagnosis and therapy: preparation, biological and clinical perspectives *Curr. Drug Metab.* **13** 1035–45
[35] Sun Y P *et al* 2006 Quantum-sized carbon dots for bright and colorful photoluminescence *J. Am. Chem. Soc.* **128** 7756–7
[36] Liu C, Zhang F, Hu J, Gao W and Zhang M 2021 A mini review on pH-sensitive photoluminescence in carbon nanodots *Front. Chem.* **8** 605028
[37] Ray S C, Saha A, Jana N R and Sarkar R 2009 Supporting information fluorescent carbon nanoparticle: synthesis, characterization and bio-imaging application *J. Phys. Chem.* C **113** 118546–51
[38] Siddique A B *et al* 2018 Amorphous carbon dots and their remarkable ability to detect 2,4,6-trinitrophenol *Sci. Rep.* **8** 9770

[39] Zheng X T, Ananthanarayanan A, Luo K Q and Chen P 2015 Glowing graphene quantum dots and carbon dots: properties, syntheses, and biological applications *Small* **11** 1620–36
[40] Feng Z, Adolfsson K H, Xu Y, Fang H, Hakkarainen M and Wu M 2021 Carbon dot/polymer nanocomposites: from green synthesis to energy, environmental and biomedical applications *Sustainable Mater. Technol.* **29** e00304
[41] Jhonsi M A 2018 Carbon quantum dots for bioimaging *State of the Art in Nano-bioimaging.* ed G Morteza (London: IntechOpen)
[42] Yang S T *et al* 2009 Carbon dots as nontoxic and high-performance fluorescence imaging agents *J. Phys. Chem.* C **113** 18110–4
[43] Cheng Z, Al Zaki A, Hui J Z, Muzykantov V R and Tsourkas A 2012 Multifunctional nanoparticles: cost versus benefit of adding targeting and imaging capabilities *Science* **338** 903–10
[44] Molaei M J 2019 Carbon quantum dots and their biomedical and therapeutic applications: a review *RSC Adv.* **9** 6460–81
[45] Mehta V N, Jha S and Kailasa S K 2014 One-pot green synthesis of carbon dots by using saccharum officinarum juice for fluorescent imaging of bacteria (*Escherichia coli*) and yeast (*Saccharomyces cerevisiae*) cells *Mater. Sci. Eng.* C **38** 20–7
[46] Chen L, Hwang E and Zhang J 2018 Fluorescent nanobiosensors for sensing glucose *Sensors* **18** 1440
[47] Wang Q, Huang X, Long Y, Wang X, Zhang H, Zhu R, Liang L, Teng P and Zheng H 2013 Hollow luminescent carbon dots for drug delivery *Carbon* **59** 192-–9
[48] Srivastava S, Awasthi R, Tripathi D, Rai M K, Agarwal V, Agrawal V, Gajbhiye N S and Gupta R K 2012 Magnetic-nanoparticle-doped carbogenic nanocomposite: an effective magnetic resonance/fluorescence multimodal imaging probe *Small* **8** 1099–109
[49] Jiang K, Sun S, Zhang L, Lu Y, Wu A, Cai C and Lin H 2015 Red, green, and blue luminescence by carbon dots: full-color emission tuning and multicolor cellular imaging *Angew. Chem. Int. Ed.* **54** 5360–3
[50] Cao L *et al* 2007 Carbon dots for multiphoton bioimaging *J. Am. Chem. Soc.* **129** 11318–9
[51] Li W, Zhang Z, Kong B, Feng S, Wang J, Wang L, Yang J, Zhang F, Wu P and Zhao D 2013 Simple and green synthesis of nitrogen-doped photoluminescent carbonaceous nanospheres for bioimaging *Angew. Chem. Int. Edn* **52** 8151–5
[52] Zhang J H, Niu A, Li J, Fu J W, Xu Q and Pei D S 2016 *In vivo* characterization of hair and skin derived carbon quantum dots with high quantum yield as long-term bioprobes in zebrafish *Sci. Rep.* **6** 37860
[53] Wang Y, Meng Y, Wang S, Li C, Shi W, Chen J, Wang J and Huang R 2015 Direct solvent-derived polymer-coated nitrogen-doped carbon nanodots with high water solubility for targeted fluorescence imaging of glioma *Small* **11** 3575–81
[54] Cao X *et al* 2018 Photoluminescent cationic carbon dots as efficient non-viral delivery of plasmid SOX9 and chondrogenesis of fibroblasts *Sci. Rep.* **8** 7057
[55] Zhou J *et al* 2016 Cationic carbon quantum dots derived from alginate for gene delivery: one-step synthesis and cellular uptake *Acta Biomater.* **15** 209–19
[56] Younis M R, He G, Lin J and Huang P 2020 Recent advances on graphene quantum dots for bioimaging applications *Front. Chem.* **8** 424
[57] Dreyer D R, Park S, Bielawski C W and Ruoff R S 2010 The chemistry of graphene oxide *Chem. Soc. Rev.* **39** 228–40

[58] Zhao X, Gao W, Zhang H, Qiu X and Luo Y 2019 Graphene quantum dots in biomedical applications: recent advances and future challenges *Handbook of Nanomaterials in Analytical Chemistry* (Amsterdam: Elsevier) pp 493–505

[59] Tabish T A and Zhang S 2019 Graphene quantum dots: syntheses, properties, and biological applications *Compr. Nanosci. Nanotechnol.* **609** 335–54

[60] Jin L, Ren K, Xu Q, Hong T, Wu S, Zhang Y and Wang Z 2018 Multifunctional carbon dots for live cell staining and tissue engineering applications *Polym. Compos.* **39** 73–80

[61] Zeng Z, Yu D, He Z, Liu J, Xiao F X, Zhang Y, Wang R, Bhattacharyya D and Tan T T Y 2016 Graphene oxide quantum dots covalently functionalized PVDF membrane with significantly-enhanced bactericidal and antibiofouling performances *Sci. Rep.* **6** 20142

[62] Kang S, Kim K M, Jung K, Son Y, Mhin S, Ryu J H, Shim K B, Lee B, Han H S and Song T 2019 Graphene oxide quantum dots derived from coal for bioimaging: facile and green approach *Sci. Rep.* **9** 4101

[63] Kausar A 2021 Polymer dots and derived hybrid nanomaterials: a review *J. Plast. Film Sheeting* **37** 1–19

[64] Solhi E and Hasanzadeh M 2019 Recent advances on the biosensing and bioimaging based on polymer dots as advanced nanomaterial: analytical approaches *TrAC Trends Anal. Chem.* **118** 840–52

[65] Li Y, Zhang N, Zhao W W, Jiang D C, Xu J J and Chen H Y 2017 Polymer dots for photoelectrochemical bioanalysis *Anal. Chem.* **89** 4945–50

[66] Luo J H, Li Q, Chen S H and Yuan R 2019 Coreactant-free dual amplified electrochemiluminescent biosensor based on conjugated polymer dots for the ultrasensitive detection of microRNA *ACS Appl. Mater. Interfaces* **11** 27363–70

[67] Gao W, Wang W, Dong X and Sun Y 2020 Nitrogen-doped carbonized polymer dots: a potent scavenger and detector targeting Alzheimer's $\beta$-amyloid plaques *Small* **16** 2002804

[68] Liu C, Jin Y, Wang R, Han T, Liu X, Wang B, Huang C, Zhu S and Chen J 2020 Indole carbonized polymer dots boost full-color emission by regulating surface state *iScience* **23** 101546

[69] Yadav J 2017 Fullerene: properties, synthesis and application *Res. Rev. J. Phys.* **6** 1–6

[70] Alekseev N I, Filippov B M, Basargin I V and Sedov A I 2011 Investigation of semicommercial arc plants for fullerene-production *J. Eng. Phys. Thermophys.* **84** 1087–98

[71] Scott L T 2004 Methods for the chemical synthesis of fullerenes *Angew. Chem. Int. Edn* **23** 4994–5007

[72] Scott L T, Hashemi M M, Meyer D T and Warren H B 1991 Corannulene. A convenient new synthesis *J. Am. Chem. Soc.* **113** 7082–4

[73] Scott L T, Cheng P C, Hashemi M M, Bratcher M S, Meyer D T and Warren H B 1997 Corannulene. A three-step synthesis *J. Am. Chem. Soc.* **119** 10963–8

[74] Sharma A and Das J 2019 Small molecules derived carbon dots: synthesis and applications in sensing, catalysis, imaging, and biomedicine *J. Nanobiotechnol.* **17** 1–24

[75] Zhao C *et al* 2020 Synthesis of graphene quantum dots and their applications in drug delivery *J. Nanobiotechnol.* **18** 142

[76] Park J, Kwon B, Jeong W, Chae A, Choi Y, Park S Y and In I 2017 Microwave-assisted synthesis of fluorescent polymer dots from hyperbranched polyethylenimine and glycerol *Chem. Lett.* **46** 1463–5

[77] Yu Y, Tang P, Barnych B, Zhao C, Sun G and Ge M 2019 Design and synthesis of core–shell carbon polymer dots with highly stable fluorescence in polymeric materials *ACS Appl. Nano Mater.* **2** 6503–12

[78] Xia J, Chen S, Zou G Y, Yu Y L and Wang J H 2018 Synthesis of highly stable red-emissive carbon polymer dots by modulated polymerization: from the mechanism to application in intracellular pH imaging *Nanoscale* **10** 22484–92

[79] Goel A, Howard J B and Sande J B V 2004 Size analysis of single fullerene molecules by electron microscopy *Carbon* **42** 1907–15

[80] Campos B B, Abellán C, Zougagh M, Jimenez-Jimenez J, Rodríguez-Castellón E, Esteves da Silva J C G, Ríos A and Algarra M 2015 Fluorescent chemosensor for pyridine based on n-doped carbon dots *J. Colloid Interface Sci.* **458** 209–16

[81] Sharon M and Mewada A 2018 *Carbon Dots as Theranostic Agents* (Beverly, MA: Scivener )

[82] Peng J *et al* 2012 Graphene quantum dots derived from carbon fibers *Nano Lett.* **12** 844–9

[83] Li W, Li M, Liu Y, Pan D, Li Z, Wang L and Wu M 2018 Three minute ultrarapid microwave-assisted synthesis of bright fluorescent graphene quantum dots for live cell staining and white LEDs *ACS Appl. Nano Mater.* **1** 1623–30

[84] Klaine S J, Alvarez P J J, Batley G E, Fernandes T F, Handy R D, Lyon D Y, Mahendra S, McLaughlin M J and Lead J R 2008 Nanomaterials in the environment: behavior, fate, bioavailability, and effects *Environ. Toxicol. Chem.* **27** 1825–51

[85] Wu Z L, Liu Z X and Yuan Y H 2017 Carbon dots: materials, synthesis, properties and approaches to long-wavelength and multicolor emission *J. Mater. Chem. B.* **5** 3794–809

[86] Li P *et al* 2020 Carbon quantum dots derived from lysine and arginine simultaneously scavenge bacteria and promote tissue repair *Appl. Mater. Today.* **19** 100601

[87] Kim E S, Ahn E H, Dvir T and Kim D H 2014 Emerging nanotechnology approaches in tissue engineering and regenerative medicine *Int. J. Nanomed.* **9** 1–5

[88] Keung A J, Healy K E, Kumar S and Schaffer D V 2010 Biophysics and dynamics of natural and engineered stem cell microenvironments *Wiley Interdiscip. Rev.: Syst. Biol. Med.* **2** 49–64

[89] Shin H, Jo S and Mikos A G 2003 Biomimetic materials for tissue engineering *Biomaterials* **24** 4353–64

[90] Lin Y, Taylor S, Li H, Fernando K A S, Qu L, Wang W, Gu L, Zhou B and Sun Y P 2004 Advances toward bioapplications of carbon nanotubes *J. Mater. Chem.* 527–41

[91] Kuila T, Bose S, Mishra A K, Khanra P, Kim N H and Lee J H 2012 Chemical functionalization of graphene and its applications *Prog. Mater. Sci.* **57** 1061–105

[92] Bacakova L, Grausova L, Vacik J, Fraczek A, Blazewicz S, Kromka A, Vanecek M and Svorcik V 2007 Improved adhesion and growth of human osteoblast-like MG 63 cells on biomaterials modified with carbon nanoparticles *Diamond Relat. Mater.* **16** 2133–40

[93] Krishnan V *et al* 2015 Vortex-aligned fullerene nanowhiskers as a scaffold for orienting cell growth *ACS Appl. Mater. Interfaces* **7** 15667–73

[94] Vedhanayagam M, Raja I S, Molkenova A, Atabaev T S, Sreeram K J and Han D W 2021 Carbon dots-mediated fluorescent scaffolds: recent trends in image-guided tissue engineering applications *Int. J. Mol. Sci.* **2** 5378

[95] Travlou N A, Giannakoudakis D A, Algarra M, Labella A M, Rodríguez-Castellón E and Bandosz T J 2018 S- and N-doped carbon quantum dots: surface chemistry dependent antibacterial activity *Carbon* **1** 693–707

[96] Sun H, Gao N, Dong K, Ren J and Qu X 2014 Graphene quantum dots-band-aids used for wound disinfection *ACS Nano* **8** 6202–10

[97] Liu Q, Guo B, Rao Z, Zhang B and Gong J R 2013 Strong two-photon-induced fluorescence from photostable, biocompatible nitrogen-doped graphene quantum dots for cellular and deep-tissue imaging *Nano Lett.* **13** 2436–41

[98] Han G *et al* 2019 Membrane-penetrating carbon quantum dots for imaging nucleic acid structures in live organisms *Angew. Chem., Int. Ed.* **58** 7087–91

[99] Huang Y C, Chen C P, Wu P J, Kuo S Y and Chan Y H 2014 Coumarin dye-embedded semiconducting polymer dots for ratiometric sensing of fluoride ions in aqueous solution and bio-imaging in cells *J. Mater. Chem.* B **2** 6188–91

[100] Yan C, Ren Y, Sun X, Jin L, Liu X, Chen H, Wang K, Yu M and Zhao Y 2020 Photoluminescent functionalized carbon quantum dots loaded electroactive silk fibroin/PLA nanofibrous bioactive scaffolds for cardiac tissue engineering *J. Photochem. Photobiol.* B **202** 111680

[101] Pal P, Das B, Dadhich P, Achar A and Dhara S 2017 Carbon nanodot impregnated fluorescent nanofibers for: *in vivo* monitoring and accelerating full-thickness wound healing *J. Mater. Chem.* B **5** 6645–56

[102] Omidi M, Yadegari A and Tayebi L 2017 Wound dressing application of pH-sensitive carbon dots/chitosan hydrogel *RSC Adv.* **7** 10638–45

[103] Dehghani A, Ardekani S M, Hassan M and Gomes V G 2018 Collagen derived carbon quantum dots for cell imaging in 3D scaffolds via two-photon spectroscopy *Carbon* **131** 238–45

[104] Hsieh F Y, Shrestha L K, Ariga K and Hsu S H 2017 Neural differentiation on aligned fullerene C60 nanowhiskers *Chem. Commun.* **53** 11024–7

[105] Miyazawa K, Kuwasaki Y, Obayashi A and Kuwabara M 2002 C60 nanowhiskers formed by the liquid-liquid interfacial precipitation method *J. Mater. Res.* **17** 83–8

[106] Zheng F F, Zhang P H, Xi Y, Chen J J, Li L L and Zhu J J 2015 Aptamer/graphene quantum dots nanocomposite capped fluorescent mesoporous silica nanoparticles for intracellular drug delivery and real-time monitoring of drug release *Anal. Chem.* **87** 11739–45

[107] Alifu N, Sun Z, Zebibula A, Zhu Z, Zhao X, Wu C, Wang Y and Qian J 2017 Deep-red polymer dots with bright two-photon fluorescence and high biocompatibility for *in vivo* mouse brain imaging *Opt. Commun.* **399** 120–6

# Chapter 12

# Antibiofilm potential of zero-dimensional carbon nanomaterials

**Charles Oluwaseun Adetunji, John Tsado Mathew, Kshitij RB Singh, Abel Inobeme, Olugbemi T Olaniyan, Jay Singh and Ravindra Pratap Singh**

Zero-dimensional carbon nanomaterial are identified as a sustainable antimicrobial agent that can act against several pathogenic micro-organisms, as these materials possess unique physicochemical features that enable them to penetrate the cell wall of pathogenic micro-organisms. Bacterial biofilms have been recognized as complex surface-attached groups of bacteria joined together, in particular by self-produced polymer matrices that include extracellular DNAs, polysaccharides, and secreted proteins. Biofilm development has been established to have an adverse influence on marine industries, healthcare, drinking water distribution systems, food, and many other fields. Therefore, this study intends to provide detailed information on the relevance of zero-dimensional carbon nanomaterials and their antibiofilm potential.

## 12.1 Introduction

It has been found that 40%–80% of bacterial cells present on Earth carry the potential to develop biofilms [1]. Biofilm development is known to cause many problems, for example in the food industries, the occurrence of pathogenic bacteria developing biofilms, most significantly inside processing facilities, results in food spoilage and could also affect the health of consumers [2, 3]. Similarly, in medical centers, biofilms could persist on the surface of medical devices and the tissues of patients, resulting in infections [4]. Biofilms constitute a complex surface that includes groups of micro-organisms joined together by self-generated matrices that include extracellular DNAs, polysaccharides, and secreted proteins [5]. A biofilm can be a combination of numerous species or be made up of individual microbial species of yeast, bacteria, filamentous fungi, archaea, protozoa, and algae that have a great capacity to attach on abiotic and biotic surfaces [6]. Moreover, biofilm formation is among the most critical issues threatening the stability of both

doi:10.1088/978-0-7503-4048-9ch12

off-shore and on-shore marine constructions. Nanotechnology's rapid advancement has aided in developing new ideas in various fields. Because of their mechanical strength, and electrical and biological properties, carbon-based nanostructures (CNSs) have developed as intriguing substrates for a wide range of uses. CNSs include carbon nanotubes (CNTs), fullerene, diamond-like carbon (DLC), and graphene, which have been shown to exhibit broad-spectrum antibacterial activity against micro-organisms. The exact mechanism which governs the antimicrobial activity of CNSs must be comprehended so as to ensure the safe and efficient incorporation of these structural elements as antibacterial agents into biomaterials, although the unraveling the individual and synergic effects of the electrical, chemical, and physical properties of CNSs on cells is challenging. The current developments in this area are covered in this chapter, emphasizing the contact of distinct families of carbon nanomaterials and microbes to assess their bactericidal ability [7].

In therapeutics, microbial biofilms are a significant issue. They invade and destroy medical equipment and implantations since they frequently promote the growth of multidrug-resistant bacteria. Bacterial adhesion to the substrate and the generation of extrinsic polymeric compounds are the first steps in the production of extracellular polymer substrate (EPS). EPS serves as a structural framework for bacterial cell division. The EPS coatings may serve as a barrier, inhibiting medicines from reaching biofilm-associated microbes. Moreover, the appearance of multidrug-resistant infectious bacterial communities has been a serious health issue in the last decade. Therefore, nanoparticles (NPs) are increasingly employed to target bacteria as an alternative to antibiotics. Nanotechnology could be efficiently utilized in the treatment of bacterial infections. Many works in the literature have mentioned the novel use of various bionanomaterials in inhibiting biofilms [9–11], in which metal nanomaterials are among the most commonly used, such as silver, titanium, copper, zinc, etc. In addition to these nanocomposites, metallic nanotubes and bioinspired nanomaterials have also been established to display anticorrosive and antifouling properties. Antimicrobial adhesives for medicinal materials and implantable devices to disease and promote wound healing, bacterial monitoring devices to create microbial diagnostics, antibiotic delivery systems to treat disease, and antimicrobial antibiotics to regulate infectious diseases are examples of nanotechnology applications. Although our understanding of the antimicrobial processes of NPs is still lacking, oxidative stress induction, non-oxidative processes, and metal ion release are some of the widely accepted processes that NPs cause in bacterial cells. Moreover, generating resistance would require many simultaneous gene alterations in the same bacterial cell because of the numerous methods of action of NPs against germs; hence, for bacterial cells, it is difficult to become resistant to NPs. The bactericidal effects of NPs against bacteria and the components involved are discussed in [12]. Fouling of maritime structures is considered a primary target for the development of antibiofilm materials. The marine industry suffers significant financial losses due to biofouling and biofilm. Whenever the surface of a ship is affected by fouling, it is believed that an additional 10% of fuel is consumed. However, hazardous compounds are frequently utilized in the control and avoidance of biofouling and biofilms on marine structures and these constitute a serious threat

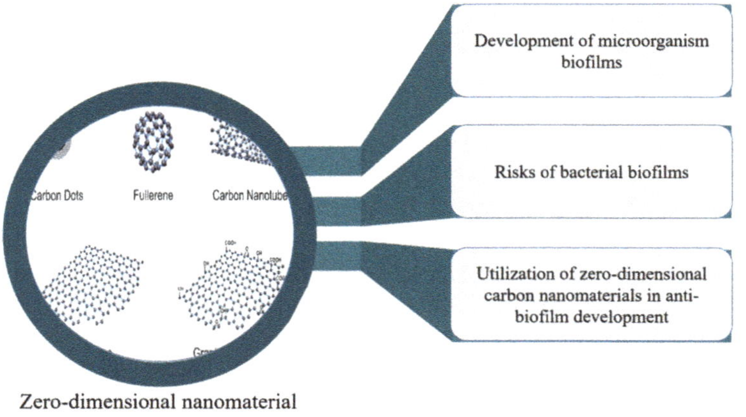


**Figure 12.1.** A schematic representation of the areas discussed in the chapter.

to marine ecosystems and are closely controlled under global trade agreements. Therefore, the development of biomaterials to manage the fouling, corrosion, and biofilms of marine structures has become necessary. This chapter provides detailed information on the relevance of zero-dimensional carbon nanomaterials and their antibiofilm potential. Moreover, this chapter also focuses on the harmful effects of biofilm formation on the healthcare system, food industry, and drinking water system, and will also lay an essential foundation for enhancing further research on the prevention of biofilm development. Figure 12.1 illustrates the main areas discussed in this chapter.

## 12.2 The development of micro-organism biofilms

Micro-organism biofilms are responsible for the majority of device-related illnesses, and they continue to pose a problem in modern medical practice. Therefore, it has become necessary to prevent biofilm formation by using innovative antimicrobial substances and by surface functionalization of equipment. However, effective prevention treatments are still highly restricted due to high biotoxicity and low biocompatibility. Due to the presence of the biofilm's EPS and the antimicrobial resistance in biofilm-growing bacteria, most traditional antibacterial medicines or antibiofilm medicines present particular challenges. Moreover, the therapy becomes more complex because it must be acceptable for use on equipment accompanied by host cells in different circumstances. The formation of bacterial biofilm involves five stages: the reversible attachment phase, in which bacteria non-specifically attach themselves to the surface; the irreversible attachment phase, where a relationship between bacterial cells is initiated through bacterial adhesion, such as lipopolysaccharide and fimbriae; the generation of extracellular polymeric substances by the inhabitant bacterial cells; the biofilm maturation phase, where the bacterial cell begins to reproduce and release signaling molecules that could detect the occurrence of many bacteria, thereby leading to the development of biofilms and micro-colonies; and the detachment dispersal phase, which involves bacterial cells separating from biofilms and returning to a self-regulating planktonic lifestyle, as illustrated in figure 12.2 [13].

Although various nanomaterials have already been studied extensively for prohibiting bacterial growth on medical equipment, their uses in treating bacterial biofilms are yet to be explored thoroughly, and inhibiting the formation of microbial biofilms and destroying the EPS are very difficult. One study showed successful EPS destruction and also destruction of a pathogenic biofilm, and supplied an in-depth characterization of the nanomaterials and biofilm elimination, followed by an analysis of key issues in this field and recommendations for future research [14]. Pathogenic micro-organisms have wreaked havoc in healthcare services worldwide. Biofilms are thought to be responsible for over 60% of systemic infections, and the bacterial ability to develop biofilms helps them exhibit exceptional drug resistance. However, a breakthrough was achieved by the introduction of zero-dimensional nanomaterials. They possess unique physical and chemical characteristics and a high specific surface area to volume ratio, which helps treat biofilm-associated diseases. The research involves distinct forms of zero-dimensional nanomaterials whose antibiofilm action is currently being studied. However, the effectiveness of zero-dimensional nanomaterials cannot be described solely by examining biofilm reduction or disruption rates. Various other parameters, such as the bacterial strain, type of nanoparticle, measurement methodologies, biofilm culture environment, and NP concentration, must be considered [15]. This chapter focuses on recent studies on the synthesis, characterization, and assessment of nanoparticles (NPs) for biofilm disease diagnosis and treatment.

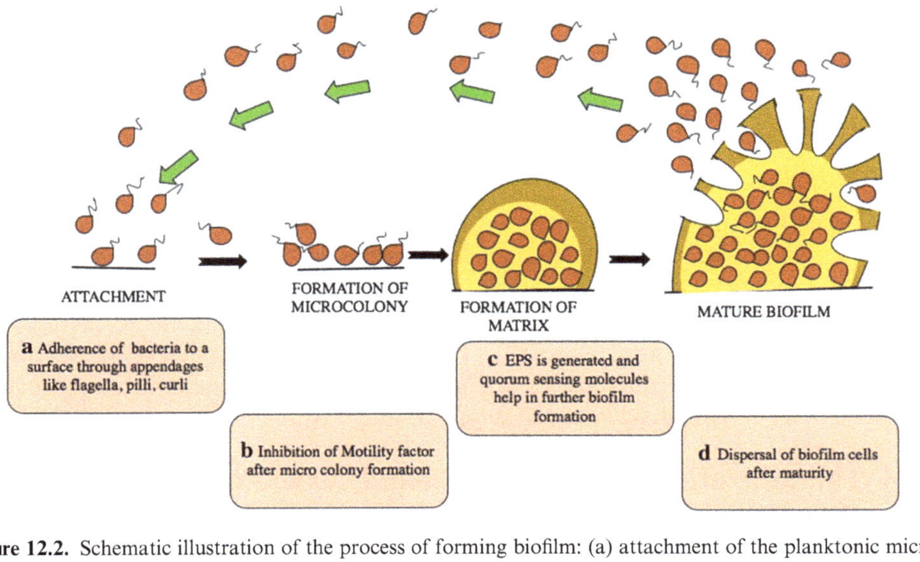


**Figure 12.2.** Schematic illustration of the process of forming biofilm: (a) attachment of the planktonic micro-organisms to the surface, (b) leading to the inhibition of motility factor, (c) consequently generating EPS with quorum sensing, and (d) finally resulting in the development of mature biofilm and later the release of mature cells. (Reproduced with permission from [13]. Copyright 2020 Springer Nature.)

## 12.3 The risks of bacterial biofilms

### 12.3.1 Healthcare issues

Biofilms can develop on any living or dead tissue [16]. Moreover, they can also develop on the surface of biomedical devices, such as cerebrospinal fluid (CSF) shunts, pacemakers, contact lenses, prosthetic heart valves, etc [17]. It has been observed that gram-positive and gram-negative bacteria can attach to a device's surface and develop a biofilm. Typical examples of bacteria that have biofilm capability include *Pseudomonas aeruginosa*, *Staphylococcus aureus*, and *Staphylococcus epidermidis* [18].

Further, bacteria can also grow in healthcare based water distribution systems. For example, it has been observed that *P. aeruginosa* can develop biofilms, in particular on the interior of the metal pipes of a hospital water system [19]. Moreover, most bacteria can produce several diseases, such as infective endocarditis, osteomyelitis, periodontitis, and chronic wounds [20]. At the same time, *P. aeruginosa* biofilm could result in pulmonary infections, particularly in patients suffering from cystic fibrosis [21].

### 12.3.2 Food safety and the food industry

The food industry faces a significant loss every year due to the growth of biofilm on surfaces, both in contact and not in contact with foods [22]. It has been stated that biofilm is responsible for 60% of foodborne outbreaks [23]. Further, several studies have reported that biofilms could present many challenges and risks to the food industry and constitute a threat to food safety [2]. The development of biofilm could spoil foods that could serve as sources of contaminants to the public [2]. Typical examples of biofilms producing foodborne spoilage and pathogens include *Listeria monocytogenes* [2], Salmonella spp., *Escherichia coli* 0157:H7 [24], *Pseudomonas* spp. [25], *Vibrio parahaemolyticus* [26], *Clostridium perfringens* [24], *Campylobacter jejuni* [24], *Bacillus* spp. [2], *S. aureus* [27], *Shewanella putrefaciens* [28], *Cronobacter* spp. [24], and *Geobacillus stearothermophilus* [29].

### 12.3.3 Drinking water distribution systems

Several scientists have documented biofilm occurrence in various drinking water distribution systems [30, 31]. Biofilm has been reported to pose numerous challenges in drinking water distribution systems. The consumption of highly contaminated water, particularly with pathogenic biofilms, could be related to waterborne outbreaks and human infections [32]. Typical examples of pathogens that could produce biofilm in water include *Klebsiella pneumonia*, *Mycobacteria*, *P. aeruginosa*, *Campylobacter jejuni*, *Aeromonas hydrophila*, and *Legionella pneumophila* [33].

## 12.4 The utilization of zero-dimensional carbon nanomaterials in antibiofilm development

With the increase in the application of antibiotics, the generation of drug-resistant micro-organisms has also sped up, and these 'superbugs' are continually changing and becoming progressively more difficult to treat. As a result, innovative and effective control strategies with little danger of producing drug-resistant pathogens are desperately required. The zero-dimensional materials are considered innovative due to their excellent drug-releasing/loading capacities [34]. The spread of lethal viruses and infections has become a huge concern worldwide. Initiatives have been devoted to battling the spread of deadly pathogens through the air, human interaction, and contaminated objects, which has led to an increase in the study of antibacterial and antiviral compounds. The invention of antimicrobial nanoparticles has provided significant advantages in combating the risks posed by surface and airborne viruses, particularly during the COVID-19 pandemic [35].

Distinctive electrochemical properties have helped carbon to be utilized as an electrochemical sensing surface. Furthermore, recent advancements in material design and fabrication, particularly nanomaterials, have resulted in highly reliable electrochemical sensing devices with improved analytical performance. Similarly, due to the presence of their exceptional features, carbon nanotubes (CNTs) are among the most investigated zero-dimensional nanomaterials. The capacity of carbon nanotubes to enhance the electrochemical sensitivity of key biomolecules and stimulate electron transfer events in proteins has become particularly important in electroanalysis. CNTs' exceptional susceptibility to affect the surface resistance caused by adsorbates allows them to be used in extremely sensitive nanoscale sensing. An experiment was performed in which multi-walled carbon nanotubes (MWCNTs) were surface grown and anchored vertically, and were then utilized as antibiofilm agents against human pathogens, namely *Klebsiella oxytoca*, *Pseudomonas aeruginosa*, and *Staphylococcus epidermidis*, in medical and technical areas. It is well known that MWCNTs exhibit high anti-adhesive properties that increase with their increase in length which, along with their vertical alignment, aids in preventing biofilm formation by bacteria. Therefore, this novel technique can be further utilized to avoid biofilm development on different medical devices, sensors, or other technical areas. A schematic illustration of the experiment is presented in figure 12.3 [36]. Moreover, CNT-modified electrodes have shown their ability to act as attachments for biomolecules, including nucleic acids, and their ability to reduce

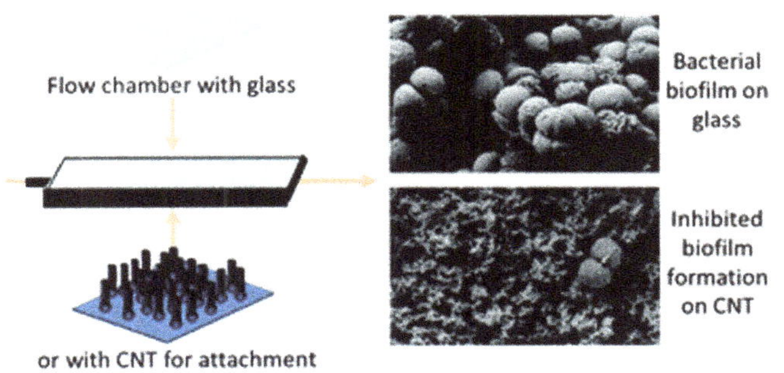


**Figure 12.3.** Vertical arrangement of multi-walled carbon nanotubes (MWCNTs) on a glass surface and its role in inhibiting the development of biofilm. (Reproduced with permission from [36]. Copyright 2016 Royal Society of Chemistry.)

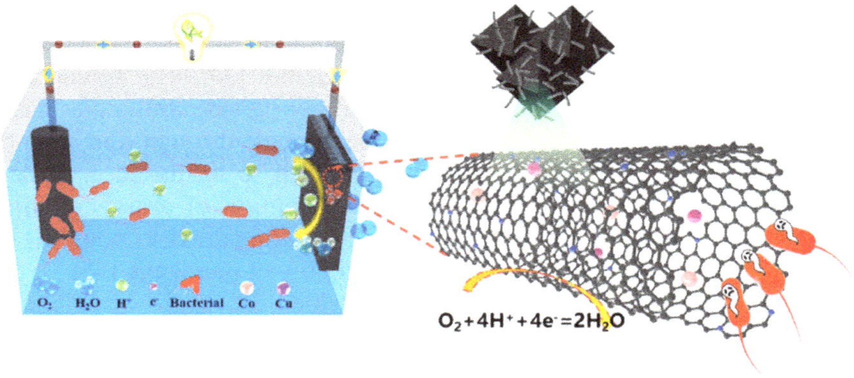


**Figure 12.4.** Copper-based metal–organic frameworks (MOFs) co-doped with cobalt and nitrogen (CuCo@NCNTs) as a cathode, used as a microbial fuel cell and also prevents the development of bacterial biofilm on its surface. (Reproduced with permission from [39]. Copyright 2014 Elsevier.)

surface fouling. As a result, CNTs are a popular choice among researchers as a foundation for various electrochemical sensors [37]. Many zero-dimensional nanoparticles have been studied to inhibit biofilm production by bacteria. Further, it was found that pristine-MWCNT (p-MWCNT)-based surfaces can be applied on medical devices. Similarly, p-SWCNTs can easily damage a deposited layer of bacteria, as well as reducing its viability by 80%. Therefore, both p-SWCNTs and p-MWCNTs are utilized for inhibiting gram-positive and gram-negative bacteria [38],

Further, it has also been observed that biofilm development on a cathode surface can enhance the mass transfer resistance that affects the diffusion of $H^+$ ions on cathodes. One of the highly common cathode degrading bacteria is *S. oneidensis.* It was found that copper-based metal–organic frameworks (MOFs) co-doped with cobalt and nitrogen nanotubes (CuCo@NCNTs) prevented biofilm formation by the bacteria. Moreover, the experiment also reported that these CuCo@NCNTs exhibited various active sites of Cu–N and Co–N, which helped them exhibit high conductivity. They were therefore utilized as the cathode and increased the output of microbial fuel cells, as depicted in figure 12.4 [39]. However, the overall mechanisms through which CNTs prevent biofilm formation are still not clear. However, more in-depth knowledge of the interaction of CNTs with biofilms is needed to design robust and prolonged coating materials to resist biofilm development.

## 12.5 Conclusion

This chapter has provided information on the antibiofilm potential of zero-dimensional carbon nanomaterials and the procedure involved in the application of zero-dimensional carbon nanomaterials with antimicrobial properties. Relevant information was provided on the risks of bacterial biofilms and healthcare issues, including drinking water distribution systems, food safety, and the food industry. The processes involved in the development of biofilms were also discussed.

## References

[1] Flemming H-C, Wingender J, Szewzyk U, Steinberg P, Rice S A and Kjelleberg S 2016 Biofilms: an emergent form of bacterial life *Nat. Rev. Microbiol.* **14** 563–75

[2] Galié S, García-Gutiérrez C, Miguélez E M, Villar C J and Lombó F 2018 Biofilms in the food industry: health aspects and control methods *Front. Microbiol.* **9** 898

[3] Coughlan L M, Cotter P D, Hill C and Alvarez-Ordóñez A 2016 New weapons to fight old enemies: novel strategies for the (bio)control of bacterial biofilms in the food industry *Front. Microbiol.* **7** 1641

[4] Percival S L, Suleman L, Vuotto C and Donelli G 2015 Healthcare-associated infections, medical devices and biofilms: risk, tolerance and control *J. Med. Microbiol.* **64** 323–34

[5] Tremblay Y D, Lévesque C, Segers R P and Jacques M 2013 Method to grow *Actinobacillus pleuropneumoniae* biofilm on a biotic surface *BMC Vet. Res.* **9** 213

[6] Raghupathi P K, Liu W, Sabbe K, Houf K, Burmølle M and Sørensen S J 2018 Synergistic interactions within a multispecies biofilm enhance individual species protection against grazing by a pelagic protozoan *Front. Microbiol.* **8** 2649

[7] Al-Jumaili A, Alancherry S, Bazaka K and Jacob M 2017 Review on the antimicrobial properties of carbon nanostructures *Materials* **10** 1066

[8] Adetunji C O *et al* 2021 Bionanomaterials for green bionanotechnology *Bionanomaterials* (Bristol: IOP Publishing) **10** 1–24

[9] Singh K R, Nayak V, Singh J, Singh A K and Singh R P 2021 Potentialities of bioinspired metal and metal oxide nanoparticles in biomedical sciences *RSC Adv.* **11** 24722–46

[10] Singh K R, Nayak V and Singh R P 2021 Introduction to bionanomaterials: an overview *Bionanomaterials* (Bristol: IOP Publishing) 1-1–1-21

[11] Singh P, Singh K R, Singh J, Prasad P and Singh R P 2021 Bioinspired triangular ZnO nanoclusters synthesized by *Argyreia nervosa* nascent leaf extract for the efficient electrochemical determination of vitamin C *RSC Adv.* **11** 25752–63

[12] Wang L, Hu C and Shao L 2017 The antimicrobial activity of nanoparticles: present situation and prospects for the future *Int. J. Nanomed.* **12** 1227–49

[13] Banerjee D, Shivapriya P M, Gautam P K, Misra K, Sahoo A K and Samanta S K 2020 A review on basic biology of bacterial biofilm infections and their treatments by nanotechnology-based approaches *Proc. Natl Acad. Sci. India* **B 90** 243–59

[14] Tran H M *et al* 2020 Nanomaterials for treating bacterial biofilms on implantable medical devices *Nanomaterials* **10** 2253

[15] Han C, Romero N, Fischer S, Dookran J, Berger A and Doiron A L 2017 Recent developments in the use of nanoparticles for treatment of biofilms *Nanotechnol. Rev.* **6** 383–404

[16] Alav I, Sutton J M and Rahman K M 2018 Role of bacterial efflux pumps in biofilm formation *J. Antimicrob. Chemother.* **73** 2003–20

[17] Wu H, Moser C, Wang H-Z, Høiby N and Song Z-J 2015 Strategies for combating bacterial biofilm infections *Int. J. Oral Sci.* **7** 1–7

[18] Pakharukova N *et al* 2018 Structural basis for *Acinetobacter baumannii* biofilm formation *Proc. Natl Acad. Sci.* **115** 5558–63

[19] Loveday H P, Wilson J A, Kerr K, Pitchers R, Walker J T and Browne J 2014 Association between healthcare water systems and *Pseudomonas aeruginosa* infections: a rapid systematic review *J. Hosp. Infect.* **86** 7–15

[20] Masters E A *et al* 2019 Evolving concepts in bone infection: redefining 'biofilm', 'acute vs chronic osteomyelitis', 'the immune proteome' and 'local antibiotic therapy' *Bone Res.* **7** 20

[21] Rabin N, Zheng Y, Opoku-Temeng C, Du Y, Bonsu E and Sintim H O 2015 Biofilm formation mechanisms and targets for developing antibiofilm agents *Future Med. Chem.* **7** 493–512
[22] Kumar C G and Anand S 1998 Significance of microbial biofilms in food industry: a review *Int. J. Food Microbiol.* **42** 9–27
[23] Han N, Mizan M F R, Jahid I K and Ha S-D 2016 Biofilm formation by *Vibrio parahaemolyticus* on food and food contact surfaces increases with rise in temperature *Food Control.* **70** 161–6
[24] Lelieveld H *et al* (ed) 2016 *Handbook of Hygiene Control in the Food Industry* (Amsterdam: Elsevier)
[25] Rajmohan S, Dodd C E R and Waites W M 2002 Enzymes from isolates of *Pseudomonas fluorescens* involved in food spoilage *J. Appl. Microbiol.* **93** 205–13
[26] Yeung P S M and Boor K J 2004 Epidemiology, pathogenesis, and prevention of foodborne *Vibrio parahaemolyticus* infections *Foodborne Pathog. Dis.* **1** 74–88
[27] Argudín M Á, Mendoza M C and Rodicio M 2010 Food poisoning and *Staphylococcus aureus* enterotoxins *Toxins* **2** 1751–73
[28] Bagge D, Hjelm M, Johansen C, Huber I and Gram L 2001 *Shewanella putrefaciens* adhesion and biofilm formation on food processing surfaces *Appl. Environ. Microbiol.* **67** 2319–25
[29] Burgess S A, Flint S H, Lindsay D, Cox M P and Biggs P J 2017 Insights into the *Geobacillus stearothermophilus* species based on phylogenomic principles *BMC Microbiol.* **17** 140
[30] Liu S, Gunawan C, Barraud N, Rice S A, Harry E J and Amal R 2016 Understanding, monitoring, and controlling biofilm growth in drinking water distribution systems *Environ. Sci. Technol.* **50** 8954–76
[31] Mahapatra A 2015 Study of biofilm in bacteria from water pipelines *J. Clin. Diagn. Res.* **9** DC09–DC11
[32] Prest E I, Hammes F, van Loosdrecht M C M and Vrouwenvelder J S 2016 Biological stability of drinking water: controlling factors, methods, and challenges *Front. Microbiol.* **7** 45
[33] Chan S, Pullerits K, Keucken A, Persson K M, Paul C J and Rådström P 2019 Bacterial release from pipe biofilm in a full-scale drinking water distribution system *NPJ Biofilms Microbiomes.* **5** 9
[34] Yeh Y-C, Huang T-H, Yang S-C, Chen C-C and Fang J-Y 2020 Nano-based drug delivery or targeting to eradicate bacteria for infection mitigation: a review of recent advances *Front. Chem.* **8** 286
[35] Shah K W and Huseien G F 2020 Inorganic nanomaterials for fighting surface and airborne pathogens and viruses *Nano Express.* **1** 032003
[36] Malek I, Schaber C F, Heinlein T, Schneider J J, Gorb S N and Schmitz R A 2016 Vertically aligned multi walled carbon nanotubes prevent biofilm formation of medically relevant bacteria *J. Mater. Chem.* B **4** 5228–35
[37] Power A C, Gorey B, Chandra S and Chapman J 2018 Carbon nanomaterials and their application to electrochemical sensors: a review *Nanotechnol. Rev.* **7** 19–41
[38] Kang S, Pinault M, Pfefferle L D and Elimelech M 2007 Single-walled carbon nanotubes exhibit strong antimicrobial activity *Langmuir* **23** 8670–3
[39] Wang Y *et al* 2021 Bimetallic hybrids modified with carbon nanotubes as cathode catalysts for microbial fuel cell: effective oxygen reduction catalysis and inhibition of biofilm formation *J. Power Sources.* **485** 229273

# Chapter 13

# Zero-dimensional carbon nanomaterials in agriculture: from biosensors to photosynthesis enhancement

**Eduardo E Andrada Suarez, Sergio A Cuozzo, Gonzalo Tortella, Olga Rubilar and Martín A Fernández-Baldo**

Recently, carbon dots have attracted significant interest in agriculture due to their chemical properties and other attributes such as photoluminescence reaction and photoinduced electron transfer. These qualities allow potential applications related to bioimaging, biosensors, catalysis, and photovoltaic conversion. Indeed, carbon quantum dots (CQDs) and graphene quantum dots (GQDs) are widely used in monitoring pesticides, mycotoxins, metal ions, and environmental parameters (moisture and oxygen). Furthermore, these nanomaterials have positively affected crops by increasing photosynthesis and nutritional assimilation in plants, promoting their growth and resistance to abiotic and biotic stresses.

## 13.1 Introduction

Nanotechnology is considered one of the most studied disciplines in the last century due to the characteristics of nanomaterials, which provide excellent versatility for the development of applications in countless fields. These nanomaterials have progressed in various sectors due to their size and structure dependent physico-chemical properties and have become fascinating materials in the technological world [1]. The agricultural sector is currently facing the need to improve production systems to cope with demographic growth and the demand for sustainable methods for managing natural resources and reducing climate change. For this reason, agriculture 4.0 is presented as a green revolution with science-based tools and has adopted new technologies such as robotics, genetic engineering, artificial intelligence, and nanotechnology, among others [2, 3]. One of these approaches began with developing and implementing products based on nanomaterials to respond to the demand for environmentally safe methods and high-quality standards for food production [4, 5]. It is essential to highlight that a wide range of products that include different nanoparticles have been developed to replace the current harmful and less effective practices. This approach allows establishing ecological tools and methods that respond to the sustainability objectives proposed by the United Nations [6, 7]. At present, the study of carbon nanoparticles has acquired great interest, as is the case for carbon quantum dots, due to their properties that provide great potential and versatility in the development of applications and give them advantages over other similar nanoparticles used to date [8]. Within these groups there are many materials, such carbon nanotubes, many of which were reported in multiple applications in agriculture [9, 10]. This chapter will emphasize the knowledge and technological innovations related to agriculture that were developed from zero-dimensional nanomaterials, mainly carbon quantum dots (CQDs) and graphene quantum dots (GQDs).

## 13.2 Carbon quantum dots: a green alternative

CQDs are recently studied nanoparticles belonging to the carbon family, classified as zero-dimensional nanomaterials. Their importance is based on their optical, electrochemical, and biocompatibility properties, and competitive cost, which are critical for developing applications [11, 12]. In general, carbon allotropes are usually classified according to their nanoscale range into zero-dimensional (0D), one-dimensional (1D), two-dimensional (2D), and three-dimensional (3D) [13, 14]. This scheme is based on the Pokropivny classification according to the dimensionality and surface concept [15]. CQDs were discovered in 2004 when fluorescence was detected from a mixture of nanoparticles during a single-walled carbon nanotube electrophoretic assay [16]. Since then, many types of research have been carried out, and the properties discovered to date have been reviewed, which allowed opening new fields of technological development. It has been reported that some of the advantages of carbon dots are their low toxicity, good dispersibility in water, and high conductivity [8, 17–19]. Currently, these nanoparticles can be produced through two different strategies, called the top-down and bottom-up methods [11, 20]. The former consists of physical or chemical procedures that generally utilize carbon material sources such as graphite, but have the disadvantages of requiring strong synthesis conditions, longer times, and the precursors tend to be more expensive. Bottom-up techniques can involve various organic substances as a starting point for synthesis, such as polymers, organic acids, and carbohydrates, which are subjected to processes that require specific temperature and irradiation conditions [21, 22]. Some of these bottom-up methods are of interest in the scientific world because they allow carbon quantum dots to be obtained cleanly and are therefore are ecological and economical. Therefore they respond consistently to the fundamental principles of green chemistry because the use of hazardous substances is greatly limited during the production of these nanomaterials [23, 24]. Together with biological methods, this strategy has been proposed and implemented to place nanotechnology on the path to sustainability and take advantage of its competitive characteristics [25–27]. As reported, great efforts are currently being made to develop green fabrication methodologies for CQDs and GQDs derived from renewable sources, because they have the advantages of low cost, simplicity, and sustainability in production [28]. Table 13.1 summarizes various plant-based sources as alternatives for the production of carbon dots, including fruits, vegetables, and their derivatives such as juices and peel [29–34], spices [35], vegetables, plant leaves, and derivatives [36–46], and forestry or agricultural biomass [47, 48]. As CQDs can be obtained from any plant-based carbon source, there has been interest in testing agricultural by-products, which are generated in large quantities, are economically profitable, and, therefore, allow products with added value to be obtained [47, 49]. Some of these by-products are sugarcane bagasse [50, 51], corncob residues [48], lignocellulosic residues [52, 53], among others [54, 55].

## 13.3 Nanosensors: applications in agriculture and related areas

### 13.3.1 Developments based on CQDs and GQDs for sensing substances used in agriculture

Carbon nanoparticles and other bio-nanomaterials are emerging platforms that are being used widely for the development of detection devices and probes. Currently, residual pesticides represent a problem in the agricultural sector, even more so with the paradigm shift towards sustainable agriculture [56, 57]. Likewise, studies have increased to include CQDs and GQDs in new technologies to determine pesticides and other substances related to agribusiness and food. Zhao *et al* evaluated innovative developments in enzyme and non-enzyme biosensors constructed from carbon nanomaterials for pesticide detection, including carbon quantum dots [58]. These biosensors are being developed as a technology for food safety, for detecting, quantifying, or semi-quantifying dangerous substances derived from agricultural activity, such as potentially toxic pesticides or mycotoxins [59, 60]. CQD and GQD based biosensors for pesticide detection were classified according to their electron transfer properties into electro-chemiluminescent, fluorescent, and photoluminescent biosensors [58].

Pesticides influence the fluorescence intensity of CDs, so this process depends on the nature of the interaction. In this way, some pesticides can increase fluorescence. In contrast, others quench it through static or dynamic extinction or across different phenomena with contact, such as photoinduced electron transfer, or without contact, such as the inner filter effect (IFE), chemiluminescence resonance energy transfer, and fluorescence (CRET and FRET) [61]. Also, sensors based on CQDs can show variable sensitivity so that their detection limit can range from the nanomolar to femtomolar ranges [62].

Heteroatom-doped carbon dots have become a valuable tool since the doping process improves the photoluminescent characteristics. Combined with their biocompatibility and the other beneficial properties of carbon dots, this results in promising tools for multiple applications and an efficient alternative to traditional metal quantum dots (tQDs). Bottom-up methods are usually better to produce these nanoparticles since they can be adapted easily for doping carbon dots. In addition,

**Table 13.1.** Comparison of the synthesis of carbon dots from various plant-based sources, agro-industrial waste, and food.

| Carbon source | Nanoparticle | Size (nm) | Application | Synthesis methods | Ref. |
|---|---|---|---|---|---|
| Fruit (*Elaeagnus angustifolia*) | CQDs | 3.5 | Tartrazine detection | Hydrothermal | Ghereghlou *et al* [40] |
| Lotus root | N-CQDs | 9.4 | $Hg^{2+}$ detection/cell imaging | Microwave | Gu *et al* [38] |
| Peach gum exudates (*Prunus persica*) | N-CQDs | 2–5 | $Au^{3+}$ detection | Hydrothermal carbonization | Liao *et al* [37] |
| Sugarcane juice (*Saccharum officinarum*) | CQDs | 3 | Cell imaging | Hydrothermal | Mehta *et al* [45] |
| Peel (*Trapa bispinosa*) | CQDs | 5–10 | Cell imaging | Thermal oxidation | Mewada *et al* [34] |
| Algal bloom mixture | CQDs | 8 | Biomarkers | Microwave thermolysis | Ramanan *et al* [36] |
| Coriander leaves | CQDs | 2.4 | $Fe^{3+}$ detection | Hydrothermal | Sachdev *et al* [44] |
| Mulberry leaves (*Morus alba* L.) | CQDs | 2–4 | Cell imaging | Hydrothermal | Shao *et al* [43] |
| Spices | CQDs | 3.1–4.3 | Bioanalytical/clinical applications | Hydrothermal | Vasimalai *et al* [35] |
| *Aloe vera* leaves | CQDs | 5 | Tartrazine detection | Hydrothermal | Xu *et al* [46] |
| Bamboo leaves | N-S-CQDs | 11 | Nitrophenol compounds detection | Calcination | Yang *et al* [39] |
| Cardamom leaves (*Elettaria cardamomum*) | CQDs | | Dyes degradation | Ultrasonication | Zaib *et al* [41] |
| Sugarcane bagasse (*Saccharum officinarum*) | CQDs/GQDs | | | Chemical oxidation | Baweja *et al* [51] |
| Food waste (turtle shell) | CQDs | 2.6 | Multisignal coding/ anticounterfeiting | Carbonization | Guo *et al* [55] |
| Waste biomass (hemicellulose/xilane) | CQDs | 7 | | Hydrothermal carbonization | Liang *et al* [47] |
| Lignocellulose (pulp and paper industry) | CQDs | 17.5 | | Microwave | Rodriguez *et al* [53] |
| Lignocellulose (aminated alkali lignin) | N-CQDs | 4–10 | Cell imaging/$Fe^{3+}$ detection | Carbonization | Shi *et al* [52] |
| Sugarcane bagasse (*Saccharum officinarum*) | CQDs | 4.1 | | Chemical oxidation | Thambiraj *et al* [50] |
| Corncob residues | N-CQDs/E-N-CQDs | 9.3 | $Fe^{3+}$ detection | Hydrothermal | Zhang *et al* [48] |
| Fruits juice (kiwi, avocado, and pear) | CQDs | 3.9–4.4 | | Hydrothermal | Dias *et al* [31] |
| Citrus fruit peels | CQDs | 4.6 | Cell imaging | Sand bath strategy | Gudimella *et al* [32] |
| Rose-heart radish | N-CQDs | 3.6 | Cell imaging/$Fe^{3+}$ detection | Hydrothermal | Liu *et al* [30] |
| Spinach leaves | CQDs | 5.6 | $Fe^{3+}$ detection | Hydrothermal | Ran *et al* [42] |
| Apple juice | CQDs | 4.5 | Cell imaging | Hydrothermal | Metha *et al* [29] |
| Residues (*Lolium perenne*/malt bagasse) | CQDs | | Cell imaging/$Cr^{6+}$ detection | Hydrothermal/microwave | Afonso *et al* [54] |
| Garlic | N-S-CQDs | 11 | Cell imaging/radical scavenging | Hydrothermal | Zhao *et al* [33] |

some strategies are generally low-cost and environmentally friendly, such as hydrothermal synthesis [63].

For example, a fluorescent probe was designed using nitrogen-modified carbon dots (N-CQDs) to detect atrazine. It was obtained by hydrothermal treatment of o-phenylene diamine, showing high selectivity to determine the presence of this xenobiotic in aqueous solutions, with high sensitivity (up to 3 pM) and good performance in agricultural samples even at low concentrations [64]. Nanocomposites are also being developed to detect metobromuron, another herbicide used in many crops such as beans, soybeans, tomatoes, tobacco, and others. This electrochemical sensor consists of a core of gold nanoparticles that includes a molecularly imprinted polymer, an intermediate layer of nitrogen-doped CQDs, and an outer layer of silver nanoparticles [65]. Insecticides used in crops can also be effectively tested, such as ethion, using a detection system built with graphene quantum dots associated with nitrogen and sulfur heteroatoms in the presence of $Hg^{2+}$ [66]. A nanoprobe based on fluorescent GQDs and the quenching mechanism with silver nanoparticles (AgNp) was also developed to analyze parathion. The system had good selectivity and sensitivity in food and water samples and emitted a double response signal (fluorescence and ultraviolet–visible) [67]. A photoelectrochemical biosensor was also created for other organophosphate pesticides. The biosensor was prepared by taking nanoparticles of indium tin oxide, titanium dioxide ($TiO_2$), and N-CQDs [68]. Recently, multiple applications based on zero-dimensional carbon nanoparticles have been reported to detect a large spectrum of chemical substances used in agriculture [61, 69, 70].

### 13.3.2 Carbon quantum dot based nanosensors applied to food safety

CQDs and GQDs have other versatile applications, such as in the area of food, in terms of testing objects or environments, nutrient analysis, finding dangerous substances, and food microbiology [60]. An example is a method for the sensitive and specific detection of *Escherichia coli* with green fluorescence CQDs combined with magnetic nanoparticles. The nanosensor was tested in contaminated milk, and the carbon dots were synthesized from orange peel, which shows the potential of these technologies using sustainable and low-cost materials [71]. A system combining a CQD–chitosan nanocomposite and ultrasound to improve the food quality of processed cucumber was also reported. The results indicated that the treatment markedly inhibited the bacterial and fungal count after 15 days of storage, showing that carbon dots can also be used to improve food preservation [72]. In the same way, a fluorescence immunoassay with carbon dots and cadmium quantum dots was developed to detect concentrations up to 220 ng $mL^{-1}$ of *Citrus tristeza* virus in citrus plants [73]. In this regard, CQDs and GQDs have also been used to develop mycotoxin detection methods in crops and food. For example, an ultrasensitive immunosensor was developed using carbon dots doped with $MnO_2$ nanosheets to detect aflatoxin B1 in food [74]. Similarly, a photoelectrochemical aptasensor was based on zinc oxide–nitrogen-doped GQDs to evidence zearalenone in cereal crops with mildew [75]. In most cases, the introduction of carbon nanoparticles made it possible to obtain detection methods of great precision and sensitivity.

### 13.3.3 Environmental applications for agriculture

Other nanosensors have also been developed that could be applied effectively in the field of agriculture, such as a micro-sensor made with GQDs and tested as a platform for measuring soil moisture. This system has a varied range of responses to moisture content for silt loam soils and clay soils in the short time of three minutes. In addition, the sensor is stable for approximately five months or more, with minimal changes in resistance over time [76]. Similarly, another capacitive detector has been developed using graphene nanodots for humidity measurements with great sensitivity and shorter response–recovery times [77].

Regarding the photoluminescence properties of carbon dots, there is still no precise mechanism explaining the redshift phenomenon. Some studies have proposed that the size and surface characteristics of the particles could be the key to this process [78–80]. This property allows obtaining nanoparticles with different emissions in the UV–visible spectrum that can be useful for building detection systems related to the excitation of nanoparticles. By associating this behavior with other strategies, such as dyes and colorimetric visual tests, sensors could be developed to measure pH [81, 82]. In a study, nitrogen-doped GQDs were obtained with a high quantum yield and showed a sensitive behavior to pH and behaved in a stable way against ionic strength. Likewise, the fluorescence intensity presented a direct proportional relationship to pH in a wide range from 2 to 9. Therefore, it was determined that the graphene nanodots could function as fluorescent pH probes to measure the fluctuations of intracellular environments and water-soluble substances. It was also concluded that they could serve as a reliable parameter since the results of measurements in water are comparable to a conventional pH meter [83].

## 13.4 The influence of CQDs and GQDs on plants

### 13.4.1 Impact on plant physiology and general growth

Several carbon nanomaterials, including carbon dots, have demonstrated their potential applications in seed germination, root permeability, photosynthesis, and growth enhancement, thus improving overall plant performance [10]. Figure 13.1 outlines some of the known effects of carbon dots on plants.

Water-soluble carbon dots have been shown to increase wheat growth in the presence or absence of light. This study proposed that carbon dots enter the root cells through the plasmodesmata, passing through the apoplastic and symplastic cells to the cortex, where the endodermis allows the passage of water molecules. From there, they would follow the xylem pathway thus reaching the entire plant [84]. Other studies later confirmed the same entry route for nanoparticles [85]. Similarly, it was possible to monitor the absorption of nutrients (N, P, K) using carbon dots of 2 nm derived from bee pollen in *Brassica parachinensis* L. It was observed that the CQDs follow the pathway through the vascular bundles and enter the cells by translocation, depending on the plasmodesma size exclusion limit. It was also

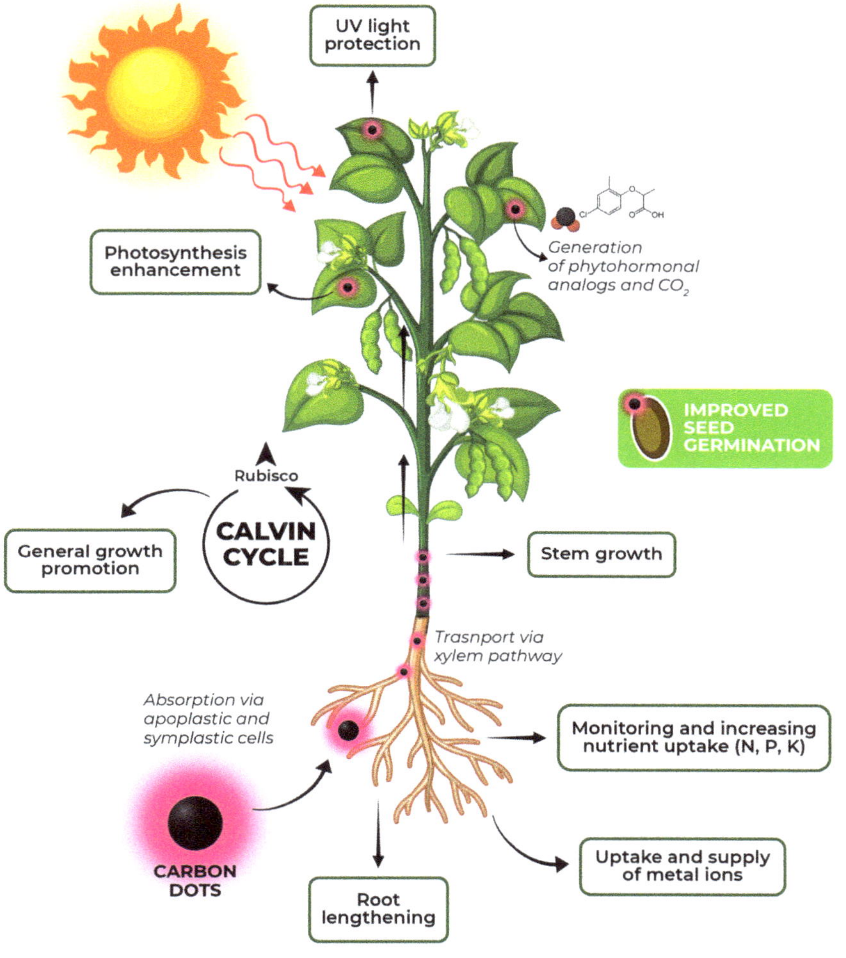


**Figure 13.1.** Diagram of some effects of carbon dots in the development of a plant.

highlighted that the potassium absorption is significantly increased, reducing the accumulation of nitrates simultaneously and effectively improving the growth of the vegetable [86]. Subsequently, the uptake of N-doped CQDs in tobacco cell lines could be monitored in real-time through fluorescence or confocal microscopy, revealing that nanodots enter cells by endocytosis with negligible cytotoxicity [87].

Studies using carbon dots with different oxygen content showed that the nanoparticles penetrate all parts of the rice plant, and they can be degraded to form phytohormonal analogs and carbon dioxide. These nanoparticles improved plant growth throughout the Calvin cycle, evidenced by increased RuBisCO activity [88]. The same author later corroborated that ~5 nm degradable carbon quantum dots could indeed promote dicot growth and improve grain production by 20%. In addition to the generation of carbohydrates derived from carbon dots, the increase in RuBisCO activity, and therefore plant growth, it was also found that carbon nanoparticles accelerate seed germination and improve the absorption and supply of metal ions [89]. They also allowed the development of stems and roots, demonstrating a positive increase in biomass [84, 89, 90]. Interestingly, CQDs could have low toxicity because they can be metabolized into the plant as carbon dioxide to improve intracellular components and could also have an antioxidant-protective function by absorbing in the UV spectrum [91, 92]. In this way, CQDs and GQDs lead to a new strategy to reduce fertilizer application and increase yields in modern agriculture.

### 13.4.2 Advances in improving photosynthesis

Photosynthesis is a process that is naturally determined by the presence of chlorophyll, accessory pigments, and the wavelength that they capture. One of the proposed strategies to increase the efficiency of absorbed photons is to maximize the capture of solar energy by expanding the range of use towards the ultraviolet spectrum. Li *et al* observed that CQDs improved photosynthesis when were conjugated with chloroplasts transforming UV radiation into photosynthesis radiation. Therefore, significant results were achieved in rice plant growth [93]. In another specific approach, double emission CQDs showed strong light uptake in the ultraviolet spectrum and intense blue/red light projection in water, matching the chloroplast's absorption spectrum. Consequently, an increase in the energy efficiency of *in vitro* photosystems was observed in romaine lettuce plants [94]. In another study, high quantum yield far-red carbon dots were obtained to test romaine lettuce seedlings and mung bean sprouts. These convert ultraviolet A light and emit an appropriate wavelength that photosynthetic pigments can absorb. As a result, an increase in electron transfer efficiency was observed in *in vitro* and *in vivo* photosystems [95]. The effect of near-red fluorescent CQDs can be understood from the Emerson effect, which explains the increase in the photosynthesis rate when chloroplasts are irradiated simultaneously with wavelengths corresponding to the red and far-red [96].

The application of CQDs can directly influence the photosynthetic pigments, as could be observed for CQDs co-doped with magnesium and nitrogen that, when applied in rice plants, positively regulated the production of chlorophyll A and B, increasing their content. Furthermore, the Mg also changed the fluorescence of the carbon dots to red, which was absorbed by the chloroplasts. Therefore, an increase in photosynthetic activity of 109.54% and an electron transfer rate of 104.48% was observed [93].

CQDs functionalization with glucose has also been reported to improve uptake, photoprotection, and pigment production, increasing the yield of *Triticum aestivum* (dwarf bread wheat) by 18% [97].

### 13.4.3 Resistance to stress caused by biotic and abiotic factors

CQDs and GQDs have good solubility in water, making them feasible to use together with the nutrients that plants require in their growing medium [84]. It has been reported that CQDs improve seeds' moisture levels and promote their germination [90] and resistance to drought by promoting water absorption [98].

Nutrient absorption is also a viable way to decrease abiotic stress, as observed in rice plants, by improving nutrient assimilation and antioxidant capacity after using CQDs. This resistance against saline stress and light stress maintained a prolonged effect over time [75]. It was also reported that proline-associated carbon dots could function as priming agents in plants and increase resistance to saline stress [99].

Due to climate change, drought is a relevant issue on the world agenda, in particular where access to water is difficult and directly determines food production [100, 101]. Some studies have used CQDs on peanut plants which caused an increase in water stress resistance and improved crop productivity [102]. Other nitrogen and sulfur co-doped CQDs also demonstrated the ability to enhance antioxidant metabolism and increase water stress resistance by regulating the expression of aquaporin genes [98].

Other approaches have been tested to improve antioxidant capacity in plants and remove phenol using N-doped CQDs as a platform to immobilize peroxidase. The incorporation of the nanoparticles enhanced the activity and stability of the method compared to the free enzyme [103]. Passivated carbon dots with a polymer have also been tested, which conferred a high antioxidant capacity and improved the oxidative damage generated by saline stress in Italian lettuce plants [104].

Heavy metals are other abiotic factors that can negatively affect plant growth. Some studies have explored the ability of CQDs to mitigate cadmium-induced toxicity, alleviating damage and improving the phytoremediation capacity [103]. It was also shown that carbon nanoparticles could adsorb these metals on their surface and induce the regulation of genes related to glutathione to reduce oxidative stress and improve the physiological response in *Triticum* and *Citrus maxima* [105, 106].

Regarding biotic stress, infections by phytopathogens can generate drastic reductions in crop productivity. In a trial with rice plants, it was reported that CQDs entered the cell nucleus and increased the expression of the thionin gene (Os06g32600), which improved the resistance to disease. In addition, rice plant leaves pre-cultivated with carbon dots and inoculated with *Rhizoctonia solani* (sheath blight) showed high resistance to infection five days later, compared to the control (without carbon dots) where the leaves were affected [88].

### 13.4.4 Assessment of phytotoxicity

Regarding the phytotoxicity of CQDs and GQDs, contrasting results have been presented. As mentioned above, many plants can assimilate CQDs and become metabolic intermediaries for enhancing growth, such as biocompatible and biodegradable molecules [88]. However, the capacity for absorption and metabolism of carbon nanoparticles is not always associated with positive effects. On the one hand, some works reported that toxicity might depend on the applied concentration. For example, in mung bean plants, CQD concentrations between 0 and 100 ppm stimulated growth, while at a concentrations four times higher, they showed the opposite effect [85].

On the other hand, Zhang *et al* proposed that the toxicity of graphene quantum dots is strongly related to the functionalization of its surface and the generation of oxidative radicals. The carboxylated GQDs and the hydroxylated GQDs produced oxidative damage, affecting photosynthesis, the biomass, and the regulation of nutrients in lettuce, compared to the amino GQDs that turned out to be less harmful [110]. Although the mechanisms of toxicity still appear to be unclear, there is variable evidence of the negative impact on plants. For example, in recent studies, *Arabidopsis thaliana* seedlings showed the decreased division of meristematic cells at the tips of primary roots when treated with nitrogen-doped carbon nanoparticles. This effect could be due to the deregulation of genes involved in the synthesis of auxins and carriers that allow growth [108]. Similarly, graphene quantum dots showed a negative impact on mung bean (*Vigna radiata* L.) and tomato (*Solanum lycopersicum* L.) plants' growth; however, the germination of the seeds of both plants was not affected [109]. In another example, *Chlorella vulgaris* generated oxidative stress, evidenced by increased superoxide dismutase when exposed to CQDs, which led to the interruption of photosynthesis [107].

## 13.5 Conclusion and prospects

CQDs and GQDs are nanoparticles that have attracted the attention of scientific minds worldwide mainly for their particular and beneficial properties. Due to this, a range of innovative applications is being developed to respond to multiple needs and improve the performance of existing technologies, as is the case in modern medicine and agriculture. The creative processes in the latter seem to follow the path of sustainability, which results in a necessary condition to obtain competitive products and services that adapt to the economic and environmental scenarios that currently prevail. Although there is currently exponential growth in research related to the application of these nanoparticles and their technological transfer, it is imperative to broaden our understanding of them in crucial aspects. Studies should be carried out that explicitly postulate theories about the mechanisms that govern the behavior of these carbon nanomaterials and that allow us to elucidate their properties, and therefore their full potential. Due to their versatility and beneficial properties, the technological development and replacement of traditional materials in the future seem to be very broad and promising. Given their wide applicability in different fields, some of which significantly impact humans, such as food and health, it is essential to carry out studies on nanotoxicity since there is still no consensus on the biocompatibility of carbon dots. These studies would allow knowledge of their short- and long-term effects on living beings and the environment. In parallel, governments must create laws that adequately manage these innovative materials to ensure safe technologies and prevent their inappropriate application.

## Acknowledgments

The authors want to thank CONICET, Universidad Nacional de Tucumán (Argentina), Universidad Nacional de San Luis (Argentina), and Universidad de La Frontera (Temuco, Chile) for providing free access to databases and for the technical advice received. We also extend special thanks to the graphic designer Mariana Manzanares for producing the graphics and images for this chapter.

## References


[1] Wang D, Behura S, Karim F, Kazi A, Wangila G and Wang S 2021 Nanoscience and nanotechnology for food and agroforestry *ES Food Agrofor.* **3** 1–3

[2] De C M, Vats A and Biel A 2018 Agriculture 4.0: The Future *World Government Summit Report*

[3] Klerkx L and Rose D 2020 Dealing with the game-changing technologies of agriculture 4.0: how do we manage diversity and responsibility in food system transition pathways? *Glob. Food. Sec* **24** 100347

[4] Singh Sekhon B 2014 Nanotechnology in agri-food production: an overview *Nanotechnol. Sci. Appl.* **7** 31–53

[5] Baker S, Volova T, Prudnikova S V, Satish S and Prasad M N N 2017 Nanoagroparticles emerging trends and future prospect in modern agriculture system *Environ. Toxicol. Pharmacol.* **53** 10–7

[6] Singh R P, Handa R and Manchanda G 2021 Nanoparticles in sustainable agriculture: an emerging opportunity *J. Control. Release* **329** 1234–48

[7] United Nations 2015 Transforming our World: The 2030 Agenda for Sustainable Development *Sustainable Development Goals Knowledge Platform*

[8] Liu Y, Huang H, Cao W, Mao B, Liu Y and Kang Z 2020 Advances in carbon dots: from the perspective of traditional quantum dots *Mater. Chem. Front.* **4** 1586–613

[9] Abd-Elsalam K A 2019 Carbon nanomaterials: 30 years of research in agroecosystems *Carbon Nanomaterials for Agri-food and Environmental Applications* (Amsterdam: Elsevier) pp 1–18

[10] Aacharya R and Chhipa H 2019 Nanocarbon fertilizers: implications of carbon nanomaterials in sustainable agriculture production *Carbon Nanomaterials for Agri-food and Environmental Applications* (Amsterdam: Elsevier) pp 297–321

[11] Zheng X T, Ananthanarayanan A, Luo K Q and Chen P 2015 Glowing graphene quantum dots and carbon dots: properties, syntheses, and biological applications *Small* **11** 1620–36

[12] Baker S N and Baker G A 2010 Luminescent carbon nanodots: emergent nanolights *Angew Chem.—Int. Ed.* **49** 6726–44

[13] Tiwari J N, Tiwari R N and Kim K S 2012 Zero-dimensional, one-dimensional, two-dimensional and three-dimensional nanostructured materials for advanced electrochemical energy devices *Prog. Mater. Sci.* **57** 724–803

[14] Mostofizadeh A, Li Y, Song B and Huang Y 2011 Synthesis, properties, and applications of low-dimensional carbon-related nanomaterials *J. Nanomater.* **2011** 685081

[15] Pokropivny V V and Skorokhod V V 2007 Classification of nanostructures by dimensionality and concept of surface forms engineering in nanomaterial science *Mater. Sci. Eng.* C **27** 990–3

[16] Xu X *et al* 2004 Electrophoretic analysis and purification of fluorescent single-walled carbon nanotube fragments *J. Am. Chem. Soc.* **126** 12736–7

[17] Li H *et al* 2010 Water-soluble fluorescent carbon quantum dots and photocatalyst design *Angew Chem.* **122** 4532–6

[18] Chong Y *et al* 2014 The *in vitro* and *in vivo* toxicity of graphene quantum dots *Biomaterials* **35** 5041–8

[19] Puvvada N, Kumar B N P, Konar S, Kalita H, Mandal M and Pathak A 2012 Synthesis of biocompatible multicolor luminescent carbon dots for bioimaging applications *Sci. Technol. Adv. Mater.* **13** 4

[20] Li Z, Wang L, Li Y, Feng Y and Feng W 2019 Frontiers in carbon dots: design, properties and applications *Mater. Chem. Front.* **3** 2571–601
[21] Wang Y and Hu A 2014 Carbon quantum dots: synthesis, properties and applications *J. Mater. Chem.* C **2** 6921–39
[22] Yao B, Huang H, Liu Y and Kang Z 2019 Carbon dots: a small conundrum *Trends Chem.* **1** 235–46
[23] Anastas P T and Warner J C 1998 *Green Chemistry Frontiers* **640** 1998
[24] Anastas P and Eghbali N 2010 Green chemistry: principles and practice *Chem. Soc. Rev.* **39** 301–12
[25] Albrecht M A, Evans C W and Raston C L 2006 Green chemistry and the health implications of nanoparticles *Green Chem.* **8** 417–32
[26] Parveen K, Banse V and Ledwani L 2016 Green synthesis of nanoparticles: their advantages and disadvantages *AIP Conf. Proc.* **1724** 020048
[27] Ijaz I, Gilani E, Nazir A and Bukhari A 2020 Detail review on chemical, physical and green synthesis, classification, characterizations and applications of nanoparticles *Green Chem. Lett. Rev.* **13** 59–81
[28] Iravani S and Varma R S 2020 Green synthesis, biomedical and biotechnological applications of carbon and graphene quantum dots. A review *Environ. Chem. Lett.* **18** 703–27
[29] Mehta V N, Jha S, Basu H, Singhal R K and Kailasa S K 2015 One-step hydrothermal approach to fabricate carbon dots from apple juice for imaging of mycobacterium and fungal cells *Sensors Actuators* B **213** 434–43
[30] Liu W, Diao H, Chang H, Wang H, Li T and Wei W 2017 Green synthesis of carbon dots from rose-heart radish and application for $Fe^{3+}$ detection and cell imaging *Sensors Actuators* B **241** 190–8
[31] Dias C *et al* 2019 Biocompatibility and bioimaging potential of fruit-based carbon dots *Nanomaterials* **9** 199
[32] Gudimella K K *et al* 2021 Sand bath assisted green synthesis of carbon dots from citrus fruit peels for free radical scavenging and cell imaging *Colloids Surf.* B **197** 111362
[33] Zhao S *et al* 2015 Green synthesis of bifunctional fluorescent carbon dots from garlic for cellular imaging and free radical scavenging *ACS Appl. Mater. Interfaces* **7** 17054–60
[34] Mewada A *et al* 2013 Green synthesis of biocompatible carbon dots using aqueous extract of *Trapa bispinosa* peel *Mater. Sci. Eng.* C **33** 2914–7
[35] Vasimalai N *et al* 2018 Green synthesis of fluorescent carbon dots from spices for *in vitro* imaging and tumour cell growth inhibition *Beilstein J. Nanotechnol.* **9** 530–44
[36] Ramanan V, Thiyagarajan S K, Raji K, Suresh R, Sekar R and Ramamurthy P 2016 Outright green synthesis of fluorescent carbon dots from eutrophic algal blooms for *in vitro* imaging *ACS Sustain. Chem. Eng.* **4** 4724–31
[37] Liao J, Cheng Z and Zhou L 2016 Nitrogen-doping enhanced fluorescent carbon dots: green synthesis and their applications for bioimaging and label-free detection of $Au^{3+}$ ions *ACS Sustain. Chem. Eng.* **4** 3053–61
[38] Gu D, Shang S, Yu Q and Shen J 2016 Green synthesis of nitrogen-doped carbon dots from lotus root for Hg(II) ions detection and cell imaging *Appl. Surf. Sci.* **390** 38–42
[39] Yang X, Wang D, Luo N, Feng M, Peng X and Liao X 2020 Green synthesis of fluorescent N,S-carbon dots from bamboo leaf and the interaction with nitrophenol compounds *Spectrochim. Acta* A **239** 118462

[40] Ghereghlou M, Esmaeili A A and Darroudi M 2021 Green synthesis of fluorescent carbon dots from *Elaeagnus angustifolia* and its application as tartrazine sensor *J. Fluoresc.* **31** 185–93

[41] Zaib M, Akhtar A, Maqsood F and Shahzadi T 2021 Green synthesis of carbon dots and their application as photocatalyst in dye degradation studies *Arab. J. Sci. Eng.* **46** 437–46

[42] Ran Y *et al* 2020 Green synthesis of fluorescent carbon dots using chloroplast dispersions as precursors and application for $Fe^{3+}$ ion sensing *Luminescence* **35** 870–6

[43] Shao Y *et al* 2020 Green synthesis of multifunctional fluorescent carbon dots from mulberry leaves (*Morus alba* L.) residues for simultaneous intracellular imaging and drug delivery *J. Nanopart. Res* **22** 229

[44] Sachdev A and Gopinath P 2015 Green synthesis of multifunctional carbon dots from coriander leaves and their potential application as antioxidants, sensors and bioimaging agents *Analyst* **140** 4260–9

[45] Mehta V N, Jha S and Kailasa S K 2014 One-pot green synthesis of carbon dots by using *Saccharum officinarum* juice for fluorescent imaging of bacteria (*Escherichia coli*) and yeast (*Saccharomyces cerevisiae*) cells *Mater. Sci. Eng.* C **38** 20–7

[46] Xu H, Yang X, Li G, Zhao C and Liao X 2015 Green synthesis of fluorescent carbon dots for selective detection of tartrazine in food samples *J. Agric. Food Chem.* **63** 6707–14

[47] Liang Z, Zeng L, Cao X, Wang Q, Wang X and Sun R 2014 Sustainable carbon quantum dots from forestry and agricultural biomass with amplified photoluminescence by simple $NH_4OH$ passivation *J. Mater. Chem.* C **2** 9760–6

[48] Zhang L, Wang Y, Liu W, Ni Y and Hou Q 2019 Corncob residues as carbon quantum dots sources and their application in detection of metal ions *Ind. Crops. Prod* **133** 18–25

[49] Shojaei T R *et al* 2018 Applications of nanotechnology and carbon nanoparticles in agriculture *Synthesis, Technology and Applications of Carbon Nanomaterials* (Amsterdam: Elsevier) pp 247–77

[50] Thambiraj S and Shankaran R 2016 Green synthesis of highly fluorescent carbon quantum dots from sugarcane bagasse pulp *Appl. Surf. Sci.* **390** 435–43

[51] Baweja H and Jeet K 2019 Economical and green synthesis of graphene and carbon quantum dots from agricultural waste *Mater. Res. Express* **6** 0850g8

[52] Shi Y *et al* 2019 Synthesis of N-doped carbon quantum dots from bio-waste lignin for selective irons detection and cellular imaging *Int. J. Biol. Macromol* **128** 537–45

[53] Rodríguez-Padrón D *et al* 2018 Catalyzed microwave-assisted preparation of carbon quantum dots from lignocellulosic residues *ACS Sustain. Chem. Eng.* **6** 7200–5

[54] Afonso A J G, Aquino F T, Dalmônico G M L, Nascimento M V, Wrasse E and de Aguiar K M F R 2021 Green synthesis of carbon nanodots from agro-industrial residues *Carbon Lett.* **32** 131–41

[55] Guo J *et al* 2020 Green synthesis of carbon dots toward anti-counterfeiting *ACS Sustain. Chem. Eng.* **8** 1566–72

[56] Carvalho F P 2017 Pesticides, environment, and food safety *Food Energy Secur.* **6** 48–60

[57] Boudh S and Singh J S 2019 Pesticide contamination: environmental problems and remediation strategies *Emerging and Eco-Friendly Approaches for Waste Management* ed R N Bharagava and P Chowdhary (Singapore: Springer) pp 245–69

[58] Zhao F, Wu J, Ying Y, She Y, Wang J and Ping J 2018 Carbon nanomaterial-enabled pesticide biosensors: design strategy, biosensing mechanism, and practical application *TrAC—Trends Anal. Chem.* **106** 62–83

[59] Shi X *et al* 2019 Review on carbon dots in food safety applications *Talanta* **194** 809–21

[60] Qu J H, Wei Q and Sun D W 2018 Carbon dots: principles and their applications in food quality and safety detection *Crit. Rev. Food Sci. Nutr.* **58** 2466–75

[61] Chu H W, Unnikrishnan B, Anand A, Lin Y W and Huang C C 2020 Carbon quantum dots for the detection of antibiotics and pesticides *J. Food Drug. Anal.* **28** 539–57

[62] Molaei M J 2020 Principles, mechanisms, and application of carbon quantum dots in sensors: a review *Anal. Methods* **12** 1266–87

[63] Xu Q *et al* 2016 Heteroatom-doped carbon dots: synthesis, characterization, properties, photoluminescence mechanism and biological applications *J. Mater. Chem.* **B 4** 7204–19

[64] Mohapatra S, Bera M K and Das R K 2018 Rapid 'turn-on' detection of atrazine using highly luminescent N-doped carbon quantum dot *Sensors Actuators* B **263** 459–68

[65] Kıran T R, Yola M L and Atar N 2019 Electrochemical sensor based on Au@nitrogen-doped carbon quantum dots@Ag core–shell composite including molecular imprinted polymer for metobromuron recognition *J. Electrochem. Soc.* **166** H691–7

[66] Nemati F, Hosseini M, Zare-Dorabei R and Ganjali M 2018 Sensitive recognition of ethion in food samples using turn-on fluorescence N and S co-doped graphene quantum dots *Anal. Methods* **10** 1760–6

[67] Li Y, Chen S, Lin D, Chen Z and Qiu P 2020 A dual-mode nanoprobe for the determination of parathion methyl based on graphene quantum dots modified silver nanoparticles *Anal. Bioanal. Chem.* **412** 5583–91

[68] Cheng W *et al* 2019 The visible light-driven and self-powered photoelectrochemical biosensor for organophosphate pesticides detection based on nitrogen doped carbon quantum dots for the signal amplification *Electrochim. Acta* **296** 627–36

[69] Carneiro S V *et al* 2019 Sensing strategy based on carbon quantum dots obtained from riboflavin for the identification of pesticides *Sensors Actuators* B **301** 127149

[70] Nair R V, Thomas R T, Mohamed A P and Pillai S 2020 Fluorescent turn-off sensor based on sulphur-doped graphene quantum dots in colloidal and film forms for the ultrasensitive detection of carbamate pesticides *Microchem. J.* **157** 104971

[71] Hu X *et al* 2021 Green one-step synthesis of carbon quantum dots from orange peel for fluorescent detection of *Escherichia coli* in milk *Food Chem.* **339** 127775

[72] Fan K, Zhang M and Chen H 2020 Effect of ultrasound treatment combined with carbon dots coating on the microbial and physicochemical quality of fresh-cut cucumber *Food Bioprocess Technol.* **13** 648–60

[73] Shojaei T R *et al* 2016 Fluorometric immunoassay for detecting the plant virus *Citrus tristeza* using carbon nanoparticles acting as quenchers and antibodies labeled with CdTe quantum dots *Microchim. Acta* **183** 2277–87

[74] Lin Y, Zhou Q, Tang D, Niessner R and Knopp D 2017 Signal-on photoelectrochemical immunoassay for aflatoxin B1 based on enzymatic product-etching $MnO_2$ nanosheets for dissociation of carbon dots *Anal. Chem.* **89** 5637–45

[75] Luo L, Liu X, Ma S, Li L and You T 2020 Quantification of zearalenone in mildewing cereal crops using an innovative photoelectrochemical aptamer sensing strategy based on ZnO-NGQDs composites *Food Chem.* **322** 126778

[76] Kalita H, Palaparthy V S, Baghini M S and Aslam M 2020 Electrochemical synthesis of graphene quantum dots from graphene oxide at room temperature and its soil moisture sensing properties *Carbon* **165** 9–17

[77] Chaudhary P, Maurya D K, Yadav S, Pandey A, Tripathi R K and Yadav B C 2021 Ultrafast responsive humidity sensor based on roasted gram derived carbon quantum dots: experimental and theoretical study *Sensors Actuators* B **329** 129116
[78] Nie H *et al* 2014 Carbon dots with continuously tunable full-color emission and their application in ratiometric pH sensing *Chem. Mater.* **26** 3104–12
[79] Ding H, Yu S B, Wei J S and Xiong H M 2016 Full-color light-emitting carbon dots with a surface-state-controlled luminescence mechanism *ACS Nano* **10** 484–91
[80] Jiang K *et al* 2015 Red, green, and blue luminescence by carbon dots: full-color emission tuning and multicolor cellular imaging *Angew Chem.—Int. Ed.* **54** 5360–3
[81] Kong W, Wu H, Ye Z, Li R, Xu T and Zhang B 2014 Optical properties of pH-sensitive carbon-dots with different modifications *J. Lumin.* **148** 238–42
[82] Lei C W, Hsieh M L and Liu W R 2019 A facile approach to synthesize carbon quantum dots with pH-dependent properties *Dye Pigment* **169** 73–80
[83] Wu Z L, Gao M X, Wang T T, Wan X Y, Zheng L L and Huang C Z 2014 A general quantitative pH sensor developed with dicyandiamide N-doped high quantum yield graphene quantum dots *Nanoscale* **6** 3868–74
[84] Tripathi S and Sarkar S 2015 Influence of water soluble carbon dots on the growth of wheat plant *Appl. Nanosci.* **5** 609–16
[85] Li W *et al* 2016 Phytotoxicity, uptake, and translocation of fluorescent carbon dots in mung bean plants *ACS Appl. Mater. Interfaces* **8** 19939–45
[86] Zheng Y *et al* 2017 Pollen derived blue fluorescent carbon dots for bioimaging and monitoring of nitrogen, phosphorus and potassium uptake in *Brassica parachinensis* L *RSC Adv.* **7** 33459–65
[87] Loukanov A, Mladenova P, Toshev S, Karailiev A, Ustinovich E and Nakabayashi S 2018 Real time monitoring and quantification of uptake carbon nanodots in eukaryotic cells *Microsc. Res. Tech.* **81** 1541–7
[88] Li H *et al* 2018 Impacts of carbon dots on rice plants: boosting the growth and improving the disease resistance *ACS Appl. Bio. Mater* **1** 663–72
[89] Li H *et al* 2019 Enhanced RuBisCO activity and promoted dicotyledons growth with degradable carbon dots *Nano Res.* **12** 1585–93
[90] Wang H *et al* 2018 Carbon dots promote the growth and photosynthesis of mung bean sprouts *Carbon* **136** 94–102
[91] Zhang M *et al* 2018 Pristine carbon dots boost the growth of *Chlorella vulgaris* by enhancing photosynthesis *ACS Appl. Bio. Mater* **1** 894–902
[92] Xue R, Fu L, Dong S, Yang H and Zhou D 2020 Promoting *Chlorella* photosynthesis and bioresource production using directionally prepared carbon dots with tunable emission *J. Colloid. Interface Sci.* **569** 195–203
[93] Li Y *et al* 2021 Carbon dots as light converter for plant photosynthesis: augmenting light coverage and quantum yield effect *J. Hazard Mater* **410** 124534
[94] Li W *et al* 2018 Enhanced biological photosynthetic efficiency using light-harvesting engineering with dual-emissive carbon dots *Adv. Funct. Mater.* **28** 1804004
[95] Li D *et al* 2020 Far-red carbon dots as efficient light-harvesting agents for enhanced photosynthesis *ACS Appl. Mater. Interfaces* **12** 21009–19
[96] Emerson R and Rabinowitch E 1960 Red drop and role of auxiliary pigments in photosynthesis *Plant Physiol.* **35** 477–85

[97] Swift T A *et al* 2021 Photosynthesis and crop productivity are enhanced by glucose-functionalised carbon dots *New Phytol.* **229** 783–90

[98] Kou E *et al* 2021 Nitrogen and sulfur co-doped carbon dots enhance drought resistance in tomato and mung beans *ACS Appl. Bio. Mater* **4** 6093–102

[99] Gohari G *et al* 2021 Enhanced tolerance to salinity stress in grapevine plants through application of carbon quantum dots functionalized by proline *Environ. Sci. Pollut. Res* **28** 42877–90

[100] Lesk C, Rowhani P and Ramankutty N 2016 Influence of extreme weather disasters on global crop production *Nature* **529** 84–7

[101] Ngcamu B S and Chari F 2020 Drought influences on food insecurity in Africa: a systematic literature review *Int. J. Environ. Res. Public Health* **17** 5897

[102] Su L X *et al* 2018 Carbon nanodots for enhancing the stress resistance of peanut plants *ACS Omega* **3** 17770–7

[103] Chen Q *et al* 2020 Enhanced bioaccumulation efficiency and tolerance for Cd (II) in *Arabidopsis thaliana* by amphoteric nitrogen-doped carbon dots *Ecotoxicol. Environ. Saf.* **190** 110108

[104] Li Y *et al* 2021 *Salvia miltiorrhiza*-derived carbon dots as scavengers of reactive oxygen species for reducing oxidative damage of plants *ACS Appl. Nano Mater.* **4** 113–20

[105] Xiao L, Guo H, Wang S, Li J, Wang Y and Xing B 2019 Carbon dots alleviate the toxicity of cadmium ions ($Cd^{2+}$) toward wheat seedlings *Environ. Sci. Nano* **6** 1493–506

[106] Li J *et al* 2019 Applications of carbon quantum dots to alleviate $Cd^{2+}$ phytotoxicity in *Citrus maxima* seedlings *Chemosphere* **236** 124385

[107] Zhang M *et al* 2019 Biotoxicity of degradable carbon dots towards microalgae: *Chlorella vulgaris Environ. Sci. Nano* **6** 3316–23

[108] Yan X *et al* 2021 Response of primary root to nitrogen-doped carbon dots in *Arabidopsis thaliana*: alterations in auxin level and cell division activity *Environ. Sci. Nano* **8** 1352–63

[109] Feng P, Geng B, Cheng Z, Liao X, Pan D and Huang J 2019 Graphene quantum dots-induced physiological and biochemical responses in mung bean and tomato seedlings *Rev. Bras. Bot.* **42** 29–41

[110] Zhang P, Wu X, Guo Z, Yang X, Hu X and Lynch 2021 Stress response and nutrient homeostasis in lettuce (*Lactuca sativa*) exposed to graphene quantum dots are modulated by particle surface functionalization *Adv. Biol.* **5** 2000778

# Chapter 14

# Broad spectrum utility of zero-dimensional carbon nanomaterials for agricultural applications

**Elaine Gabutin Mission**

This chapter discusses the broad and recent developments in the utilization of zero-dimensional carbon materials (ZCNs), such as fullerenes, carbon quantum dots, and graphene quantum dots, in particular, as applied in agriculture. A brief description of the ZCNs is provided and the growing trend to synthesize them from natural and biological precursors is discussed. Specific applications of ZCNs as plant growth regulators and fertilizers, as early detectors and biomarkers, and as environmental remediates are described. A discussion of the negative aspects of ZCN utilization is also included. Finally, a summary and future perspectives regarding the drawbacks and concerns of ZCNs are provided.

## 14.1 Introduction

Nanomaterials have demonstrated versatility in terms of application and development, thus their exploitation and exploration have been reported in an exponentially increasing number of research articles, patents, and public releases. To date, nanomaterials have found applications in advanced materials for catalysis, electronics, biosensing, optics, nanomedicine, bioimaging, energy storage, and agriculture [1–3]. What facilitated such enormous possibilities for nanomaterials is their unique combination of properties, including optical, thermal, structural, electrical, and chemical features that endow them with superior characteristics and performance that are not evident in their bulk counterparts [4–6]. Nanomaterials have been tested and developed as nanofertilizers, nanopesticides, and nanodelivery systems to influence plant growth and increase agricultural production, mainly due to their size, which allows penetration into the roots and plant system [7–9]. In fact, several metal based, carbon based, and polymer based nanomaterials in pure and composite forms have been reported to elicit positive responses from plants during uptake [10–12].

doi:10.1088/978-0-7503-4048-9ch14

However, due to the tendency of nanomaterials to bioaccumulate and exhibit phytotoxicity, their agricultural application has become quite controversial [13].

Thus, attention has been diverted into carbon based nanomaterials (CNM) due to their perceived biocompatibility and non-toxicity. CNMs have been studied widely due to their low mass production cost, low inherent toxicity, and multifunctional surface modification [14–18]. These excellent properties propelled CNMs as a promising candidate over other types of nanomaterials in a broad range of biological applications, particularly in imaging, therapeutics, bio-chemical sensing, and controlled drug delivery [16, 19–24]. Aside from the usual encapsulation technique, carbon nanotubes have been used as a carrier of the anticancer agent doxorubicin, which was attached to folic acid and hyaluronan as a ligand [21]. Keihan and co-workers have summarized several works on the utilization of CNMs as scaffolds for tissue engineering, tissue regeneration, and implantation [25]. CNMs also have potential in theranostics, or the combined effect of therapeutics and diagnostics. In this case, the CNMs can be introduced into target cells and can be used to image the area (*in vivo* imaging) and, at the same time, provide the desired active molecules that can induce healing or repair of the damaged cells in that area. This is particularly of interest for rehabilitative medicine, for those who have met with accidents, and cancer treatment [16]. These early positive contributions of CNMs in the field of biomedicine have implied their potential biocompatibility and the possibility of application to ecological systems in the hope of revitalizing food and agricultural systems. The introduction of CNMs into a plant system was reported to affect plant physiological phenomena and metabolic activities in various ways, including those that contribute towards plant growth and productivity. Some of these include enhanced seed germination, greater seedling growth, taller shoots and longer roots, bigger and greener leaves, more flowers and fruits, faster flowering and fruit production, improved photosynthesis rates, and developed defense against pests [26–35]. Not all reports were positive, such as the case of the effect of carbon nanotubes in the tomato plant, which caused a delay in the period at which early growth was observed [31]. It was interesting, however, that the measured salicylic acid in the tomatoes was higher than in those that were not exposed to carbon nanotubes, which may indicate that the presence of nanomaterials can trigger changes in the chemical constituents of the plant. These chemical changes maybe a signal of stress and any change in plant growth and mechanisms could then suggest that carbon nanotubes can acts as aggressors or are perceived by the plant as a foreign material. These chemical changes can then have repercussions regarding the taste and nutrient level of the fruits and may affect food quality.

Aside from plant growth, CNMs were also found to be crucial in conserving soil structure through the slow or timed release of nutrients, enhanced water and nutrient retention capacity, and by maintaining the soil's pH balance [36–38]. Previous works have identified the mode by which CNMs enter into the plant system, which is by entry from the roots into the main plant body through the symplastic (a selectively permeable route by which water and mineral movement is governed by osmosis through the cytoplasm) or apoplastic (a fully permeable route by which water and minerals move via passive diffusion in an upward flow direction through the cell wall) pathways. From thereon, transport of CNMs throughout the plant body is facilitated by the xylem vessels, reaching all parts of the plant. As such, CNMs can provide an ecofriendly and sustainable solution towards secondary production from agriculture, first through the reduction of the amount of fertilizer being utilized in fertigation systems (a method by which fertilizer is incorporated within the irrigation water through a drip system) and second through the controlled/timed release of CNMs as next-generation fertilizers and pesticides. CNMs have demonstrated enormous potential as plant growth stimulants or regulators, as insect or pest control nanosensors or active agents, as pesticide adsorbents, and in nutrient encapsulation in the agricultural industries [39–42]. The most popular carbon allotropes in recent years have comprised low-dimensional carbon forms that include fullerenes ($C_{60}$), carbon nanotubes (CNTs), carbon black, carbon onions, carbon nanohorns, carbon nanodots, and graphene. CNTs, graphene, and fullerene have been employed as nanocarriers for active molecules and compounds aimed at pest and insect control and replenishment of nutrients [43–46].

Since nanomaterial properties were found to be better and more promising with further size reduction, researchers and scientists have been continually reducing their size from the bulk to molecular to atom scale. There are two major factors that cause nanomaterial properties to substantially deviate from other materials: a high relative surface area and quantum effects. Manipulation or control of these factors can alter or boost nanomaterial properties, including their reactivity, inherent strength, and electrical features. Therefore, a newly synthesized nanostructure can be imparted with a novel set of properties distinct from those of higher-dimensional materials by altering and adjusting the nanomaterial size and fabricating them within the zero-dimensional spectrum [47, 48]. This possibility paved the way for the introduction of zero-dimensional carbon nanomaterials (ZCNs) into the carbon allotrope portfolio.

Pokropivny and Skorokhod classified materials into four categories according to their dimensions, hence the adoption of the 0D, 1D, 2D, and 3D conventions for nanoparticles [49]. For a quick differentiation between the four aforementioned classes, a zero-dimensional (0D) nanoparticle has none of its dimensions larger than 100 nm. In one-dimensional or two-dimensional nanomaterials (1D and 2D, respectively), one dimension can be beyond the nanoscale. Meanwhile, three-dimensional nanomaterials (3D) are not restricted in the nanoscale regime in any dimension. Despite having zero dimensions, 0D materials also exist in various shapes, such as solid and hollowed spheres, nanoclusters, core–shell structures, and onions. The 0D CNMs have fascinating properties encompassing exceptional optical and electrical properties, high quantum yield, low toxicity, and superior photostability, that have enabled their utilization as solar cells, supercapacitors, nanosensors, electrical devices, light emitting diodes, batteries, transistors, and lasers [50–54]. More recent 0D applications included advancements in biomedicine (where 0D nanomaterials acts as nanocarriers or encapsulate active pharmaceutical compounds), theranostics (which involves cells and tissue imaging of cells and *in situ* delivery of therapeutic molecules for tissue repair and treatment of disease), and tissue engineering (where carbon nanomaterials are used as a scaffold for bone regeneration) [22, 23, 47, 55].

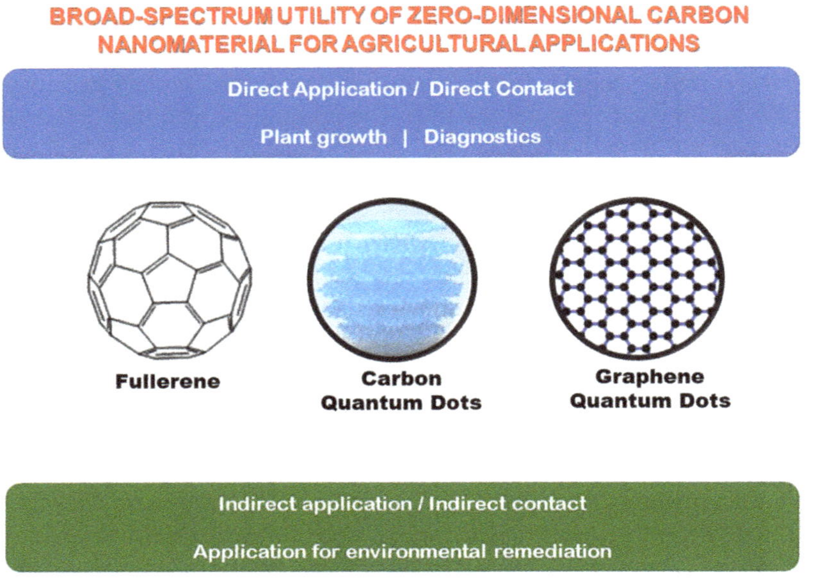


**Figure 14.1.** Broad spectrum utility of zero-dimensional carbon nanomaterials for agricultural applications as presented in this chapter.

The interesting properties and the ability for modification, together with the remarkable achievements in the field of medicine, opened more opportunities for ZCNs to be exploited in other fields, such as food production and agriculture. In this chapter, the broad spectrum application of zero-dimensional carbon nanomaterials, particularly fullerenes, graphene quantum dots, and carbon quantum dots, is discussed. First, an overview of the development, synthesis, and application of 0D carbon nanoparticles is provided. Next, the latest trends in the synthesis and properties of ZCNs are summarized. Then, the applications of ZCNs in the field of agriculture, either directly or indirectly, are discussed within the framework illustrated in figure 14.1. The negative and positive effects of ZCNs are also indicated. Finally, a summary and future perspective is provided.

## 14.2 Zero-dimensional carbon nanomaterials (ZCNs)

ZCNs are not limited to fullerenes, graphene quantum dots, and carbon quantum dots, however, these three are the most widely reported and the rest are known as their derivatives.

### 14.2.1 Fullerenes

Fullerenes are the spherical and caged molecular allotropes of carbon. Their existence was first revealed in the 1985 by researchers from Rice University through laser spectroscopy. Fullerenes were named after Richard Buckminster Fuller, an architect/designer/engineer recognized for designing and building geodesic dome structures that bear a resemblance to spherical fullerenes in appearance. As spherical fullerene also resembles a soccer ball, it earned the monicker ‘buckyball’. About a decade after its first report, the Nobel Prize in Chemistry was jointly awarded to Robert F Curl, Jr, Richard E Smalley, and Sir Harold W Kroto ‘for their discovery of fullerenes’ [56]. Fullerenes are commonly illustrated as polyhedral structures consisting of pentagons and hexagons, whose vertices are the carbon atoms [57]. Typically, they are expressed by the chemical formula $C_{20+n}$, where $n$ is any integer. In fullerenes, the carbon atoms typically have $sp^2$-hybridization and are interconnected with covalent bonds. Let us take $C_{60}$ as an example. $C_{60}$ is the most popular and most common fullerene. It is composed of about five to six $sp^2$-hybrid carbon rings, that assume a truncated icosahedron. It has a diameter of about 0.7 nm [58]. High resolution transmission electron microscopy (HRTEM) has made possible the accurate measurement of the diameters of some other fullerenes, such as $C_{36}$, $C_{58}$, $C_{90}$, and $C_{130}$, with corresponding diameters of 0.52, 0.685, 0.86, and 1.03 nm, respectively.

At the beginning of its study, the basic constituent for fullerene synthesis was carbon vapor. In 1985 Kroto *et al* used carbon in a laser vaporization set-up in an inert atmosphere to synthesize fullerenes. Meanwhile, German physicist Wolfgang Krätschmer and American professor Donald R Huffman evaporated graphite electrodes in a helium atmosphere in 1990. Then came the electric arc reactor that employed two graphite electrodes which yielded soot. The soot was allowed to condense on the reactor’s cold surface and was then collected via extraction using a variety of organic solvents including toluene, benzene, and xylene. Once the solvents were evaporated, a black residue or condensate remained, comprising about 10%–15% of $C_{60}$ and $C_{70}$ fullerenes, with traces of higher order fullerenes. Another synthesis approach involved the use of microwave plasma in a low-pressure argon atmosphere. Fullerenes were also obtained by the pyrolysis of naphthalene at very high temperatures of 1000 °C. The dominant fraction is more often a $C_{60}$ in a mixture containing $C_{70}$, depending on the synthesis conditions.

One of fullerene’s interesting properties is its ability to withstand high pressures of over 3000 atm and revert to its initial shape once the pressure is released. $C_{60}$ species have a dielectric constant of ~4 and possesses an effective bulk modulus of 668 GPa, making them even harder than diamond [59]. They also possess a high refraction index, structural stability, a large molecular volume, high electron affinities, large surface-to-volume ratios, and superconductivity. C60 is insoluble or minimally soluble in water, thus the necessity to utilize organic solvents such as toluene, 1-chloronaphthalene, benzene, p-xylene, and carbon disulfide to dissolve it. Other molecules interact with $C_{60}$ due to its molecular $\pi$-electron system and $C_{60}$ can be incorporated into a wide variety of polymers via the formation of donor–acceptor or covalent bonds. With these characteristics, fullerenes are used in a diverse range of applications, including electronics, biology, and medicine. As fullerenes have good affinity with various organic molecules, they have also been used to construct various biosensors.

### 14.2.2 Carbon quantum dots

Carbon quantum dots (CQDs), also known as carbon dots (CDs), are a class of carbon nanomaterials that are less than 10 nm in size and were discovered in 2004 while CNTs were being purified. CQDs are attracting attention for their fluorescent properties (chemiluminescence (CL) and electrochemiluminescence (ECL)). They also exhibit optical properties such as strong optical absorption manifesting in the UV range (mid-300 nm to 700 nm) that can extend to some portions of the visible spectrum, bright photoluminescence, and excellent light stability (resistance to light decomposition, photobleaching, and blinking). For this reason, it is common for CDs to be used in the fields of energy conversion, bioimaging, optoelectronics, drug delivery, photocatalysis, and biosensing. The quantum yields (QYs) reported vary greatly and the highest values achieved were over 90%.

CQDs mainly consist of carbon, hydrogen, and oxygen in varying amounts depending on the synthesis procedure. Their surface presents complex oxygenated functionalities, including carboxyls and hydroxyls, that aid the optical properties and water-dispersability of CQDs. The presence of these functionalities have been the foremost reason for the viability of CQDs as drug nanocarriers and nanodelivery vehicles. The active compound or drug molecule can be conjugated through the free functional groups. For example, CQDs have been loaded with anticancer agents, such doxorubicin [21]. Tailoring the relative proportion and combination of oxygenated functional groups by virtue of doping or surface functionalization can further address weaknesses in terms of the reactivity and biocompatibility of CQDs. In contrast to GQDs, CDs have lower amount of crystalline $sp^2$ carbon (mainly amorphous in nature and with more surface defects), hence they have poorer crystallinity [60]. However, similarly to GQDs, CQDs can be synthesized and functionalized quickly and easily.

All CDs can emit either single or multiple long wavelength light even under a single excitation wavelength owing to their surface functionalities and sizes. For example, a CD with 400–710 nm and a QY of 6%–20% can emit photoluminescence colors of blue to red. In addition, CDs with blue (435 nm, QY = 17.6%), green (535 nm, QY = 4.8%), and red (604 nm, QY = 26.1%) emissions are also possible. Lastly, blue to red emissions can be obtained from 440 to 625 nm CDs at a QY of 8.5%–35% [61].

### 14.2.3 Graphene quantum dots

Graphene quantum dots (GQDs) are a newly identified type of 0D graphene nanomaterial arising as small fragments of graphene. GQDs are characterized by atom-thin graphitized planes (usually one or two layers, not exceeding 2 nm in thickness) and small transverse dimensions (<10 nm in general) [62]. They was synthesized by Ponomarenko and Geim in 2008 based on the work done by Xu *et al* on carbon dots, thus they contains characteristics that are derivative from graphene and carbon dots [63, 64]. Due to the limitations of graphene, one of which is being a zero-bandgap semiconductor resulting in low absorptivity and the absence of optical luminescence, its electronic applications are limited. For this reason, the bandgap was introduced via quantum confinement in all three directions, which exists in 0D materials. Therefore, by synthesizing graphene as a 0D nanoparticles, photoluminescence (PL) can be harnessed [52, 65].

Contrary to CQDs, with a core of amorphous $sp^3$-hybrid forms of carbon, GQDs are crystalline and have a 'molecule-like' character instead of being colloidal in nature. GQDs can exist in different sizes but of the same 3.4 eV bandgap energy, in contrast to the 0 eV or close to 0 eV bandgap energy of pristine graphene, which is why they are called quantum dots instead [59]. Despite this difference, GQDs have identical photoluminescence and similar emission properties to CQDs.

There is also an abundance of recent studies that have reported and demonstrated that the PL excitation and emission wavelength of GQDs can be modified via adjustment of their dimensions, morphologies, or dopant atoms [66–70]. Furthermore, the edges of GQDs are rich in oxygen-rich functional groups, which enables their effective water solubility and biocompatibility [71]. In fact, GQDs have been found to be safe for both *in vitro* and *in vivo* use [55, 72–74]. GQDs have been synthesized via electrolysis of graphite rods, pyrolysis, hydrothermal strategy, chemical exfoliation of multiwalled carbon nanotubes, and oxidation of carbon fibers [66, 68, 75–77].

Due to their tunable PL properties, oxygenated functionalities, and biocompatibility, GQDs have been utilized widely for bioimaging and biosensing applications for the detection of biomarkers and early diagnosis of major diseases, such as lung cancer and Alzheimer's disease [68, 78–80].

Cui *et al* [76] developed a sensitive and promising magnetic fluorescent biosensor based on GQDs that can be used for quantitative detection of circulating tumor cells (CTCs), which can lead to early diagnosis of cancer. The biosensor also contained $Fe_3O_4$ as donors and $MoS_2$ nanosheets as quenchers and could detect tumor cells in as short a time as 15 min. Moreover, the captured tumor cells can also be separated and captured if further experimental study is desired.

GQDs have a strong sorption capacity for heavy metals compared to other substrates due to the large amount of surface groups, such as carboxyl, hydroxyl, and amine they contain. The same factor enables their excellent water solubility, which provides convenience in developing composites with other materials without the need for phase separation.

## 14.3 Preparation and synthesis of ZCNs

ZCNs share the same methods of synthesis as any other nanomaterial synthesis. The two main approaches include the top-down approach (where nanoscale materials are produced by subdividing bulk materials) and the bottom-up approach (where nanomaterials are produced from precursors). The major representative techniques are summarized in table 14.1. The differences lie in the sensitivity and precision to produce a desired target size. Furthermore, an additional step of filtration may be necessary to homogenize the particles and filter out-of-range nanoparticles [81]. It is worth noting that ZCN synthesis is a lengthy and tedious process and a huge consumer of chemicals, and is sometimes performed in sophisticated reactors, yet some methods generate only a few grams of high quality product that would only be suitable for characterization studies. Therefore, a variety of techniques were developed to (i) synthesize ZCNs in bigger batches for large-scale applications, (ii) develop homogeneous ZCNs with homogeneous and well-preserved properties, (iii) modify ZCN properties and make them more susceptible for reactions depending on the desired application or modification, and (iv) decrease the cost of production.

**Table 14.1.** Synthesis routes and representative techniques for the synthesis of zero-dimensional carbon nanomaterials.

| Synthesis route | Representative techniques | References |
|---|---|---|
| **Physical** | • Evaporation<br>• Lithography<br>• Sputtering<br>• Phase condensation, hot and cold plasma spray<br>• Pyrolysis<br>• Inert gas phase condensation<br>• Pulsed laser ablation method<br>• Sonochemical reduction<br>• Thermal evaporation method | [5, 18, 36, 38, 109–112] |
| **Chemical** | • Lyotropic<br>• Liquid crystal templates<br>• Electrochemical deposition<br>• Electrodeless deposition<br>• Hydrothermal<br>• Solvothermal techniques<br>• Laser chemical vapor deposition<br>• Laser pyrolysis<br>• Chemical vapor deposition<br>• Solvent exchange | [2, 5, 60, 104, 110, 113–115] |
| **Biological** | • Microbial derived materials | [95] |

The main drawback of the well-established approaches for ZCN synthesis is that the process, as well as the product, become expensive. Thus, in order to obtain a cost-effective material, the precursor should be lower in terms of cost. A growing trend recently has been to synthesize ZCNs from natural and biological precursors. Recently, the utilization of industrial [82], biological, or agricultural waste [19] has been explored, also in support of a circular economy framework and sustainable development goals. Figure 14.2 shows a simplified diagram to visualize the conventional approach for carbon nanotube synthesis in comparison to carbon quantum dots. Aside from the difference in the feedstock, the wide temperature gap is also striking. Thus, a cheap and widely available feedstock, moderate temperature range, and less or absence of solvent would put CQDs in a better position than other ZCNs. For example, GQDs were prepared by the pyrolysis of citric acid [82]. Then, Ding and co-workers utilized lignin as a precursor for GQD preparation through a two-step method that involves an oxidized cleaving step of alkali lignin molecules followed by an aromatic refusion step [83]. Surendran and colleagues reported the synthesis of CQDs from the agro waste of *Ananas comosus* by hydrothermal treatment. Other reported precursors with varying quantum sizes (between 2 and 5 nm) and yields (between 5%–10%) include sugar cane juice, ascorbic acid, strawberry juice,

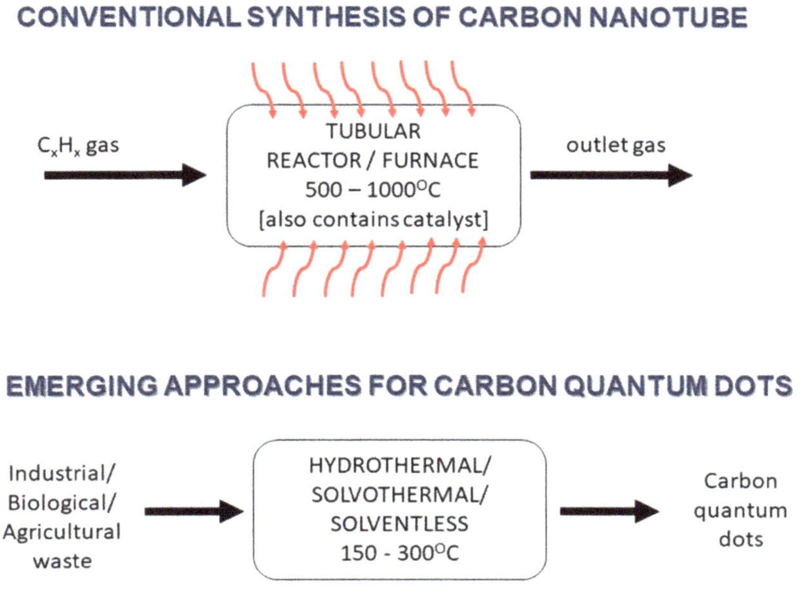


**Figure 14.2.** Simplified diagram of the conventional synthesis of carbon nanotubes and emerging approaches for carbon quantum dot synthesis.

apple juice, chicken egg, pomelo peel, watermelon peel, and pineapple peel [84–91]. The fish scales of crucian carp were also explored as a precursor via hydrothermal reaction for CQDs synthesis. The as-prepared CDs exhibited strong fluorescent emissions at 430 nm with a relative quantum yield of 6.9% [92]. Rapeseed pollen has also been used as precursor for CDs via a hydrothermal method [93]. Broccoli was also shown as a potential industrial source of CQDs after hydrothermal treatment followed by carbonization [94]. Wang *et al* utilized cyanobacteria as a carbon source for CQDs. They utilized a hydrothermal treatment resulting in CQDs with a mean diameter of 2.48 nm and a QY of 9.24%. The absence of chemical solvent in the process rendered low cytotoxic CQDs [95]. The main considerations for success in CQD synthesis are the excitation wavelength and the quantum yield. These two values are largely tunable by the selection of the precursor or starting carbon material, synthesis method (solvothermal, hydrothermal or solventless reactions), and reaction time and temperature. In a novel large-scale one-step hydrothermal method, Li and co-workers [96] reported a kilogram production of CQD from poplar leaves. The poplar leaves were simply heated to 200 °C for 24 h in pressurized water, providing roughly 1.5 kg of CQDs in one pot. After drying, the yielded brownish-yellow CQDs with a crystalline hexagonal structure were analyzed and found to reach a photoluminescent quantum yield of 10.64%. This promising research has shown that we are closer to actually producing CQDs at an industrial scale.

Despite the notable progress in synthesis approaches for ZCNs, much work needs to be done to fully unlock the promising potential of these intriguing materials. Based on several reported works, it can be seen that the proper inspection and reporting of the dimensions and their correlation to respective properties is still a weak area.

Although the emergence of various fabrication methods has made the synthesis of intricate ZCNs possible, no specific or precise atomic structure has been reported, which limits the in-depth study of the relationship between the structure and performance, and the precise control of performance and the sensing mechanism.

## 14.4 Agricultural applications of ZCNs

The applications of zero-dimensional nanoparticles in agriculture can be summarized into two broad aspects: direct application through plant growth and soil biota effects, and indirect application involving environmental remediation. The main advantages of ZCNs over other nanomaterials are their perceived lower production costs and the wider availability of precursors. In addition, having zero dimensionality is associated with chemical inertness.

### 14.4.1 Plant growth/fertilizer

Having a carbon structure, ZCNs are perceived as suitable for agricultural applications. In fact, the utilization of ZCNs has not been as widely explored in agriculture as in medicine, sensors, environmental science, packaging, and food processing. The literature also does not provide a consensus as to whether ZCNs are truly beneficial as they have shown both positive and negative impacts when in direct contact with the plant under study. There are several parameters that are considered to induce a response from the plant, including the plant type, age and species, ZCN type and concentration, growing conditions (simulated environment versus actual environmental conditions), and introduction of fertilizers. These reports are presented and discussed in the following.

For example, GQDs can act as smart treatment-delivery systems in regulating plant growth and can pass through cell walls easily since the size of the GQDs is smaller than the size of cell walls [97]. GQDs 5 nm in size were used to improve the growth rate of coriander and garlic plants from seeds [97]. It was observed that the addition of GQDs resulted in: (i) a greater number of flowers; (ii) a more intense green leaf color; (iii) greater average root length; and (iv) heavier roots (a more robust root system) for the coriander. On the other hand, there was no significant difference measured in the height of the plants between the GQD-treated and untreated ones. The lengths of the leaves and roots and the mass of the roots of the garlic plants were observed to be higher with the GQDs compared to the untreated ones, while the numbers of leaves was not affected by the treatment. The effects of the addition of GQDs were clearly demonstrated, although the reason for these effects was not explored in detail.

The work of Zheng *et al* [93] showed a more traditional approach of incorporating CDs into the hydrophonic solution used to grow *Lactuca sativa* L. lettuce. The CDs were sourced from rapeseed pollen and had a mean diameter of 5.2 nm. At a concentration of 30 mg $L^{-1}$ solution, the lettuce plant size and leaf area almost doubled compared to that the untreated ones. Furthermore, the nutrient content, such as ascorbic acid, sugar, and protein, were not affected.

Since carbon nanoparticles have the tendency to aggregate in the presence of water, other organic materials, and agitation, derivatives of the original form have been explored. For example, the water soluble daughter product of fullerene $C_{60}$, known as fullerols $C_{60}(OH)_{24}$ [98] have been found to induce the phytomedicinal properties of bitter melon. The effect of fullerols, having hydrodynamic diameters of around 1.5 ± 0.2 nm and 5.0 ± 0.7 nm, on biomass yield, plant water content, and phytomedicine content was studied by Kole and co-workers [99]. They reported that bitter melon seeds treated with fullerols resulted in an increase in biomass yield, water content, fruit length, fruit number, and fruit weight by 54%, 24%, 20%, 59%, and 70%, respectively, leading to an overall fruit yield increase of 112% to 128%. Major phytomedicine content, namely, cucurbitacin-B (74%), lycopene (82%), charantin (20%), and insulin (91%), were also observed to have increased, with percentages are indicated in the parenthesis. While the results were indeed promising, there was no reporting of any negative effects observed nor of the possibility of accumulation in the plant environment. Fullerol was also investigated in use against the saprophytic soil fungus *Aspergillus flavus*, that is capable of producing aflatoxins, inflicts diseases and colonizes crops. It was found that fullerol, with an 8 nm diameter and a zeta potential of −33 mV, can affect mycelial growth and alter the biosynthetic pathways of the *A. flavus*, indicating that its mineralization in soil could threaten mycotoxigenic fungi [98].

There have been few attempts to study the uptake, translocation, and transmission of carbon nanoparticles in plants. Lin *et al* [100] demonstrated that the $C_{70}$ fullerene, with an about 1.19 nm hydrodynamic diameter, can be accumulated and compartmentalized in various parts of rice plants. The C70 was found to be prevalent in the vascular system of the stem, indicating that the nanoparticles follow the same transport system as water and nutrients. Their potential accumulation and clogging along the plant's transport system could be the potential reason as to why some findings documented negative effects of carbon nanoparticles towards plant growth. At high concentrations, $C_{70}$ aggregation within the vascular system and in plant tissues and cells is expected to interfere with nutrients and water uptake, and hinder plant development. The main concern in this work was the potential that carbon nanoparticles could lead to plant gene expression and affect biological pathways. In fact, it was observed that flowering was delayed for 1 month in the presence of $C_{70}$. In addition, the absorbed nanoparticles of first generation plants can also be 'passed on' to a second generation plant derived from first generation seeds.

Aside from potential accumulation in plants, De La Torre-Roche and co-workers [26] found out that $C_{60}$ fullerenes curbed the weathered chlordane uptake of corn and zucchini, but amplified the uptake of soybean and tomato. Such behavior was contrasted in the co-exposure to multiwalled carbon nanotubes, wherein all four plants decreased chlordane uptake between 21%–80% depending on the concentration.

### 14.4.2 Early detection/diagnostics

The photoluminescence property of CQDs was harnessed in a study by Kang *et al* and they observed zebrafish husbandry and embryo harvesting for *in vivo* multicolor imaging [101]. In this work, CQDs were injected into the zebrafish embryos and larvae by soaking, which revealed that different parts contained different amounts of CQDs depending on tissue affinity. The team also reported that the livability of embryos as well the growth development of the larvae was not affected, hinting at the biocompatibility of CQDs. This study could be used to monitor the growth of individual species or the early detection and diagnosis of any impending abnormality, particularly in propagating fish for food purposes. While the initial results revealed that the CQDs do not interfere in the normal growth and development of the zebrafish models, the widespread utilization of such an approach should be cautioned, particularly as environmental factors affect the nanoparticle properties.

If zebrafish have shown a positive response towards CQDs, juvenile largemouth bass reacted negatively towards $C_{60}$. [102] In the study by Eva Oberdorster, it was found that $C_{60}$ actually improved the water quality of the tank where the fish were, which she hypothesized could be due to its bactericidal properties. However, it was revealed that the $C_{60}$ could preferentially localize and translocate into the brains of the fish, where it led to the depletion of total glutathione levels and oxidative stress. This study revealed that $C_{60}$ exposure could hamper normal metabolic processes in living organisms and that further studies *in vivo* should be carried out to avoid massive damage to environmental systems.

CQDs of 3 nm mean diameter capable of emitting blue color were also synthesized from the juice of *Saccharum officinarum* and were utilized in the cell imaging of bacteria and yeast, represented by *Escherichia coli* and *Saccharomyces cerevisiae*, respectively [86]. It was shown that CQDs have the versatility to penetrate even the simplest molecules and can therefore be useful in visualizing and monitoring real-time biomolecular interactions which could pave the way to a better understanding of biological systems.

### 14.4.3 Environmental remediation

Owing to the structure and surface properties of ZCNs, they have fast become an attractive component in biosensors. Their small size facilitates rapid reactions and interactions with compounds of interest, which translate to quicker responses, faster detection, shorter lag or waiting times, higher accuracy, and repeatable measurements. ZCNs have been particularly applied to highly toxic and complex compounds as discussed in the following examples.

Bellesteros *et al* studied the potential of $C_{60}$ as a sorbent in a column for analytical gas chromatography and flame atomic absorption spectrometry. The compounds used in the study included organic compounds such as *N*-methylcarbamates, phenols, polycyclic aromatic hydrocarbons, and amines; and organometallic compounds including two metalocenes and two organoleads in aqueous solutions. It was found that while fullerene adsorbs many types of organic substances as mentioned earlier, the recorded efficiencies were never higher than 60% and this was highly dependent on the nature of the substance being analyzed. Therefore, there are much better adsorbents for these compound types other than $C_{60}$. Meanwhile, the adsorption of organometallic compounds is facilitated by the formation of neutral

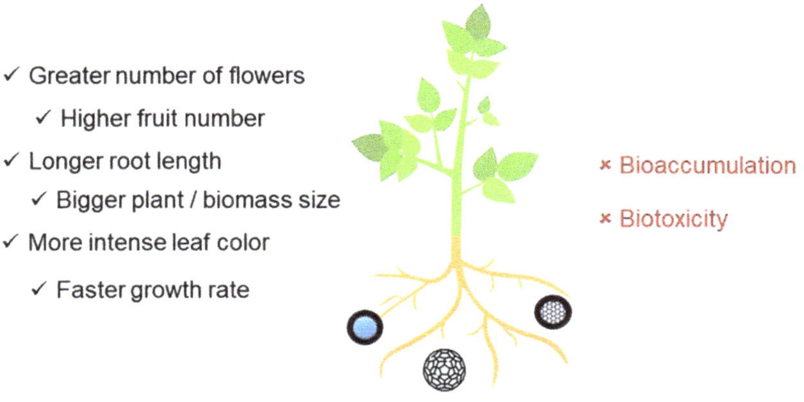


**Figure 14.3.** The observed effects of zero-dimensional carbon nanomaterials on a plant.

complexes or chelates by using classical reagents such as pyrrolidinedithiocarbamate or diethyldithiocarbamate. In this case, $C_{60}$ was found to perform better than RP-C and silica gel 100. This study revealed that $C_{60}$ has a limited range of application as a sorbent and may not necessarily be a promising choice for highly contaminated solutions [103].

In a study by Ma and Wang [104], it was shown that 2–15 mg $L^{-1}$ $C_{60}$ did not result in any acute toxicity in cottonwood (*Populus deltoids*) but instead enhanced its ability for trichloroethylene uptake in a batch hydroponic system. In such case, the $C_{60}$ did not necessarily act as the main reason for trichloroethylene removal, but rather improved the existing conditions for phytoremediation. This shows that $C_{60}$ reacts with any substance present in its immediate environment and interactions should therefore be evaluated.

In a report by Yang and co-workers, CQDs were shown to outperform $C_{60}$ in coordinating with electronic polymers in order to improve photon harvesting ability. This study intends to improve the viability of CQDs and $C_{60}$ as electron donors/acceptors in redox reactions, which is fundamental for nanosensors in environmental applications [105]. Figure 14.3 illustrates the positive and negative impacts of ZCNs on plants. It can be seen that the observed positive effects reported in the literature are much more prominent than the negative effects. For these reasons, ZCNs are widely regarded as a promising approach despite potential drawbacks arising from our limited knowledge of them.

## 14.5 Drawbacks and setbacks

As already discussed in an earlier section, there is no definite consensus regarding the effectiveness of ZCNs. Therefore, it is worth discussing illustrative studies aimed at demonstrating the key concerns in the utilization of ZCNs.

For example, Canesi *et al* [106] studied the *in vitro* effects of $C_{60}$ fullerene and other metal based nanoparticles on the bivalve mussel (*Mytilus galloprovincialis* Lam.). These bivalves have the ability to filter and internalize contaminants via endocytosis and phagocytosis at both the micro- and nanoscale range. Exposure to engineered nanoparticles may then affect or impair the key physiological and biological functions of these organisms. While a toxic effect was not evident as a result of short-term exposure, their findings revealed that they could induce stress in *Mytilus* that could lead to inflammatory failure and, therefore, loss of viability. Furthermore, the engulfed nanoparticles formed aggregates/agglomerates which may thus indicate more complications as to how the impacts of ZCNs are evaluated. ZCNs in their original form may not necessarily induce acute toxicity, but the occurrence of agglomeration or aggregation may pose a higher risk since the ZCN properties have already been altered.

In another study by Sumi *et al* [107], *Anabas testudineus* fish were exposed to two sub-lethal concentrations of 5 mg $L^{-1}$ (one-tenth of LC50–96 h) and 10 mg $L^{-1}$ (one-fifth of LC 50–96 h) of fullerene $C_{60}$ for short-term (24, 48, 72, and 96 h) and long-term (7, 15, 30, and 60 days) durations. No mortality was observed throughout the study but the exposure led to drastic changes in their behavior, i.e. the fish preferred to stay at the bottom of the tank all the time but tended to re-surface to engulf atmospheric air. Upon dissecting and observing the fish, morphological changes in their body included the occurrence of a slight hemorrhage on the body surface, increased descaling and mucous secretion, and the presence of black aggregates in their intestines. Hematological changes observed include a significant reduction in the erythrocyte and leukocyte counts and hemoglobin concentration. Clearly, this study shows that some species may respond abnormally to ZCNs, and that repeated exposure to sub-lethal amounts could be detrimental.

Yao and his team [108] demonstrated that CQDs do not have a universal impact on organisms, but rather each organism at a specific trophic level will react differently towards it. The study involved the analysis of the toxicity of CQDs to zebrafish (*Danio rerio*), zooplankton (*Daphnia magna*), and phytoplankton (*Scenedesmus obliquus*). CQDs were not substantially toxic to *D. rerio* up to 200 mg $L^{-1}$ but such levels were proven to be detrimental to *D. magna*, with mortality and immobility recorded at a 48 h EC50 value and LC50 value of 97.5 and 160.3 mg $L^{-1}$, respectively. In the same manner, photosynthesis and nutrition absorption were inhibited by CQDs in *S. obliquus* and algal growth was inhibited with a 96 h EC50 value of 74.8 mg $L^{-1}$. This means that *S. obliquus*, representing the lowest trophic level in this study, was most sensitive to and affected by CQDs exposure.

ZCN-based biosensors play an important role in the *in vivo* bioassay. However, their application is limited due to the complexity of the *in vivo* environment. Non-specific components in the biological body affect the accuracy of detection. Furthermore, currently most of the ZCN-based biosensors can only be used for *in vitro* testing and laboratory experiments. A large number of studies on disease diagnosis have been conducted, but most of them are limited to human serum, cells, and animal models (mice). The lack of long-term biotoxicity evaluation and biodegradation of these materials further limits their clinical application.

Overall, key concerns involve the tendency of synthesized ZCNs to aggregate and alter their properties, which is often overlooked in the assessment of acute or environmental toxicity. Once its properties have changed and no longer serve their original purpose, ZCN behavior in environmental compartments or biological systems could pose hazards or health risks. It was shown that the lowest level in the trophic system would be at risk, which could be magnified as the issue climbs to higher trophic levels. In addition, if the effect was on the behavior and disruption of physiological systems it may be overlooked easily since this would take time to be noticed. Also, further studies on the responses of agricultural plants and aquatic organisms should be explored. If the plants or animals consider ZCNs as an abiotic threat similar to pathogens or external stress factors, there is a possibility that their chemical response would lead to the formation of other compounds that would affect their taste or composition. This could eventually affect their usage as food and they may even develop mechanisms to induce chemicals that could be poisonous or detrimental to consumers in the food chain.

## 14.6 Summary and future perspectives

Recent progress on ZCNs for agricultural applications has been highlighted in this chapter. It should be noted that there is an increasing interest in ZCNs due to their interesting and exciting properties that are not reflected by their bulk counterparts. These properties also paved the way for ZCNs to be exploited almost in every possible field of application, in order to fulfill the ever-changing needs arising from industrial and regulatory requirements, as well as to unravel their behavior along the way. Also, their widespread application is due to the success of applying one-dimensional and two-dimensional nanostructures towards a variety of applications including energy devices, medical supplies, and biological systems. As many carbon based materials have been previously reported to bring about unprecedented revolution in the field of agriculture, including nanofertilizer delivery, plant growth, and pest control, there was a high expectation that ZCNs would also deliver. However, there is a lack of studies on zero-dimensional materials. Therefore, this opportunity opens up new avenues to be explored:

1. With various conflicting reports regarding the effectiveness of ZCNs, having shown both a negative and positive influence towards observed plant physiological activities, it is necessary widen the scope of study. The results of environmental studies should be assessed since abiotic factors also influence actual ZCN performance.
2. *In vivo* results are not conclusive, potentially due to the lack of a systematic or holistic approach. Most studies claiming biocompatibility were performed under well-defined and controlled conditions, which discounts the effects of abiotic factors on ZCN properties. Therefore, protocols and definitions should be standardized and operating conditions should be benchmarked against existing universally accepted standard procedure.
3. It was observed in the materials reviewed for this chapter that most ZCN studies are still in their initial stages. For example, the current synthesis and fabrication methods still lack control of the structure of materials at the atomic level. Moreover, the fluorescence quenching mechanism of ZCNs is not yet well understood. Comprehensive reviews are still lacking in ZCNs

unlike their nanoscale 1D and 2D counterparts, therefore, there is a need to provide an in-depth analysis of findings on these topics.

4. It should be noted that commercialization is quite far from reality due to the high cost of materials, lengthy processes, and the need for purification. Irrespective of the innovation in fabrication approaches, large-scale synthesis of environmentally benign ZCNs at an affordable cost is still far from reality. The carbon precursors used for the synthesis of carbon nanomaterials (CNs) play a pivotal role in feasible large-scale production. It was observed that more researchers have been favoring CQDs due to their ease of synthesis, as they can be synthesized cheaply from biomass or waste materials. In the future, the accuracy in synthesis, fabrication, and assembly, as well as the corresponding optical and electrical properties of ZCNs should be studied, so as to prepare products or devices that are more relevant for actual clinical applications.

## Acknowledgments

The author would like to acknowledge the European Union's Horizon 2020 research and innovation programme under the Marie Skłodowska-Curie Actions grant agreement No. 894088 (SUBLUBE) for her fellowship.

## References

[1] Quitain A T, Sumigawa Y, Mission E G, Sasaki M, Assabumrungrat S and Kida T 2018 Graphene oxide and microwave synergism for efficient esterification of fatty acids *Energy Fuels* **32** 3599–607

[2] Cui L, Ren X, Wang J and Sun M 2020 Synthesis of homogeneous carbon quantum dots by ultrafast dual-beam pulsed laser ablation for bioimaging *Mater Today Nano* **12** 100091

[3] Khodakovskaya M V, De Silva K, Biris A S, Dervishi E and Villagarcia H 2012 Carbon nanotubes induce growth enhancement of tobacco cells *ACS Nano.* **6** 2128–35

[4] Qiu L *et al* 2020 A review of recent advances in thermophysical properties at the nanoscale: from solid state to colloids *Phys. Rep.* **843** 1–81

[5] Baig N, Kammakakam I, Falath W and Kammakakam I 2021 Nanomaterials: a review of synthesis methods, properties, recent progress, and challenges *Mater Adv.* **2** 1821–71

[6] Jeevanandam J, Barhoum A, Chan Y S, Dufresne A and Danquah M K 2018 Review on nanoparticles and nanostructured materials: history, sources, toxicity and regulations *Beilstein J. Nanotechnol.* **9** 1050–74

[7] Anjum S, Anjum I, Hano C and Kousar S 2019 Advances in nanomaterials as novel elicitors of pharmacologically active plant specialized metabolites: current status and future outlooks *RSC Adv.* **9** 40404–23

[8] Bai D P, Lin X Y, Huang Y F and Zhang X F 2018 Theranostics aspects of various nanoparticles in veterinary medicine *Int. J. Mol. Sci.* **19** 3299

[9] Chavan S and Nadanathangam V 2019 Effects of nanoparticles on plant growth-promoting bacteria in Indian agricultural soil *Agronomy* **9** 140

[10] Lang C, Mission E G, Ahmad Fuaad A A-H and Shaalan M 2021 Nanoparticle tools to improve and advance precision practices in the agrifoods sector towards sustainability—a review *J. Clean. Prod.* **293** 126063

[11] Kumar V, Sachdev D, Pasricha R, Maheshwari P H and Taneja N K 2018 Zinc-supported multiwalled carbon nanotube nanocomposite: a synergism to micronutrient release and a smart distributor to promote the growth of onion seeds in arid conditions *ACS Appl. Mater. Interfaces* **10** 36733–45

[12] Zhang R, Bai X, Shao J, Chen A, Wu H and Luo S 2020 Effects of zero-valent iron nanoparticles and quinclorac coexposure on the growth and antioxidant system of rice (*Oryza sativa* L.) *Ecotoxicol. Environ. Saf.* **203** 111054

[13] Paramo L A, Feregrino-Pérez A A, Guevara R, Mendoza S and Esquivel K 2020 Nanoparticles in agroindustry: applications, toxicity, challenges, and trends *Nanomaterials* **10** 1654

[14] Noamani S, Niroomand S, Rastgar M and Sadrzadeh M 2019 Carbon-based polymer nanocomposite membranes for oily wastewater treatment *NPJ Clean Water* **2** 20

[15] Zhang Y, Wu M, Wu M, Zhu J and Zhang X 2018 Multifunctional carbon-based nanomaterials: applications in biomolecular imaging and therapy *ACS Omega* **3** 9126–45

[16] Patel K D, Singh R K and Kim H W 2019 Carbon-based nanomaterials as an emerging platform for theranostics *Mater. Horizons* **6** 434–69

[17] Nasir S, Hussein M Z, Zainal Z and Yusof N A 2018 Carbon-based nanomaterials/allotropes: a glimpse of their synthesis, properties and some applications *Materials* **11** 1–24

[18] Lopes J L, Martins M J, Nogueira H I S, Estrada A C and Trindade T 2021 Carbon-based heterogeneous photocatalysts for water cleaning technologies: a review *Environ. Chem. Lett.* **19** 643–68

[19] Pan M, Yin Z, Liu K, Du X, Liu H and Wang S 2019 Carbon-based nanomaterials in sensors for food safety *Nanomaterials* **9** 1330

[20] Hassanpour S *et al* 2021 Carbon based nanomaterials for the detection of narrow therapeutic index pharmaceuticals *Talanta* **221** 121610

[21] Pardo J, Peng Z and Leblanc R M 2018 Cancer targeting and drug delivery using carbon-based quantum dots and nanotubes *Molecules* **23** 378

[22] Panwar N *et al* 2019 Nanocarbons for biology and medicine: sensing, imaging, and drug delivery *Chem. Rev.* **119** 9559–656

[23] Peng Z, Zhao T, Zhou Y, Li S, Li J and Leblanc R M 2020 Bone tissue engineering via carbon-based nanomaterials *Adv. Healthc. Mater.* **9** 1901495

[24] Maiti D, Tong X, Mou X and Yang K 2019 Carbon-based nanomaterials for biomedical applications: a recent study *Front. Pharmacol.* **9** 1401

[25] Eivazzadeh-Keihan R *et al* 2019 Carbon based nanomaterials for tissue engineering of bone: building new bone on small black scaffolds: a review *J. Adv. Res.* **18** 185–201

[26] De La Torre-Roche R *et al* 2013 Multiwalled carbon nanotubes and $C_{60}$ fullerenes differentially impact the accumulation of weathered pesticides in four agricultural plants *Environ. Sci. Technol.* **47** 12539–47

[27] Jiao J *et al* 2016 The role of graphene oxide on tobacco root growth and its preliminary mechanism *J. Nanosci. Nanotechnol.* **16** 12449–54

[28] Mondal A, Basu R, Das S and Nandy P 2011 Beneficial role of carbon nanotubes on mustard plant growth: an agricultural prospect *J. Nanopart. Res.* **13** 4519–28

[29] Samadi S, Asgari Lajayer B, Moghiseh E and Rodríguez-Couto S 2021 Effect of carbon nanomaterials on cell toxicity, biomass production, nutritional and active compound accumulation in plants *Environ. Technol. Innov.* **21** 101323

[30] Chhipa H 2017 Nanofertilizers and nanopesticides for agriculture *Environ. Chem. Lett.* **15** 15–22

[31] Jordan J T *et al* 2020 Carbon nanotubes affect early growth, flowering time and phytohormones in tomato *Chemosphere* **256** 127042

[32] Gohari G *et al* 2020 Modified multiwall carbon nanotubes display either phytotoxic or growth promoting and stress protecting activity in *Ocimum basilicum* L. in a concentration-dependent manner *Chemosphere* **249** 126171

[33] Khodakovskaya M *et al* 2009 Erratum: Carbon nanotubes are able to penetrate plant seed coat and dramatically affect seed germination and plant growth *ACS Nano* **3** 3221–7

[34] Joshi A, Kaur S, Dharamvir K, Nayyar H and Verma G 2018 Multi-walled carbon nanotubes applied through seed-priming influence early germination, root hair, growth and yield of bread wheat (*Triticum aestivum* L.) *J. Sci. Food Agric.* **98** 3148–60

[35] Yuan Z, Zhang Z, Wang X, Li L, Cai K and Han H 2017 Novel impacts of functionalized multi-walled carbon nanotubes in plants: promotion of nodulation and nitrogenase activity in the rhizobium-legume system *Nanoscale* **9** 9921–37

[36] Elemike E E, Uzoh I M, Onwudiwe D C and Babalola O O 2019 The role of nanotechnology in the fortification of plant nutrients and improvement of crop production *Appl. Sci.* **9** 499

[37] Sonkar S K, Roy M, Babar D G and Sarkar S 2012 Water soluble carbon nano-onions from wood wool as growth promoters for gram plants *Nanoscale* **4** 7670–5

[38] Lateef A *et al* 2019 Synthesis and characterization of environmental friendly corncob biochar based nano-composite—a potential slow release nano-fertilizer for sustainable agriculture *Environ. Nanotechnol. Monit. Manage* **11** 100212

[39] Wang Y *et al* 2021 A new temperature-responsive controlled-release pesticide formulation—poly(*N*-isopropylacrylamide) modified graphene oxide as the nanocarrier for lambda-cyhalothrin delivery and their application in pesticide transportation *Colloids Surf.* A **612** 125987

[40] Abdullayev E and Lvov Y 2011 Halloysite clay nanotubes for controlled release of protective agents *J. Nanosci. Nanotechnol.* **11** 10007–26

[41] Li T, Gao B, Tong Z, Yang Y and Li Y 2019 Chitosan and graphene oxide nanocomposites as coatings for controlled-release fertilizer *Water Air Soil Pollut.* **230** 146

[42] Shang Y, Hasan K, Ahammed G J, Li M and Yin H 2019 Applications of nanotechnology in plant growth and crop protection: a review *Molecules* **24** 2558

[43] Andelkovic I B, Kabiri S, Tavakkoli E, Kirby J K, McLaughlin M J and Losic D 2018 Graphene oxide-Fe(III) composite containing phosphate—a novel slow release fertilizer for improved agriculture management *J. Clean. Prod.* **185** 97–104

[44] Troncoso O P and Torres F G 2020 Bacterial cellulose–graphene based nanocomposites *Int. J. Mol. Sci.* **21** 6532

[45] Bertolino V, Cavallaro G, Milioto S and Lazzara G 2020 Polysaccharides/halloysite nanotubes for smart bionanocomposite materials *Carbohydr. Polym.* **245** 116502

[46] Anumary A, Ashokkumar M, Thanikaivelan P and Ajayan P M 2017 Hybrid composites using natural polymer blends and carbon nanostructures: preparation, characterization, and applications *Recent Developments in Polymer Macro, Micro and Nano Blends: Preparation and Characterisation* (Amsterdam: Elsevier) pp 57–74

[47] Wang Z, Hu T, Liang R and Wei M 2020 Application of zero-dimensional nanomaterials in biosensing *Front. Chem.* **8** 320

[48] Goh P S, Wong K C and Ismail A F 2020 Nanocomposite membranes for liquid and gas separations from the perspective of nanostructure dimensions *Membranes* **10** 297
[49] Pokropivny V V and Skorokhod V V 2008 New dimensionality classifications of nanostructures *Physica* E **40** 2521–5
[50] Xu Q, Li W, Ding L, Yang W, Xiao H and Ong W J 2019 Function-driven engineering of 1D carbon nanotubes and 0D carbon dots: mechanism, properties and applications *Nanoscale* **11** 1475–504
[51] Li Y, Liao K and Lee S 2014 A comparative study of antioxidant-loaded carbon nanoparticles as drug delivery vehicles *RSC Adv.* **4** 56992–7
[52] Sun H, Wu L, Wei W and Qu X 2013 Recent advances in graphene quantum dots for sensing *Mater. Today* **16** 433–42
[53] Liu Q *et al* 2020 Graphene quantum dots for energy storage and conversion: from fabrication to applications *Mater. Chem. Front.* **4** 421–36
[54] Zheng X T, Ananthanarayanan A, Luo K Q and Chen P 2015 Glowing graphene quantum dots and carbon dots: properties, syntheses, and biological applications *Small* **11** 1620–36
[55] Fan Z, Li S, Yuan F and Fan L 2015 Fluorescent graphene quantum dots for biosensing and bioimaging *RSC Adv.* **5** 19773–89
[56] Ananthaiah R 1997 Discovery of fullerenes *Resonance* **2** 68–73
[57] Schwerdtfeger P, Wirz L N and Avery J 2015 The topology of fullerenes *Wiley Interdiscip. Rev. Comput. Mol. Sci.* **5** 96–145
[58] Goel A, Howard J B and Sande J B V 2004 Size analysis of single fullerene molecules by electron microscopy *Carbon* **42** 1907–15
[59] Jani M, Arcos-Pareja J A and Ni M 2020 Engineered zero-dimensional fullerene/carbon dots-polymer based nanocomposite membranes for wastewater treatment *Molecules* **25** 4934
[60] Li H, Kang Z, Liu Y and Lee S T 2012 Carbon nanodots: synthesis, properties and applications *J. Mater. Chem.* **22** 24230–53
[61] Wang H *et al* 2017 Excitation wavelength independent visible color emission of carbon dots *Nanoscale* **9** 1909–15
[62] Bacon M, Bradley S J and Nann T 2014 Graphene quantum dots *Part. Part. Syst. Charact.* **31** 415–28
[63] Ponomarenko L A *et al* 2008 Chaotic Dirac billiard in graphene quantum dots *Science* **320** 356–8
[64] Xu X *et al* 2004 Electrophoretic analysis and purification of fluorescent single-walled carbon nanotube fragments *J. Am. Chem. Soc.* **126** 12736–7
[65] Tian P, Tang L, Teng K S and Lau S P 2018 Graphene quantum dots from chemistry to applications *Mater Today Chem.* **10** 221–58
[66] Lee N E, Lee S Y, Lim H S, Yoo S H and Cho S O 2020 A novel route to high-quality graphene quantum dots by hydrogen-assisted pyrolysis of silicon carbide *Nanomaterials* **10** 277
[67] Qu D *et al* 2014 Formation mechanism and optimization of highly luminescent N-doped graphene quantum dots *Sci. Rep.* **4** 5294
[68] Jin K *et al* 2018 Preparation of highly fluorescent sulfur doped graphene quantum dots for live cell imaging *J. Lumin.* **197** 147–52
[69] Kim S, Seo J K, Park J H, Song Y, Meng Y S and Heller M J 2017 White-light emission of blue-luminescent graphene quantum dots by europium (III) complex incorporation *Carbon* **124** 479–85

[70] Wang Z *et al* 2016 Large-scale and controllable synthesis of graphene quantum dots from rice husk biomass: a comprehensive utilization strategy *ACS Appl. Mater. Interfaces* **8** 1434–9

[71] Wang S, Cole I S, Zhao D and Li Q 2016 The dual roles of functional groups in the photoluminescence of graphene quantum dots *Nanoscale* **8** 7449–58

[72] Guo R, Zhou S, Li Y, Li X, Fan L and Voelcker N H 2015 Rhodamine-functionalized graphene quantum dots for detection of $Fe^{3+}$ in cancer stem cells *ACS Appl. Mater. Interfaces* **7** 23958–66

[73] Chong Y *et al* 2016 Crossover between anti- and pro-oxidant activities of graphene quantum dots in the absence or presence of light *ACS Nano* **10** 8690–9

[74] Tan X, Li Y, Li X, Zhou S, Fan L and Yang S 2015 Electrochemical synthesis of small-sized red fluorescent graphene quantum dots as a bioimaging platform *Chem. Commun.* **51** 2544–6

[75] Peng J *et al* 2012 Graphene quantum dots derived from carbon fibers *Nano Lett.* **12** 844–9

[76] Cui F *et al* 2019 A novel magnetic fluorescent biosensor based on graphene quantum dots for rapid, efficient, and sensitive separation and detection of circulating tumor cells *Anal. Bioanal. Chem.* **411** 985–95

[77] Wei S, Zhang R, Liu Y, Ding H and Zhang Y L 2016 Graphene quantum dots prepared from chemical exfoliation of multiwall carbon nanotubes: an efficient photocatalyst promoter *Catal. Commun* **74** 104–9

[78] Shehab M, Ebrahim S and Soliman M 2017 Graphene quantum dots prepared from glucose as optical sensor for glucose *J. Lumin.* **184** 110–6

[79] Mars A, Hamami M, Bechnak L, Patra D and Raouafi N 2018 Curcumin-graphene quantum dots for dual mode sensing platform: electrochemical and fluorescence detection of APOe4, responsible of Alzheimer's disease *Anal. Chim. Acta* **1036** 141–6

[80] Tabish T A, Hayat H, Abbas A and Narayan R J 2021 Graphene quantum dot-based electrochemical biosensing for early cancer detection *Curr. Opin. Electrochem.* **30** 100786

[81] Cheng R, Yu C, Zhen Z, Tang S and Ou S 2020 Understanding the selective-sensing mechanism of lysine by fluorescent nanosensors based on graphene quantum dots *Spectrochim. Acta* A **242** 118732

[82] Alizadeh T and Shokri M 2016 A new humidity sensor based upon graphene quantum dots prepared via carbonization of citric acid *Sensors Actuators* B **222** 728–34

[83] Ding Z, Li F, Wen J, Wang X and Sun R 2018 Gram-scale synthesis of single-crystalline graphene quantum dots derived from lignin biomass *Green Chem.* **20** 1383–90

[84] Zhou J, Sheng Z, Han H, Zou M and Li C 2012 Facile synthesis of fluorescent carbon dots using watermelon peel as a carbon source *Mater. Lett.* **66** 222–4

[85] Surendran P *et al* 2021 Fluorescent carbon quantum dots from *Ananas comosus* waste peels: a promising material for NLO behaviour, antibacterial, and antioxidant activities *Inorg. Chem. Commun.* **124** 108397

[86] Mehta V N, Jha S and Kailasa S K 2014 One-pot green synthesis of carbon dots by using *Saccharum officinarum* juice for fluorescent imaging of bacteria (*Escherichia coli*) and yeast (*Saccharomyces cerevisiae*) cells *Mater Sci. Eng.* C **38** 20–7

[87] Zhang B, Liu C Y and Liu Y 2010 A novel one-step approach to synthesize fluorescent carbon nanoparticles *Eur. J. Inorg. Chem.* **28** 4411–4

[88] Huang H *et al* 2013 One-pot green synthesis of nitrogen-doped carbon nanoparticles as fluorescent probes for mercury ions *RSC Adv.* **3** 21691–6

[89] Wang J, Wang C F and Chen S 2012 Amphiphilic egg-derived carbon dots: rapid plasma fabrication, pyrolysis process, and multicolor printing patterns *Angew Chem.—Int. Ed.* **51** 9297–301

[90] Lu W *et al* 2012 Economical, green synthesis of fluorescent carbon nanoparticles and their use as probes for sensitive and selective detection of mercury(II) ions *Anal. Chem.* **84** 5351–7

[91] Gao X, Du C, Zhuang Z and Chen W 2016 Carbon quantum dot-based nanoprobes for metal ion detection *J. Mater. Chem.* C **4** 6927–45

[92] Zhang Y, Gao Z, Yang X, Chang J, Liu Z and Jiang K 2019 Fish-scale-derived carbon dots as efficient fluorescent nanoprobes for detection of ferric ions *RSC Adv.* **9** 940–9

[93] Zheng Y *et al* 2017 Bioimaging application and growth-promoting behavior of carbon dots from pollen on hydroponically cultivated rome lettuce *ACS Omega* **2** 3958–65

[94] Arumugam N and Kim J 2018 Synthesis of carbon quantum dots from Broccoli and their ability to detect silver ions *Mater. Lett.* **219** 37–40

[95] Wang X *et al* 2019 Green preparation of fluorescent carbon quantum dots from cyanobacteria for biological imaging *Polymers* **11** 616

[96] Li W *et al* 2019 Kilogram-scale synthesis of carbon quantum dots for hydrogen evolution, sensing and bioimaging *Chin. Chem. Lett.* **30** 2323–7

[97] Chakravarty D, Erande M B and Late D J 2015 Graphene quantum dots as enhanced plant growth regulators: effects on coriander and garlic plants *J. Sci. Food Agric.* **95** 2772–8

[98] Kovač T *et al* 2018 Fullerol $C_{60}(OH)_{24}$ nanoparticles modulate aflatoxin B1 biosynthesis in *Aspergillus flavus Sci. Rep.* **8** 12855

[99] Kole C *et al* 2013 Nanobiotechnology can boost crop production and quality: first evidence from increased plant biomass, fruit yield and phytomedicine content in bitter melon (*Momordica charantia*) *BMC Biotechnol.* **13** 37

[100] Lin S *et al* 2009 Uptake, translocation, and transmission of carbon nanomaterials in rice plants *Small* **5** 1128–32

[101] Kang Y F, Li Y H, Fang Y W, Xu Y, Wei X M and Yin X B 2015 Carbon quantum dots for zebrafish fluorescence imaging *Sci. Rep.* **5** 11835

[102] Oberdörster E 2004 Manufactured nanomaterials (fullerenes, $C_{60}$) induce oxidative stress in the brain of juvenile largemouth bass *Environ. Health Perspect.* **112** 1058–62

[103] Ballesteros E, Gallego M and Valcárcel M 2000 Analytical potential of fullerene as adsorbent for organic and organometallic compounds from aqueous solutions *J. Chromatogr.* A **869** 101–10

[104] Ma X and Wang C 2010 Fullerene nanoparticles affect the fate and uptake of trichloroethylene in phytoremediation systems *Environ. Eng. Sci.* **27** 989–92

[105] Yang F *et al* 2018 Zero-dimensional carbon allotropes—carbon nanoparticles versus fullerenes in functionalization by electronic polymers for different optical and redox properties *ACS Omega* **3** 5685–91

[106] Canesi L, Ciacci C, Vallotto D, Gallo G, Marcomini A and Pojana G 2010 *In vitro* effects of suspensions of selected nanoparticles ($C_{60}$ fullerene, $TiO_2$, $SiO_2$) on *Mytilus hemocytes Aquat. Toxicol.* **96** 151–8

[107] Sumi N and Chitra K C 2019 Impact of fullerene $C_{60}$ on behavioral and hematological changes in the freshwater fish, *Anabas testudineus* (Bloch, 1792) *Appl. Nanosci.* **9** 2147–67

[108] Yao K *et al* 2018 Effects of carbon quantum dots on aquatic environments: comparison of toxicity to organisms at different trophic levels *Environ. Sci. Technol.* **52** 14445–51

[109] Fan H, Zhang M, Bhandari B and Hui Y C 2020 Food waste as a carbon source in carbon quantum dots technology and their applications in food safety detection *Trends Food Sci. Technol.* **95** 86–96

[110] Yadav V K *et al* 2021 Recent advances on properties and utility of nanomaterials generated from industrial and biological activities *Crystals* **11** 634

[111] Nair A, Haponiuk J T, Thomas S and Gopi S 2020 Natural carbon-based quantum dots and their applications in drug delivery: a review *Biomed. Pharmacother.* **132** 110834

[112] Rana S 2021 Review article toxicity induced by engineered nanoparticles in fresh water—a review of recent developments EC Pharmacolog *Toxicolog.* **9** 8

[113] Liang W, Bunker C E and Sun Y P 2020 Carbon dots: zero-dimensional carbon allotrope with unique photoinduced redox characteristics *ACS Omega* **5** 965–71

[114] Jin H, Sun Y, Sun Z, Yang M and Gui R 2021 Zero-dimensional sulfur nanomaterials: synthesis, modifications and applications *Coord. Chem. Rev.* **438** 213913

[115] Rao N, Singh R and Bashambu L 2021 Carbon-based nanomaterials: synthesis and prospective applications *Mater. Today Proc.* **44** 608–14

www.ingramcontent.com/pod-product-compliance
Ingram Content Group UK Ltd.
Pitfield, Milton Keynes, MK11 3LW, UK
UKHW051900100926
8556IPUK00005B/707

9 780750 340342